Odour Impact Assessment Handbook

Odour Impact Assessment Handbook

Editors

VINCENZO BELGIORNO, VINCENZO NADDEO
and TIZIANO ZARRA

*Sanitary Environmental Engineering Division (SEED),
Department of Civil Engineering, University of Salerno, Italy*

A John Wiley & Sons, Ltd., Publication

Contents

List of Contributors

Vincenzo Belgiorno, Sanitary Environmental Engineering Division (SEED), Department of Civil Engineering, University of Salerno, Italy

Manabendra Bhuyan, Department of Electronics and Communication Engineering, Tezpur University, India

Laura Cappelli, Politecnico di Milano, Department of Chemistry, Material and Chemical Engineering, Italy

Renato Del Rosso, Politecnico di Milano, Department of Chemistry, Material and Chemical Engineering, Italy

Licinia Dentoni, Politecnico di Milano, Department of Chemistry, Material and Chemical Engineering, Italy

Nimisha Dutta, Department of Electronics and Communication Engineering, Tezpur University, India

José M. Estrada, Department of Chemical Engineering and Environmental Technology Valladolid University, Spain

Stefano Giuliani, Sanitary Environmental Engineering Division (SEED), Department of Civil Engineering, University of Salerno, Italy

Takaya Higuchi, Graduate School of Science and Engineering, Yamaguchi University, Japan

Nikolaos Kalogerakis, Department of Environmental Engineering, Technical University of Crete, Greece

N.J.R. Kraakman, CH2M Hill, Chatswood NSW Australia and Laboratory of Biotechnology, Faculty of Applied Sciences, Delft University of Technology, Netherlands

Martin Kranert, ISWA (Institut für Siedlungswasserbau, Wassergüte- und Abfallwirtschaft) University of Stuttgart, Germany

Mihalis Lazaridis, Department of Environmental Engineering, Technical University of Crete, Greece

Raquel Lebrero, Department of Chemical Engineering and Environmental Technology Valladolid University, Spain

Jenni Lehtinen, Department of Biological and Environmental Sciences – University of Jyväskylä, Finland

Raul Muñoz, Department of Chemical Engineering and Environmental Technology Valladolid University, Spain

Vincenzo Naddeo, Sanitary Environmental Engineering Division (SEED), Department of Civil Engineering, University of Salerno, Italy

Jacques Nicolas, Department of Environmental Sciences and Management, Arlon Campus Environment, Faculty of Sciences, University of Liége, Belgium

Martin Piringer, Section Environmental Meteorology, Central Institute for Meteorology and Geodynamics, Vienna, Austria

G. Quijano, Department of Chemical Engineering and Environmental Technology, Valladolid University, Valladolid, Spain

Martin Reiser, ISWA (Institut für Siedlungswasserbau, Wassergüte- und Abfallwirtschaft), University of Stuttgart, Germany

Anne-Claude Romain, Department of Environmental Sciences and Management, Arlon Campus Environment, Faculty of Sciences, University of Liége, Belgium

Günther Schauberger, WG Environmental Health, Department for Biomedical Sciences, University of Veterinary Medicine Vienna, Austria

Selena Sironi, Politecnico di Milano, Department of Chemistry, Material and Chemical Engineering, Italy

Izabela Sówka, Institute of Environmental Protection Engineering, Wroclaw University of Technology, Poland

Kaiying Y. Wang, School of Biosystems Engineering & Food Science, Zhejiang University, China

Tiziano Zarra, Sanitary Environmental Engineering Division (SEED), Department of Civil Engineering, University of Salerno, Italy

Preface

Odour emissions are considered to be the main cause of disturbance noticed by the citizens living near some facilities. Even though a real toxicological-sanitary risk is hardly ever associated to the odour impact from sources connected to the activities of waste management and similes, due to the rarely dangerous nature of the smells as well as the generally very low concentrations, the collective imagination often associates the bad smell to conditions of 'non healthy' air. In fact, a valence higher than the one related to more dangerous contaminants, but not directly perceptible from our senses, is often attributed to them.

Odour emissions affect quality of life leading to psychological stress and symptoms such as insomnia, loss of appetite and irrational behaviour. Odours have become a priority concern for facility operators, engineers and urban planners that deal with waste and industrial treatment plants. Odour complaints can shut down facilities and prevent the expansion of existing facilities. The particular and complex nature of the many substances being dealt with in these facilities results in a 'smell impact'. The extent of this impact depends on a variety of factors, most notably these factors are time, weather conditions and the subjectivity of each individuals' perception of a specific smell. When combined, these factors have delayed progress in odour regulation.

The difficulties in resolving these controversies are also the result of an absence of universally recognized standards for objectively determining the negative effect induced. Only a few countries, in fact, have odour quality criteria or odour emission thresholds for industrial sources. On the European scale, standard EN 13725:2003 makes reference to the sampling and quantification of odours using dynamic olfactometry from sources. USA, Australia, Canada and Japan are following similar approaches.

Offensive odours are not only a direct threat for human health and welfare, but also represent a significant contribution to photochemical smog formation and particulate secondary contaminant emission. In this context, a cost-effective and environmentally friendly abatement of odours is crucial in a world increasingly concerned about sustainability and environmental preservation, but also increased life quality standards. Physical/chemical technologies have been widely used due to their low footprint, extensive experience in design and operation and rapid start-up, however, they exhibit significant environmental impacts and high operating costs. On the other hand, biotechnologies have been marketed as low-cost, environmentally friendly odour abatement methods.

The aim of this book is to provide a theoretical and practical basis for responding to the problem of sampling and measuring odours using olfactometric, analytical and mixed techniques to assess the impacts. This is in order to provide a realistic response to the demand for information from the population and from the technical-scientific world, while at the same time providing an objective instrument for the authorities managing and monitoring environment and the professional figures who are becoming more and more involved in this type of problem.

This handbook contains the work of 28 contributors from 13 different countries. The experience behind the authors stems from years of technical activity and scientific research in assessing odour environmental impact. The applicative approach that this handbook adopts in Part 8 is coherent with the numerous applications in the methods and processes field, in which the progress made in international scientific research has been integrated with the applicative limits and the need to supply results.

V. Belgiorno, V. Naddeo and T. Zarra
Fisciano, Italy (August 2012)

Glossary and Abbreviations

Acute odour effect: the effect due to short-term exposure to odours sufficiently intense to cause adverse effects.

Adaptation: the phenomena of reduced sensitivity to a stimulus after prolonged exposure. Unlike habituation this refers to a reduced physiological as opposed to psychological response to a stimuli.

Annoyance: (1) when used in relation to an odour's character or pleasantness, annoyance is akin to the hedonic rating of an odour's pleasantness. (2) When used in conjunction with population annoyance surveys, it is a function of the attitude and feelings of the community towards a source (or sources) of ongoing odour impacts.

Area Source: a surface-emitting source, which can be solid for example, the spreading of wastes, material stockpiles, surface of a biofilter, or liquid for example, storage lagoons, effluent treatment plant.

Assessor: somebody who participates in odour testing.

Delayed olfactometry: measurement of an odour with a time-lag between sampling and measurement. The odour sample is preserved in an appropriate container.

Detection Threshold: the point at which an increasing concentration of an odour sample becomes strong enough to produce a first sensation of odour in 50% of the people to whom the sample is presented. The measurement of odour concentration is based on determining the detection threshold. This is a laboratory-based test and should be conducted according to the EN13725 European standard. The odour concentration at the detection threshold is one odour unit (per cubic metre).

Diffuse Sources: sources with defined dimensions (mostly surface sources) which do not have a defined waste air flow, such as waste dumps, lagoons, fields after manure spreading, un-aerated compost piles.

Dilution factor: the dilution factor is the ratio between flow or volume after dilution and the flow or volume of the odorous gas.

Direct olfactometry: measurement of odour concentrations without any time-lag between the sampling (operation) and the measurements; equivalent to dynamic sampling or on-line olfactometry.

Dynamic dilution olfactometry (DDO): the general procedure used to establish the relative odour concentration of a gas sample. The method establishes the extent of clean air dilution required to reduce the odour strength to a level that is at the threshold of detection for a calibrated panel. The sampling of the raw gas, dilution and presentation to the panel is undertaken in a continuous manner. The backcalculated concentration of the undiluted gas sample (OU_E/m^3) represents the number of dilutions with odour-free air required to reduce the odour of the gas down to the detection threshold.

Dynamic olfactometer: a dynamic olfactometer delivers a flow of mixtures of odorous and neutral gas with known dilution factors in a common outlet.

Electronic nose (E-Nose): an electronic device that uses an array of solid-state sensors, or synthesized protein sensors, that respond to the presence of different chemical compounds. The resulting electronic signals are processed using neural network computing techniques to help produce a two-dimensional spectral pattern that is specific to a particular mix of chemical compounds. The aim is to create different spectral patterns that can identify/fingerprint specific types of odour character.

European Odour Unit OU_E/m^3: that amount of odorant(s) that, when evaporated into one cubic metre of neutral gas at standard conditions, elicits a physiological response from a panel (detection threshold) equivalent to that elicited by one European Reference Odour Mass (EROM), evaporated in 1 m^3 of neutral gas at standard conditions. One EROM is equivalent to 123 μg n-butanol.

Exposure: the dose received by a receptor, determined by the strength (concentration or intensity), time (duration and frequency) of a particular character odour.

Fugitive Releases: unintentional emissions from, for example; flanges, valves, doors, windows; that is, points which are not designated or intended as release points.

Fugitive source: any type of odour emission that cannot be readily quantified or defined. This usually refers to such sources as leaks in pipes, flanges, pump seals or structures, openings in buildings, floor spills, occasional sources such as uncovered truck loads or releases from pressure relief valves, and leaks in seals on covered tanks.

Gas Chromatography: this analytical technique is a form of chromatography that separates and detects compounds by the rate in which they move through an inert or un-reactive carrier gas such as nitrogen, helium or carbon dioxide. The time taken (residence time) to move through the glass or metal tube called a column is used to determine the type of compound present within the sample.

Habituation: a psychological term used to describe the process of decreasing behavioural response after repeated exposure to a stimulus such as odour over a prolonged period of time. This phenomena is particularly noticeable in commercial and industrial settings where occupational exposures to strong odours are no longer found offensive or even noticed by operational staff, for example, rendering plants, livestock, sewage and food processing.

Hedonic Tone: a judgement of the relative pleasantness or unpleasantness of an odour made by assessors in an odour panel. A methodology is described in VDI 3882, part 2. Odours which are more unpleasant will have a negative hedonic score whilst odours that are less unpleasant will tend towards a positive score.

Hyposmia: partial inability to detect odours (compare with anosmia).

Intensity: an assessment of odour strength based on an initial perception. This perception strength will rapidly diminish with constant exposure. The relationship between odour intensity and odour concentration depends on the specific intensity of the chemical or mixture being detected. Assessments can be made using the German method VDI 3882.

Isopleth: a line on a map connecting places registering the same amount or ratio of some geographical or meteorological phenomenon or phenomena. Commonly used to illustrate the output of odour models.

Mass Spectrometry: this is an analytical technique used to identify the chemical composition of a compound. The technique determines particles of the same type from the principle that particles with the same mass and charge will move in the same path in a vacuum when subjected to the same electric and magnetic fields. This principle of determining electronic mass and ionic charge allows the chemical composition of a sample to be determined from a database of existing compounds or unknown compounds to be detected. Three basic components make up a mass spectrometer; an ion source, a mass analyser and a detector.

Neutral gas: air or nitrogen treated in such a way that it is odourless, and which, according to panel members, does not interfere with the odour under investigation.

Odorant: a substance which stimulates a human olfactory system so that an odour is perceived.

Odour (or *Odor*): organoleptic attribute perceptible by the olfactory organ on sniffing certain volatile substances [ISO 5492].

Odour abatement efficiency: the reduction of the odour concentration or the odour flow rate due to an abatement technique, expressed as a fraction (or percentage) of the odour concentration in or the odour flow rate of the untreated gas stream

Odour annoyance survey (*Community survey*): standard survey method used to quantify the extent of population annoyance in different sectors of a community as a result of industrial odour impacts.

Odour concentration: the number of odour units in one cubic metre of gas at standard conditions. Note: odour concentration has a non-linear relationship with odour intensity.

Odour detection: to become aware of the sensation (smell) resulting from stimulation of the olfactometry system.

Odour diary: the systematic recording by individuals of odour events over a period of time at a defined location (normally a residential dwelling), including the date, time, duration, character, strength and weather conditions associated with each odour event.

Odour dose-response: the relationship derived between population annoyance and predicted odour impact concentrations, where the former is quantified via an odour annoyance survey and the latter is determined using odour emission measurement and modelling techniques.

Odour emission: the number of odour units per second discharged from a specific source.

Odour flow rate: the odour flow rate is the quantity of European odour units which crosses a given surface divided by time. It is the product of the odour concentration *cod*, the outlet velocity *v* and the outlet area *A* or the product of the odour concentration *cod* and the pertinent volume flow rate *V*. Its unit is OU_E/h (or OU_E/min or OU_E/s, respectively).

Odour intensity: the perceived strength of an odour as rated by individuals against a numerical scale, such as that contained in the German Standard VDI 3882 Odour Intensity Scale.

Odour unit (OU): one odour unit is the amount of (a mixture of) odorants present in 1 m^3 of odorous gas (under standard conditions) at the panel threshold. NOTE See also *European odour unit*.

Offensiveness: see *hedonic tone*.

Olfactometer: apparatus in which a sample of odorous gas is diluted with neutral gas in a defined ratio and presented to assessors.

Olfactometry: measurement of the response of assessors to olfactory stimuli [ISO 5492].

Olfactory: pertaining to the sense of smell [ISO 5492].

OMP: Odour Management Plan.

Panel member: an assessor who is qualified to judge samples of odorous gas, using dynamic olfactometry within the scope of this standard.

Panel: a group of panel members.

Perception: awareness of the effects of single or multiple sensory stimuli [ISO 5492].

Point source: a discrete stationary source of emission of waste gases to atmosphere through canalized ducts of defined dimension and air flow rate (e.g. chimneys, vents).

Population annoyance: a measure of the percentage of people in a community who consider themselves to be 'annoyed' or even more adversely affected by the impacts of industrial odours in their community (percentage at-least annoyed).

Quality: the totality of features and characteristics of a product or service that bear on its ability to satisfy stated or implied needs [ISO 6879].

Recognition threshold (RT): the odour concentration which has a probability of 0.5 of being recognized under the conditions of the test.

Reference odour mass (ROM): the ROM is equivalent to 123 mg of n-butanol evaporated in 1 m^3 of neutral gas.

Round: one round is the presentation of one dilution series to all assessors.

Sample: in the context of this standard, the sample is the odorous gas sample. It is an amount of gas which is assumed to be representative of the gas mass or gas flow under investigation, and which is examined for odour concentration [ISO 6879].

Specific or surface odour emission rate (SOER): the SOER per unit area of surface, which has units of odour per unit area per time (e.g. OU/m^2s or OU/m^2h).

Static dilution: dilution achieved by mixing two known volumes of gas, odorous sample and neutral gas, respectively. The rate of dilution is calculated from the volumes.

Static flux hood: an odour-sampling hood that is placed over an area source and which has a low flow-rate of neutral gas injected to allow a mixed air stream to be expelled from the hood. These devices work on the same principle as wind-tunnel sampling hoods, except that air within the static hood exhibits minimal turbulence.

Static olfactometer: a static olfactometer dilutes by mixing two known volumes of gas, odorous sample and neutral gas, respectively. The rate of dilution is calculated from the volumes.

Volume source: a source of odour emission such as a building structure from which odour diffuses from many different points.

Wind tunnel: odour-sampling wind tunnels are generally elongated hoods that are placed on to an areal odour source and have a flow-rate of neutral gas passed through in a plug-flow manner (i.e. the gas enters one end of the hood and sweeps through to the outlet end, where it is expelled for sampling). These devices generate substantially more turbulence within the hood due to the greater airflows per unit area involved.

Errata Corrige

(i) Acknowledgments

This book would not have been possible without the support, understanding, and help of all the authors which the Editors would like to acknowledge. The hard work of these authors is the good fortune of any reader of this Book. The Editors are also grateful to their many colleagues who provided valuable suggestions for corrections and improvements to the Book.

The contents under Sections 2.5 at page 24 from line 36 to 42, at page 25 from line 1 to line 2 and at page 26 from line 13 to 16 and from line 32 to 39 are reproduced by permission of K. Sucker, R. Both, M. Hangartner and G. Winneke. After the publication of the book, the Editors realized that some sentences of an unpublished manuscript sent to the Editors by Dr. G. Winneke in February 2011 for a previous collaboration, have been used, by mistake, in the drafting of section 2.5 of the book without a proper accreditation.

The contents under Sections from 3.6.1 to 3.6.3 are reproduced by permission of L. Capelli, S. Sironi and R. Del Rosso. After publication of the Book, the Editors realized that some parts of a note sent to the Editors by L. Capelli on February 2011 for a previous collaboration, have been used, by mistake, in the drafting of section 3.6 of the book without a proper accreditation. Meanwhile, the note considered unpublished by the authors, was amended, integrated, updated, completed and published by L. Capelli, S. Sironi and R. Del Rosso as new original work in the paper entitled "Odor Sampling: Techniques and Strategies for the Estimation of Odor Emission Rates from Different Source Types", 2013, Sensors (ISSN 1424-8220), Issue n. 13 – pages 938–955.

The Editors are grateful to all parties involved in these issues for the cooperation and understanding they have shown and apologize for any inconvenience caused.

Finally special thanks are given to Emma Strickland, Rebecca Stubbs and all the other people that worked on this project at the publisher, John Wiley & Sons, Ltd, for their assistance and insights in support of this book.

(ii) Page xi line 5

"Laura Cappelli" should be read "Laura Capelli"

(iii) Page 50 line 13

"*3.6 Estimation of Emission Rate*" should be read "*3.6 Estimation of Emission Rate**" where the (*) links to a footnote with the following text "*The contents under Sections from 3.6.1 to 3.6.3 are reproduced by permission of Laura Capelli, Selena Sironi and Renato Del Rosso. This material was amended, integrated, updated, completed and published by Capelli et al. as new original work in the paper entitled "Odor Sampling: Techniques and Strategies for the Estimation of Odor Emission Rates from Different Source Types", 2013, Sensors (ISSN 1424-8220), Issue n. 13 – pages 938–955. Additional information on this issue are reported in the Acknowledgments.*"

Part 1

Introduction

V. Naddeo, V. Belgiorno and T. Zarra
Sanitary Environmental Engineering Division (SEED), Department of Civil Engineering,
University of Salerno, Italy

1.1 Origin and Definition

Odour is the property of a substance, or better; a mixture of substances that depending on their concentration, are capable of stimulating the olfaction sense sufficiently to trigger a sensation of odour (Brennan, 1993; Devos *et al.*, 1990; Bertoni *et al.*, 1993). Even better, odour is a sensory response to the inhalation of air containing chemicals substances. When the sensory receptors in the nose come into contact with odorous chemicals, they send a signal to the brain, which interprets the signal as an odour. The olfactory nerve cells in humans are highly sensitive instruments, capable of detecting extremely low concentrations of a wide range of odorous chemicals. The type and amount (or intensity) of odour are both important in processing the signal sent to the brain. Most odours are a complex mixture of many odorous compounds.

Fresh or clean air is usually perceived as not containing any contaminants that could cause harm and it smells clean. Clean air may contain some chemical substances with an associated odour, but these odours will usually be perceived as pleasant, such as the smell of grass or flowers. However, not everyone likes the smell of wet grass or hay. Due to our sense of odour and our emotional response to it being synthesized by our brain, different life experiences and natural variation in the population can result in people having different sensations and emotional responses to the same odorous compounds (See Section 2.5).

Odour is a parameter that cannot be physically measured, unlike wavelength for sight or pressure oscillation frequency for hearing, nor can it be chemically determined as it is not an intrinsic characteristic of the molecule. It represents, in fact, the sensation that the substance

Odour Impact Assessment Handbook, First Edition. Edited by Vincenzo Belgiorno, Vincenzo Naddeo and Tiziano Zarra.
© 2013 John Wiley & Sons, Ltd. Published 2013 by John Wiley & Sons, Ltd.

provokes after it has been interpreted by the human olfactic system. The impossibility of physically and chemically measuring odour, the complexity of the odorants, the vast range of potentially odorous substances, the physical and psychic subjectivity of odour perception and environmental factors, together with the complexity of the olfactic system, represent a series of obstacles that render the characterization of odours and the control of olfactive pollution particularly complex (Zarra *et al.*, 2007a; Dalton, 2002).

Public opinion plays a decisive role in evaluating the extent of annoyance caused by bad odours, often leading to associating unpleasant or malodorous emissions with any industrial or sanitary installation (Bertoni *et al.*, 1993; Stuetz *et al.*, 2001). In fact, even though nuisance odours are not generally associable to harmful effects on human health, they do represent a cause of undoubted and persistent annoyance for the resident population, thus becoming an element of contention both in the case of existing plants as well as in the selection of new sites (Shusterman *et al.*, 1991; Zarra, 2007b). In this light, the impacts caused by the aesthetics of the plants and their inclusion in the landscape, the noise produced, the traffic generated and, above all, the emissions of unpleasant odours are becoming increasingly important (Zarra *et al.*, 2008b).

Over the last few years, there has been more and more technical and scientific interest in these matters thanks to the greater attention being paid to protecting the environment and human health and, above all, due to the growing number of plants located in urbanized zones (Zarra, 2007b). As a result, for some time now, attention has been drawn to the need to monitor air quality in relation to environmental odour levels. However, the particular and complex nature of the substances responsible for odour impact, their variability both over time and with respect to meteoclimatic conditions and the subjective nature of olfactic perception are factors delaying any such regulation (Park and Shin, 2001; Zarra, 2007b).

As described in the following chapters, the components that can be evaluated in order to identify an olfactic type annoyance are concentration, intensity, hedonic tone (i.e. the pleasant or unpleasant sensation obtained from an odour) and quality (association of an odour with a known natural compound). As detailed later, of these components, only the first can be determined in an objective manner, while the others are highly subjective (see Part 3).

1.2   Quantifying Odour

Dynamic olfactometry, electronic noses (e-nose) and specific chemicals can be used (with varying success) to indicate the relative amount of odorous chemicals present in the air. This and other techniques for odour sampling and measurement are described in detail in Part 3.8.

Briefly, we could distinguish between sensorial, analytical and mixed methods. Sensory analysis, carried out prevalently using dynamic olfactometry, provides precise data on odour concentration, but it does not allow to evaluate the magnitude of the disturbance to which a population is exposed, nor can it determine the effective contribution of different sources to the level of environmental odour (Jiang, 1996; Sneath, 2001). The principal causes of the uncertainty of the olfactometric method are the significant biological variability in olfactic sensitivity and its inability to detect low odour concentration. Even though the introduction of criteria for the selection and behaviour coding of the panel has notably increased the repeatability and reproducibility of the measurements, the variability associated with the use of human subjects as detectors constitutes one of the principal limitations (Koster, 1985; Zarra *et al.*, 2008b).

Analytical methods (GC-MS, colorimetric methods) allow the substances present to be screened and their concentrations identified, but they do not provide information on the odorous sensation produced by the mixture as a whole (Davoli, 2004; Zarra *et al.*, 2007b; Zarra *et al.* 2008c). The analysis methods are also heavily influenced by the sampling techniques (Gostelow *et al.*, 2001) which differ according to the type of source (areal or point, active or passive type) and the actual sampling methods (see Part 5). In order to reduce problems linked to sampling, a number of recent literary works propose the use of portable GC-MS analysers (Zarra *et al.*, 2008b; Zarra *et al.*, 2008c).

1.3   Effects of Odour

Odour exposure could cause annoyance and nuisance. A more serious effect, it may lead to feelings of nausea and headache, and other symptoms that appear to be related to stress. It has been postulated that the mechanism of 'environmental worry' helps to explain the occurrence of physiological effects in people exposed to odorous substances at concentrations much lower than might be expected to lead to actual toxic effects (see Section 2.5).

Many odorous compounds are indeed toxic at high concentrations, and in extreme cases of acute exposure toxic effects such as skin, eye or nose irritation can occur. However, such effects are most likely to occur as the result of industrial accidents, such as the rupture of tanks containing toxic compounds or severe upset conditions in chemical or combustion processes.

Repeated exposure to odour can lead to a high level of annoyance, with the receiver becoming particularly sensitive to the odour. Complaints are most likely to come from individuals who are either physiologically or psychologically sensitive to the odour, and certainly a combination of both types of sensitivity will increase the likelihood of complaint. The individual components of an odour necessary to cause an adverse reaction from people are usually present in very low concentrations; far less than will cause adverse effects on physical health or impacts on any other part of the environment.

The odour threshold values for many chemicals are several orders of magnitude less than the relative threshold limit values (TLV). This means that the chemicals can be smelled at much lower concentrations than those causing adverse effects on health. Therefore, if present in sufficient quantities, these compounds would create an odour problem at much lower concentrations than would be needed to create a public health problem.

Despite these examples, it should not be assumed that odour thresholds will always be much lower than toxicological thresholds. The potential for significant adverse effects on public health from chemicals in odorous discharges should be considered on a case-by-case basis.

There is very little information available about the physiological effects of odour nuisance on humans. However, it is known that prolonged exposure to environmental odours can generate undesirable reactions in people such as unease, irritation, discomfort, anger, depression, nausea, headaches or vomiting. In our experience, other effects reported by people subjected to environmental odours can include:

- difficulty breathing;
- frustration, stress and tearfulness;
- being woken during the night by the odour;

- odour invading the house and washing;
- reduced appetite and pleasure in eating, and difficulty preparing food;
- reduced comfort at night (the need to close bedroom windows on hot nights);
- reduced amenity due to the need;
- embarrassment when visitors experience the odours;
- reduced business due to prospective customers being affected by the odour.

All these aspects are related to odour attribute and the relative response of people, discussed in Part 2.

1.4   Odour Impact Assessment Approaches

Odour impact is defined as the alteration of air quality in terms of odours that cause nuisances. An assessment of odour impacts in the environment may need to be carried out for a variety of reasons, including:

- preparing or evaluating resource consent applications, or impact assessments, for three separate categories:
  1. renewing an existing activity,
  2. proposed modifications to an existing activity (mitigation or process change),
  3. proposed new activity.
- monitoring compliance with resource consent conditions;
- investigating odour complaints to determine if an offensive or objectionable odour is present.

The methods used to assess the odours will depend on the type of situation. A number of different techniques for odour assessment are available and discussed in Part 7. The choices of the best tools to use for an odour assessment partly depend on whether the assessment is an evaluation or a compliance issue.

Evaluation involves assessing the actual and potential effects of an activity to determine whether significant adverse environmental effects will occur. If the consent is granted, the consent holder is then required to comply with (and be able to demonstrate compliance with) any conditions imposed as part of that consent.

These two processes for evaluation and compliance are quite separate, and often the evaluation criteria are different to the criteria imposed as conditions of consent.

Assessment tools can also be classified in two categories, methods with direct measurement of odour exposures or their assessment by dispersion modelling, and respectively:

1. Odour impact assessment from exposures measurement
2. Odour impact assessment from sources

All these tools with their strengths and weaknesses are discussed in Part 7, where the criteria for choosing the best one according to the specific situation are also presented.

References

Bertoni, D., Mazzali, P., and e Vignali A. (1993) *Analisi e controllo degli odori. Quaderni di Tecniche di Protezione Ambientale n.28*. Pitagora Editrice, Bologna.

Brennan, B. (1993) Odour nuisance. *Water and Waste Treatment*, **36**, 30–33.

Dalton, P. (2002) Olfaction in Yantis, in *Handbook of Experimental Psychology*. S. Stevens (ed.), Vol. 1, Sensation and Perception, 3rd edn. John Wiley & Sons, Inc., New York, pp. 691–746.

Davoli, E. (2004) *I recenti sviluppi nella caratterizzazione dell'inquinamento olfattivo.* Tutto sugli odori, Rapporti GSISR.

Devos, M., Patte, F., Rouault, S., *et al.* (1990) *Standardized Human Olfactory Thresholds*, p. 165. Oxford University Press, New York.

Gostelow, P., Parsons, S.A. and Stuetz, R.M. (2001). Odour measurements for sewage treatment works. *Water Research*, **35** (3), 579–597.

Jiang, J.K. (1996) Concentration measurement by dynamic olfactometer. *Water Environ. Technol.*, **8**, 55–58.

Koster, E.P. (1985) *Limitations Imposed on Olfactometry Measurement by the Human Factor*. Elsevier Applied Science.

Park, J.W. and Shin, H.C. (2001) Surface Emission of Landfill Gas from Solid Waste Landfill. *Atmospheric Environment*, **35** (20), 3445–3451.

Reiser, M., Zarra, T. and Belgiorno, V. (2007) Geruchsmessung mit allen Mitteln – wie aufwendig muss die Analytik von Geruchsemissionen sein? VDI Berichte 1995, 'Gerüche in der Umwelt', 13–14 Novembre 2007, Bad Kissingen (D), ISBN: 978-3-18-091995-9.

Shusterman D., Lipscomb, J., Neutra, R., and Kenneth, S. (1991). Symptom prevalence and odour-worry interaction near hazardous waste sites. *Environmental Health Perspectives*, **94**, 25–30.

Sneath, R.W. (2001) Olfactometry and the CEN Standard prEN13725, in *Odours in Wastewater Treatment: Measurement, Modelling and Control*. R. Stuetz and B.F. Frechen (eds), pp. 130–154, IWA Publishing.

Stuetz, R. and Frechen, F.B. (2001) *Odours in Wastewater Treatment: Measurement, Modelling and Control*. IWA Publishing, ISBN 1-900222-46-9.

Zarra, T. (2007b) *Procedures for detection and modelling of odours impact from sanitary environmental engineering plants*. PhD Thesis, University of Salerno, Salerno, Italy.

Zarra, T., Naddeo, V. and Belgiorno, V. (2007a) Gestione e controllo delle emissioni odorigene da impianti di compostaggio con tecniche analitiche. *ECOMONDO 2007*, pp. 73–78, Maggioli Editore, ISBN: 978-88-387-3979-X.

Zarra, T., Naddeo, V. and Belgiorno, V. (2008a) *Tecniche analitiche per la caratterizzazione delle emissioni di odori da impianti di compostaggio di rifiuti solidi urbani*. Emissioni odorigene e Impatto olfattivo. Geva Edizioni.

Zarra, T., Naddeo, V. and Belgiorno, V. (2008c) A novel tool for estimating odour emissions of composting plants in air pollution management, in stampa su. *Global Nest International Journal*, **11** (I.4), 477–486.

Zarra, T., Naddeo, V., Belgiorno, V., *et al.* (2008b) New developments in monitoring and characterization of odour emissions – at the example of a biological waste water treatment plant. *Zeitgemäße Deponietechnik*, 2008. Oldenburg GmbH, Vol. 88, ISBN 3-486-63102-0.

Part 2

Odour Characterization and Exposure Effects

V. Naddeo, V. Belgiorno and T. Zarra
Sanitary Environmental Engineering Division (SEED), Department of Civil Engineering,
University of Salerno, Italy

2.1   Attribute Descriptors

The correlation between odorous sensations and the chemical structure of the molecules that cause them is still the subject of scientific research, and in which scientists all over the world are investing considerable resources. Nowadays, the characterization of odours is based on an accurate description of the following characteristics, known also as the characterization parameters of an odour:

- concentration;
- perceptibility or threshold;
- intensity;
- diffusibility or volatility;
- quality;
- hedonic tone.

2.1.1   Concentration

The concentration of an odour generally refers to the methods with which it is quantified. When using an analytical technique, the concentration is expressed in $\mu g\ m^{-3}$ and, as it cannot be determined with reference to the entire compound, it relates to the numerical quantification of the individual substances. The sensorial technique of dynamic olfactometry, instead, expresses concentration as OU/m^3. Particularly, a gaseous sample has a concentration of 1 OU/m3 when it is at the perception threshold, that is when at least 50% of the population perceive an odour when sniffing the sample (see Section 3.4).

2.1.2   Perceptibility or Olfactive Threshold

The concentration at which an odour is just detectable to a 'typical' human nose is referred to as the 'threshold' concentration. This concept of a threshold concentration is the basis of olfactometry in which a quantitative sensory measurement is used to define the concentration of an odour. Standardized methods for measuring and reporting the detectability or concentration of an odour sample have been defined by a European standard (EN 13725:2003). The concentration at which an odour is just detectable by a panel of selected human 'sniffers' is defined as the detection threshold and as an odour concentration of 1 European odour unit per cubic metre (1 OU$_E$/m^3 or 1 OU/m^3), (see Section 3.4).

At the detectability threshold, the concentration of an odour is so low that it is not recognizable as any specific odour at all, but the presence of some, very faint, odour can be sensed when the 'sample' odour is compared to a clean, odour-free air sample.

For a simple, single odorous compound (e.g. hydrogen sulfide), the 'amount' of odour present in an air sample can be expressed in terms of ppm, ppb or in mg m^{-3} of air. More usually, odours are very complex mixtures of compounds and the concentration of the mixture can be expressed in European odour units per cubic metre (1 OU$_E$/m^3 or 1 OU/m^3).

Relating to single odorous compound, the perceptibility or olfactive threshold represents the concentration at which a substance is capable of provoking a stimulus in human beings. It varies with differences in concentration and generally three types can be defined (Centola *et al.*, 2004):

- *perceptibility or detection threshold:* represents the concentration at which the odour is detected with certainty. The threshold of detection is also defined as the concentration at which an odour just becomes strong enough to produce a sensation of odour within the controlled conditions of an odour laboratory. This value is normally indicated with OT (odour threshold). Being dependent on the subject, this value is obviously not uniquely defined. For this reason, use is made of the terms low perceptibility threshold (the smallest value of the concentration at which the odour is detected) and high perceptibility threshold (the highest value of the concentration at which the same odour is detected – OT$_{100\%}$), in other words the perceptibility threshold interval. When not indicated, as there is a variation in sensitivity between different individuals, the OT value defined in olfactometry is a statistically derived value that represents an 'average' response from 50% of selected odour panellists (OT$_{50\%}$).
- *recognition threshold:* represents the concentration relating to an odour perceived and identified (RT). Even better, the concentration at which an odour becomes recognisable, as a specific odour, is not the same as the concentration at which it is detectable. Whilst the detection threshold is the concentration at which some odour can be sensed, a higher concentration is usually required before the odour can be recognized. The RT is generally about three times the detection threshold, although this factor may be considerably higher outside the controlled environment of a laboratory. The ability to 'discriminate' one odour from another is an important attribute when describing an odour. We rely on being able

to discriminate between odours for a whole range of reasons such as fresh and stale food, the addition of flavourings and when determining the source of an odour. This is a human ability to distinguish between odours and is important when needing to identify an odour source;

- *annoyance threshold:* represents the concentration necessary to provoke a sensation of annoyance (see Section 2.5).

The olfactive threshold is also strongly influenced by the duration of the exposure as a consequence of the adaptation conditions that may be generated. In literature, it is possible to find experimentally determined concentrations corresponding to the olfactive thresholds of many pure substances. The work published by J.H. Ruth (1986) is of particular interest in this direction, with it reporting the olfactive threshold intervals from the lowest to the highest and, where available, a description of the type of odour and its annoyance concentration. These values become difficult to evaluate when considering a mixture of different substances, in that odour intensification or masking phenomena may take place. The correlations that can derive from a combination of odorous substances are essentially those of (Centola *et al.*, 2004):

- independence: $R_{AB} \leq R_A$ or R_B;
- additivity: $R_{AB} = R_A + R_B$;
- synergism: $R_{AB} > R_A + R_B$;
- antagonism: $R_{AB} < R_A + R_B$.

where R_A and R_B represent the perceptibility threshold of two pure substances, and R_{AB} is the perception threshold of the mixture obtained when the two pure substances are combined.

2.1.3   Intensity

Odour intensity is defined as the strength of the olfactive stimulus for odorant concentration values exceeding the perceptibility threshold (McGinley *et al.*, 2002). Low concentrations of some compounds in a sample are capable of being perceived as having a high intensity even when close to threshold concentrations. These compounds are common in naturally unpleasant odours such as hydrogen sulfide (rotten eggs). The interdependence of the intensity of the olfactive sensation 'I' and the odorant concentration 'C' can be described using mathematical functions (Castano *et al.*, 1992).

According to Stevens, this relation is well represented by an exponential function (see Figure 2.1) (Stuetz *et al.*, 2001):

$$I = K_s (C - C_0)^n \text{ with } C > C_0 \qquad (2.1)$$

where:

- K_s is the Stevens constant (dependent on the substance considered);
- C_0 is the odour threshold concentration (OT);
- n is a coefficient that normally varies between 0.2 and 0.8 depending on the substance considered. Its value constitutes an important indication of the effect of an eventual dilution for odour reduction, which obviously increases as n increases. For example, for n = 0.2, a dilution of times 10 reduces the olfactive intensity by a factor of 1.6, while for n = 0.8 the same dilution causes a reduction of 6.8 (Cernuschi and Torretta, 1996).

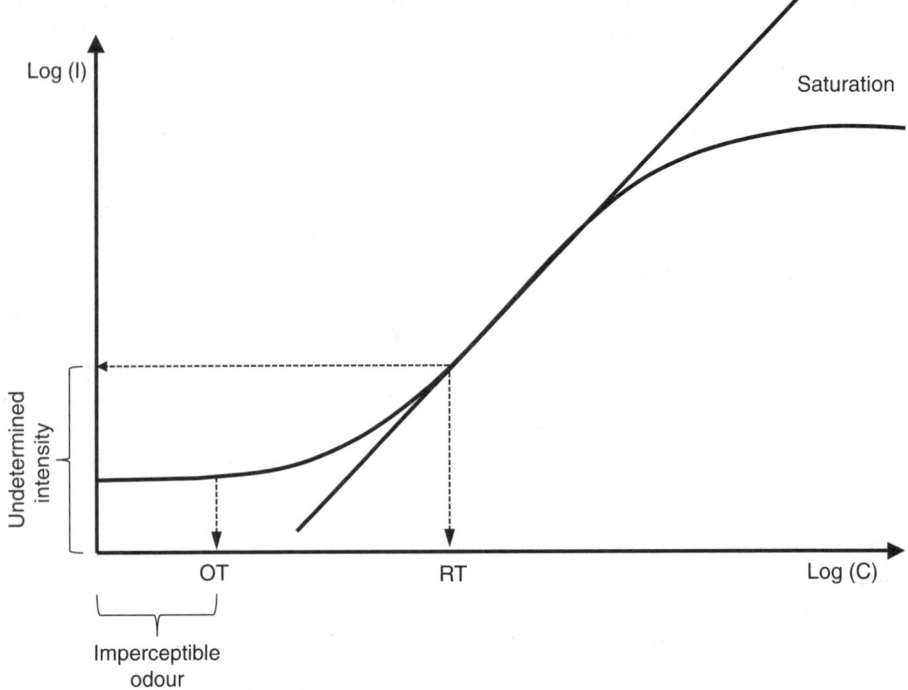

Figure 2.1 *Correlation between odour intensity (I) and concentration of odorant (C) according to Stevens (OT: odour threshold; RT: recognition threshold).*

Expressing the Stevens relation in a logarithmic form produces the following relation, which describes an increasing linear function (McGinley *et al.*, 2002):

$$\text{Log}(I) = n\,\text{Log}(C - C_0) + \text{Log}(K_s)$$

If, instead, the logarithm of the intensity is represented as a function of the dilution factor, a decreasing function is obtained, the slope of which is called persistence. From this representation, it is possible to determine how much dilution is needed in order to return a particular value of odour intensity (Centola *et al.*, 2004).

The persistence of an odour effectively represents the dose-response function (see Figure 2.2) (McGinley *et al.*, 2002).

According to Weber-Fechner, the function has a logarithmic type progression (see Figure 2.3) (Centola *et al.*, 2004):

$$I = K_w\,\text{Log}(C/C_0)$$

Where: $C > C_0$ and K_w is the Weber-Fechner constant (dependent on the substance considered).

The choice of one or other of the two formulae depends on the conditions considered. If the Weber-Fechner equation is used, representing the intensity as a function of the logarithm of the concentration, a straight line is obtained, the slope of which expresses the coefficient

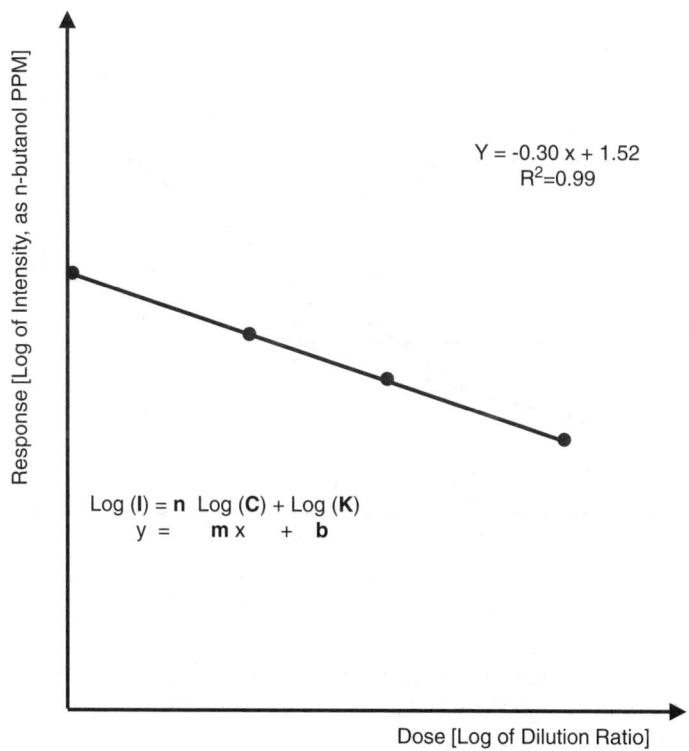

Figure 2.2   *Odour persistence for n-butanol (data from McGinley et al., 2002).*

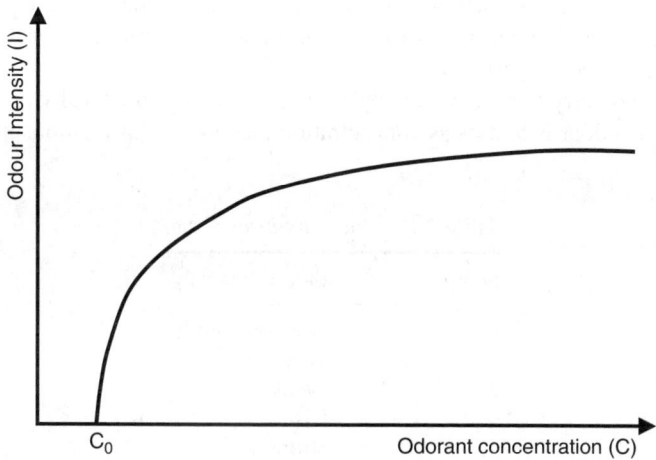

Figure 2.3   *Correlation between odour intensity (I) and odorant concentration (C) according to the Weber-Fechner model.*

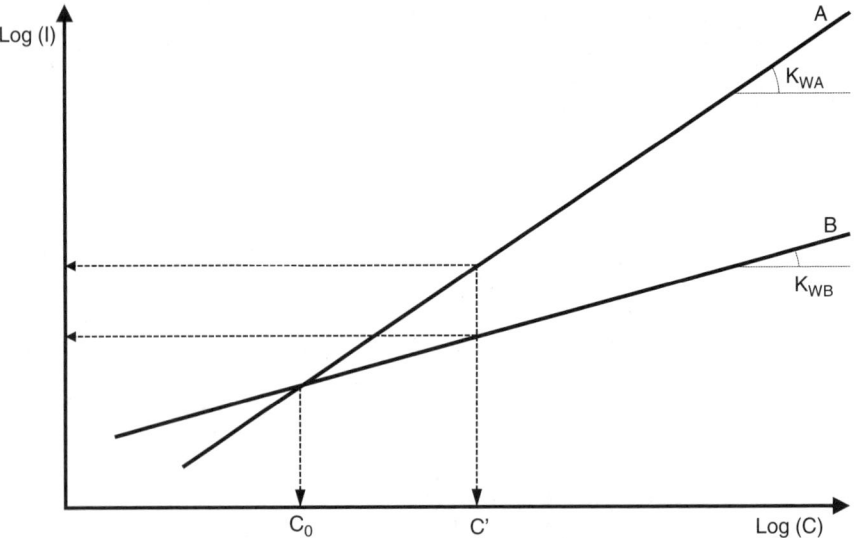

Figure 2.4   *Progress of the intensity as a function of the logarithm of the concentration.*

K_w. From this representation (see Figure 2.4), it can be seen that two different substances having the same olfactive threshold and present in the air in equal concentrations can cause an odorous sensation of very different intensity.

The intensity of the stimulus, at equal concentrations, increases as the Weber-Fechner coefficient increases (Centola *et al.*, 2004). In this case, if the logarithm of the intensity is represented as a function of the dilution factor, decreasing straight lines are obtained, and the slope of which is called persistence.

A scale of judgement (see Table 2.1) is normally used for quantifying intensity, referring to an equivalent odorous sensation of n-butanol at a known concentration (Stuetz *et al.*, 2001). In this manner, the intensity is expressed as a number to which the sensation perceived by the exposed subject corresponds.

Using a scale of very faint to extremely strong, the perceived intensity or magnitude of perception of an odour increases as concentration increases. This relationship is typically

Table 2.1   *Odour intensity scale.*

Score	Odour intensity
0	Not perceptible
1	Very weak
2	Weak
3	Distinct
4	Strong
5	Very strong
6	Extremely strong

logarithmic with concentration. However, changes in concentration do not always produce a corresponding proportional change in the odour strength as perceived by the human nose. This can be important for control purposes where an odour has a strong intensity at low concentration since even a low residual odour may cause odour problems. The method of measuring intensity is derived from the German Standard VDI 3882. Table 2.1 shows a qualitative score used by panellists for an odour sample compared to an intensity scale.

2.1.4   Diffusibility

Diffusibility is the parameter that defines the degree of volatility of odorous compounds. An odour can only be detected when a gaseous molecule manages to reach the olfactive mucus, binding itself to a receptor (olfactive cell). Volatility, therefore, is a fundamental parameter for assessing the capacity of a substance to create an odour. The diffusibility of the odour of a single substance can be evaluated by introducing a parameter called Odour Index (OI) (Centola *et al.*, 2004):

$$OI = P_{vap}/OT_{100\%}$$

where P_{vap} is the vapour tension of the substance (ppm) and $OT_{100\%}$ is the odour threshold at 100% (ppm).

Compounds with an OI of less than 10^5 are said to be slightly odorous (for example, alkanes and alcohols of low molecular weight), while odorants with higher OI values are mercaptans and sulfurs (up to 10^9) (Laraia *et al.*, 2003). Table 2.2 reports the values of the odour indices (OI) for some odorous substances.

2.1.5   Quality or Character

The quality of an odour defines its specific character. This attribute is expressed in terms of 'descriptors', for example; 'fruity', 'medical', 'fishy'. This represents an important aspect in that it allows identification of the 'type' of odour and provides, as a result, a means of 'cataloguing'. Cataloguing, however, is made difficult by the inherent subjectivity of the olfactive sensation.

The most reproducible results are obtained using the similarity evaluation technique, offering the subject a comparison term with which to associate the odour to define. For

Table 2.2   *Odour indices of some odorous substances (Lisovac and Shooter, 2003).*

Odorant substance	P_{vap} [kPa at 25°C]	$OT_{100\%}$ [ppm]	OI at 25°C
Ethanol	7.872	6000	13
Toluene	3.79	40	947
Acetone	30.8	300	1 030
2-butanone	12.6	30	4 200
Carbon disulfide	48.2	0.21	2 295 000
Dimethyl disulfide	3.82	0.0014	27 280 000
Dimethylsulfide	64.4	0.0014	460 000 000
Hydrogen sulfide	2020.0	0.0047	4 297 870 000

Table 2.3   *Quality classification according to Wise et al. (2000).*

Quality class	Description
1	fragrant
2	acid
3	burnt
4	caprylic

example, according to the Crocker and Henderson theory, an odour is evaluated by comparing it to four primary odours to which a value of between 0 and 8 is assigned, the various combinations of which allow all the others to be obtained (see Table 2.3).

Another example of odour quality classification is that proposed by Anderson through the definition of an 'odour wheel' (Figure 2.5), in which odour is divided into eight categories (floral, fruity, vegetable, earthy, offensive, fishy, chemical and medicinal).

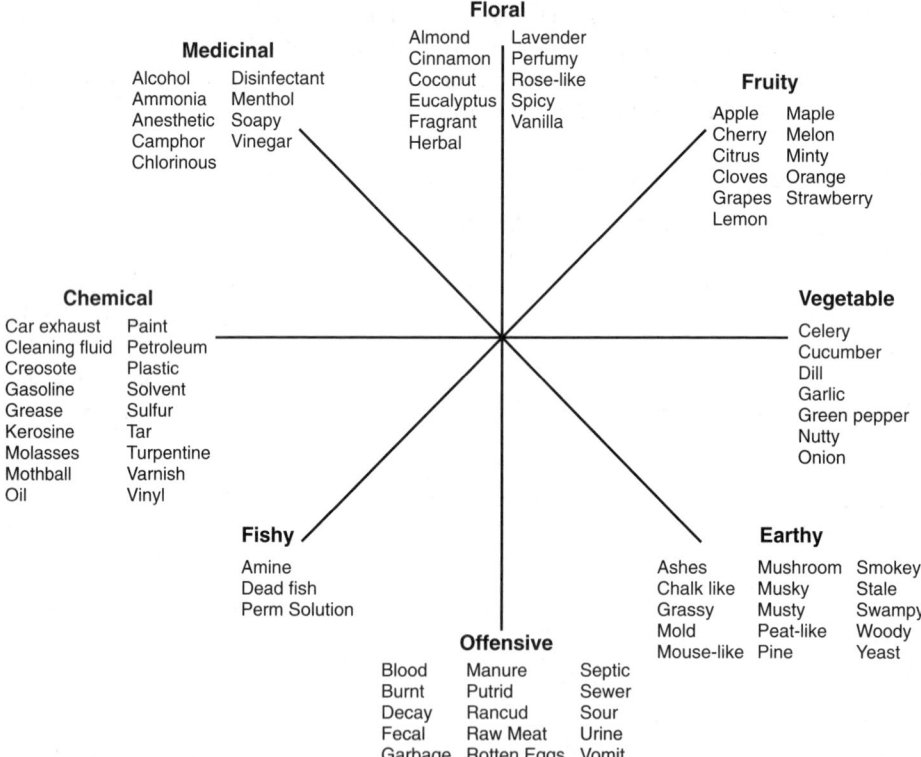

Figure 2.5   *Odour descriptors wheel (Reprinted from McGinley et al. (2002) Copyright (2002) McGinley Associates, PA).*

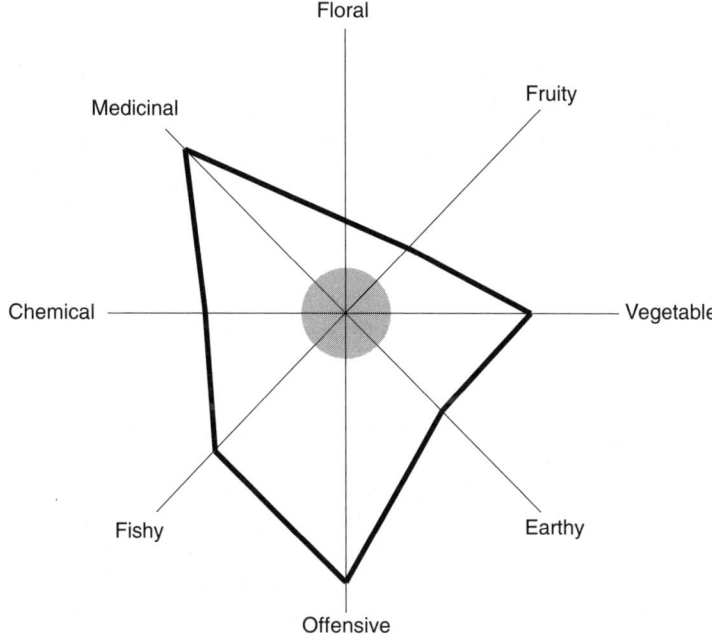

Figure 2.6   *Example of a graph describing the quality of an odour.*

By using this instrument and attributing a value from 0 to 5 to each descriptor in relation to the intensity, it is possible to obtain a spider graph that defines the quality of the odour (see Figure 2.6).

Odour quality is useful in establishing an odour source from complainants' descriptions. Alternatively, it may be possible to identify key chemical components by a description of the specific odour.

2.1.6   Hedonic Tone or Offensiveness

Hedonic tone is the parameter that defines the pleasantness or unpleasantness of an odour and is, therefore, a measure of its acceptability (Stuetz *et al.*, 2001). Importantly, the hedonic tone can be responsible for the perception leading to complaint. Here, the relative pleasantness or unpleasantness of the odour alongside the association of its source, or the context in which it is received is relevant to investigating odour complaints.

As with most odour characterization parameters, the definition of hedonic tone also involves a certain degree of subjectivity due to a number of factors such as, for example, experience or the circumstances of the individual.

The quantification of hedonic tone also makes use of the judgement scale. This judgement on the relative pleasantness or unpleasantness of an odour forms our common language when reporting unpleasant odours. Methods to make comparative judgements for such subjective reports have been established for assessors to analyse samples as part of an

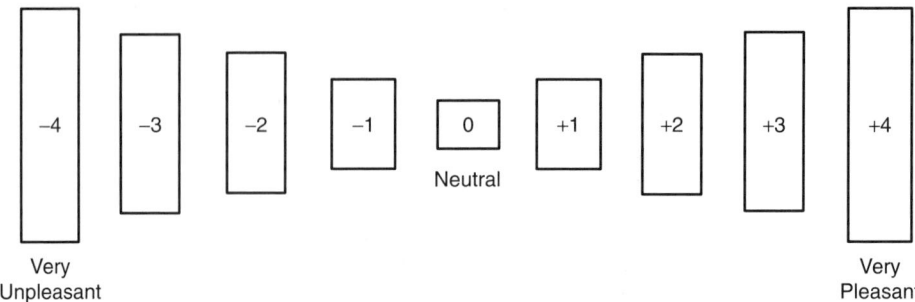

Figure 2.7   *Nine-level scale of hedonic tone.*

odour panel. A method for measuring hedonic tone is suggested next, derived from the German guideline VDI 3882.

One example, reported in Figure 2.7, uses a nine-level classification, ranging from −4 (extremely unpleasant odour) to +4 (extremely pleasant odour). It should be remembered, however, that in order to evaluate the acceptability of an odour, simply referring to its hedonic tone is not sufficient, in that even exposure to pleasant odours can alter the psychophysical equilibrium of a person and have a negative influence on his behaviour, above all if the odours are very intense and long-lasting (Nimmermark, 2011).

2.2   Chemistry and Odours

Odours in environmental engineering plants essentially originate from the degradation mechanisms of organic substances as well as from liquid to aeriform stripping. The causes of these odour formation phenomena can be natural, intrinsic or related to the effect of the treatment processes to which solid and liquid waste (as well as sewage) is subjected to in a more or less controlled manner. The principal chemical-physical properties that come into play during the formation of odours resulting from the passage from the liquid to the aeriform phase are vapour pressure and water solubility, while as far as degradation of organic substances is concerned, the phenomenon of chemical and/or biological oxidation merits particular attention.

2.2.1   Vapour Pressure

All liquids are characterized by the tendency to pass to the vapour state (evaporation), and this tendency is more pronounced the higher the temperature (i.e. the higher the kinetic energy of the constituent molecules). In an open system with a continuous supply of external heat, all the liquid will eventually be transformed into vapour. In a closed system, at a certain point a state of dynamic equilibrium is established between the quantity of liquid molecules passing to the vapour state and the quantity of vapour molecules condensing to the liquid state, in the sense that uniformity is reached in the velocities of the two processes. In this situation, the quantity of vapour overlying the liquid is the maximum compatible with the temperature conditions of the system and, consequently, it is said to be in the presence

Table 2.4   *Molecular weight and vapour pressure of some odorous compounds (Rafson, 1998).*

Odorant substance	Molecular weight [g mol^{-1}]	Vapour pressure [mmHg]
Acetaldehyde	44.05	870.0
Acetone	58.08	266.0
Benzene	78.11	95.2
Dichlorobenzene	147.00	1.2
Ethanol	46.07	50.0
Formaldehyde	30.26	3500.0
Hydrazine	32.04	14.4
Hydrogen sulfide	34.08	15 200.0
Methanol	32.04	114.0
Octane	114.23	17.0
Phenol	94.11	0.341
Styrene	104.15	7.3
Toluene	92.14	30.0

of saturated vapour. The pressure exerted by the vapour in equilibrium with its liquid is called vapour pressure or tension and is normally expressed in mmHg. Vapour pressure is normally an indication of the liquid/gas equilibrium of a pure compound. It does not, however, provide a full indication if the compound in question is dissolved in another (water). Table 2.4 shows values of vapour pressure (at 25°C) for some odorous compounds mostly present in the various environmental health engineering plants.

The vapour tension of a liquid increases as the temperature increases: when, in an open system, the vapour tension is equal to atmospheric pressure, the liquid boils. In general, therefore, the boiling point of a liquid corresponds to the temperature at which its vapour tension is equal to the pressure present on the surface of the liquid. When a liquid is made to boil at a given external pressure, its temperature can no longer increase.

The relation between temperature and vapour pressure is supplied by Antoine's equation:

$$\text{Ln}\,(P_0) = A - B/(T + C) \tag{2.2}$$

where:

- P_0 is vapour pressure (mmHg);
- T is Temperature (°C)
- A, B, C are specific constants.

2.2.2   Water Solubility

Water solubility is defined as the concentration of a compound in liquid phase in equilibrium with its pure phase. According to Henry's law, the quantity of a gas dissolved in a liquid is directly proportional to the partial pressure that the gas exerts on the surface of the liquid itself (Rafson, 1998; Stuetz *et al.*, 2001):

$$X_G = H_C\,X_L \tag{2.3}$$

Table 2.5   *Henry coefficient of the principal odorous compounds (Hvitved et al., 2001).*

Class	Odorant	Boiling point at 1 atm [°C]	Henry coefficient [atm mol^{-1}]
Sulfur compounds	Hydrogen sulfide	−59.6	563
	Methyl mercapano	6	200
	Ethyl mercaptan	35	200
	Dimethylsulfide	37	110
	Dimethyl disulfide	110	64
Nitrogen compounds	Ammonia	−33.4	0.842
	Methylamine	−6.4	0.55
	Ethylamine	17	0.55
	Dimethylamine	7	1.3
	Indole	254	–
	Skatole	265	–
Volatile fatty acids	Acids acetic acid	118	0.063
	Butyric acid	162	0.03
	Valeric acid	185	0.025
Aldehydes and ketones	Acetaldeide	21	5.88
	Butiraldeide	76	6.3
	Acetone	56	1.9
	Butanone	80	2.8

where:

- X_G is the concentration of the compound in gaseous phase (or partial pressure or vapour pressure);
- X_L is the concentration of the compound in liquid phase;
- H_C is the Henry's Law coefficient.

This state of equilibrium persists as long as the external partial pressure of the gas remains unaltered: when it decreases, the liquid is said to be in an oversaturated condition and the gas is freed, returning to the outside until a new equilibrium condition is reached.

The Henry coefficient can be expressed in various units of measurement (atmospheres/mole fraction, atmospheres/mole/m^3, etc.) and is dependent on temperature. Compounds that have a high Henry coefficient are those possessing high vapour pressure and low solubility. Generally speaking, compounds with a Henry coefficient higher than 0.01 are considered volatile; those with a Henry coefficient between 0.0001 and 0.01 are considered semivolatile; those with a Henry coefficient of less than 0.0001 are considered, nonvolatile. Table 2.5 reports the Henry coefficients for some odorous compounds.

2.2.3   Chemical and Biological Oxidation

The oxidation-reduction phenomenon occurs between a pair of substances when one of the two, which oxidizes, loses one or more electrons, while the other, which reduces, acquires one or more electrons. The reducing substance par-excellence is hydrogen, the oxidizing substance par-excellence is oxygen; a substance that oxidizes losing electrons frees energy, while a substance that reduces acquires energy. In the organic field, almost all the biological reactions can be considered as oxide-reduction reactions.

When a substance loses electrons, it generates an electric current. Consequently, substances that tend to oxidize possess a positive electric potential. The electric potential present in a system can, therefore, quantitatively represent the tendency to oxidize or be oxidized. In practice, the absolute magnitude of this potential is not measured, whereas the difference between it and the potential of a reference system is. This difference, which is expressed in millivolts and indicated by 'Eh', is called oxide-reduction potential or Redox potential (influenced by temperature and pH). As a reference system, use is made of the same electrode as that used for the potentiometric measurement of pH. The system supplied by the hydrogen electrode has mild reducing properties and an Eh of approximately +400 mV, which can therefore be considered as a neutral Redox potential.

Examples characterized by chemical type oxide-reduction include the oxidation of H_2S to sulfur or sulfate by chemical oxidizers such as air, oxygen, hydrogen peroxide, sodium hypochlorite, chlorine, potassium permanganate. The reaction speed is a function of the oxide-reduction potential of the reaction itself.

Many compounds can undergo biological oxidation (or reduction). The reactions involved are quite complex and their success depends heavily on the environmental conditions (levels of oxygen and nutrients, distribution of species, temperature, presence of food, presence of toxic elements).

There are different mathematical models that represent the kinetics of the biological reactions. One of the most well-known is the *Monod model* (Rafson, 1998):

$$R_b = (-k\, X_a\, C_1\, V)/(K_s + C_1) \qquad (2.4)$$

where:

- R_b is the rate of biodegradation (mg s^{-1});
- k represents the maximum substrate utilization (mg mgSSV^{-1} s^{-1});
- X_a is the concentration of active cells (mgSSV m^{-3});
- V is the volume of the reactor (m^{-3});
- K_S is the Monod constant (mg m^{-3}).

2.3   Odorous Compounds, Thresholds and Sources

The production activities from where unpleasant odours originate belong to a number of industrial sectors and can be schematically, though not exclusively, grouped according to the type of activity:

- *agricultural and food production sector* (livestock breeding, slaughterhouses, sludge drying, sewage and manure, oil mills, wine industry, etc.);
- *industrial sector* (chemical industry, fertilizer production plants, oil and gas refineries, foundries, plants producing plastic raw materials, paper and cardboard factories, primer and paint manufacturers, fibre or textile dyeing plants, tanneries, etc.); and

Table 2.6   *Description of the principal groups of odorous substances and principal odorous compounds (adapted from Bertoni et al., 1993; Rafson, 1998).*

Class	Description	Odorous substances
Sulfur compounds	The sulfur compounds include both inorganic compounds (such as H_2S and SO_2), and volatile organic compounds with at least one –SH group. Odorants are more suitable to indicate the presence of anaerobic conditions.	Hydrogen sulfide Carbon disulfide Dimethyl-sulfide Dimethyl-disulfide Dimethyl-trisulfide Diethyl-sulfide Ethyl-mercaptan Methyl-mercaptan Ethyl-mercaptan
Nitrogen compounds	Derive from the splitting (deamination) of the amino acids under anaerobic conditions by many bacteria. Are also formed at low pH, during the anaerobic fermentation of compounds which contain nitrogen.	Ammonia Methyl-amine Dimethyl-amine Trimethyl-amine Ethyl-amine Satolo Indole Pyridine Putrescine Cadaverine Acrylonitrile
Volatile fatty acids	Resulting from incomplete oxidation of lipids which, in conditions of oxygen deficiency, does not allow the complete oxidation to CO_2. They have, in general, a very low olfactory threshold.	Formic acid Acetic acid Propionic acid Butyric acid Valeric acid Acid isovalerianic
Ketones	Are organic compounds of empirical formula $C_nH_{2n}O$ which bear in their structure the group conventionally designated by –CO–.	Acetone Butanone 2-pentanon
Aldehydes	They are organic compounds of empirical formula $C_2H_{2n}O$ that bear in their structure conventionally designated by the group –CHO.	Acetaldehyde Formaldehyde Butyraldehyde Isobutialdeide Valeraldeide
Terpenes	Hydrocarbons of the general formula C_5H_8 that originate from the condensation of metabolic glucose to 5 carbon atoms, through the intermediate formation of benzene compounds and their hydrogenation.	α-pinene β-pinene camfene limonene γ-terpinene

Table 2.6   (Continued)

Class	Description	Odorous substances
Phenols and alcohols	Phenols are substances derived from aromatic hydrocarbons by substitution of one or more hydrogen atoms with hydroxyl OH. The alcohols are organic compounds in which a hydrogen atom is replaced by a hydroxyl group OH having the formula $C_nH_{(2n+2)}O$	Phenol Acrolein 2-Methil-2-butanol
Aromatic hydrocarbons	Are characterized from the ring hexagon of benzene C_6H_6.	Xylene Toluene Benzothiazole Benzene ethylbenzene etiltoluene 1-3-5 trimethyl-benzene 2-4-6 trimethyl-benzene 1-methyl-ethinyl-benzene

- *public utilities sector* (hazardous and non-hazardous waste treatment and recycling plants, rubbish tips, composting plants, sewage treatment works, waste to energy plants, etc.).

The substances that give rise to the diffusion of odours in areas surrounding a production plant can consist of products of an inorganic nature or volatile organic compounds (Agostinelli, 2005).

Although the described sources, in the majority of cases, produce odorants with a highly complex composition, the odorous compounds generally present in the emissions can be easily identified.

Table 2.6 reports the principal compounds, by class of substance, possessing odorous characteristics generally identifiable in the various emissions, as deduced from various research publications.

Table 2.7 instead provides a summary of the principal odorous compounds detectable at environmental health engineering plants, considered as being predominantly responsible for annoyance.

At sewage treatment works, waste disposal sites, fuel from waste plants and composting plants, the production of malodours is principally due to the anaerobiosis conditions that can be generated in some zones or phases of the treatment process, as well as during the microbial digestion of the organic matter (Koe 1989; Serra and Dugani, 1988; Stuetz *et al.*, 2001). In the case of waste to energy plants, emissions from fumes discharged from the chimney stacks, resulting from the degradation of organic substances, can be added to these causes (Azzeri, 1997).

Table 2.7 *Odorous compounds emitted by environmental plants (Stuetz et al., 2001; Frechen, 2001; Gostelow and Parsons, 2000; Gostelow et al., 2001; Azzeri, 1997).*

Source	Odorous compounds
Wastewater treatment plants	Hydrogen sulfide, mercaptans, sulfur compounds, ammonia, amines, pyridine, scatol, indole, Volatile fatty acids, aldehydes and ketones.
Landfills	Ammonia, amines, mercaptans, hydrogen sulfide, organic sulfides, disulfides, volatile fatty acids, aldehydes, ketones, aromatic hydrocarbons, limonene, alcohols, terpenes.
Composting plants	Ethanol, limonene, cadaverine, putrescine, hydrogen sulfide, methyl-mercaptan, butyric acid, acetic acid, acetaldehyde, trimethylamine, dimethyl disulfide.
Solid waste incinerators	Acetone, acetic acid, benzene, methane, hydrogen sulfide, ammonia, carbon disulfide, toluene, mercaptans.

2.4   Public Health Relevance of Odour Exposure

The study of the toxicity of odorous substances involves the investigation of the effects as a function of the concentration.

In the case of working environments, reference is usually made to the Threshold Limit Value (TLV), a parameter indicating the maximum concentration to which a worker can be exposed during his working life (8 hours per day, 5 days per week, 50 weeks per year) without experiencing any pathogenic effects. Particularly in order to analyse the potential damage generated by odorous substances, the ratio between the olfactive perceptibility threshold value (OT) and the TLV is calculated (Davoli *et al.*, 2001): substances with a ratio of less than 1 will be perceived prior to encountering their toxic effects and as such prior to causing possible harm, vice versa for the other substances (Table 2.8).

The use of TLV values, however, is not immune to some criticism. These values, in fact, are obtained assuming a healthy worker in a standard situation, while it is well-known that in a working environment, workers are exposed to various toxic substances simultaneously and continuously. It should also be pointed out that TLV values are not one hundred per cent reliable in that they need continuous updating following the discovery of new previously unknown effects. A typical example of these latter is the case of ozone: in just a few years, the limit has been reduced by one order of magnitude (Serra and Dugnani, 1998).

Another parameter normally used for characterizing the harmful effects of a substance is the Maximum Allowable Concentration (MAC), understood as being the concentration value that can never be exceeded, not even for a brief period of time. In reality, there is a great deal of confusion between MAC and TLV, as a result of which, in literature, the terms *mean MAC* and *peak MAC* have been introduced. The term mean MAC, particularly reflects the same concept as the TLV, while the definition of peak MAC remains the one previously proposed (Serra and Dugnani, 1998).

Table 2.9 reports the effects on man caused by exposure to increasing concentrations of H_2S. It is interesting to note that at concentrations approaching the lethal limits (>700 ppm)

Table 2.8 Ratio between OT and TLV for some odorous compounds (Serra and Dugnani, 1998).

Odorous compound	TLV [mg m^{-3}]	OT [mg m^{-3}]	OT/TLV
Acetaldehyde	180	1.8	0.01
Acetone	2400	240	0.1
Acetic acid	25	6.5	0.26
Acrolein	0.25	0.49	1.9
Ammonia	18	37	2.05
Benzene	32	4.5	0.14
Etilmercaptan	1.25	0.03	0.02
Phenol	19	1.18	0.06
Formaldehyde	3	1.2	0.4
Hydrogen sulfide	14	0.03	0.002
Methyl mercaptan	1	0.08	0.08
Toluene	535	5.89	0.01

H_2S produces an odour that is not entirely unpleasant, and it is actually due to this that many serious incidents have been recorded involving workers in sewerage systems.

As far as principal osmogenic substances are concerned, the main harmful effects that these have on humans are summarized next (Vincent, 2001; Davoli *et al.*, 2001):

- *mercaptans*: these are characterized by a particularly unpleasant odour that provokes intolerable gastric effects even with low exposure times. With longer exposure times, these compounds can also interfere with blood haemoglobin and consequently with the oxygen transport process, causing temporary cyanosis;
- *hydrogen sulfide*: the effects of exposure range from irritation of the eyes and respiratory tract, for concentrations between 10 and 20 ppm, up to immediate loss of consciousness and death (1000–2000 ppm). The particular and hazardous nature of this compound resides in the fact that, at particularly dangerous concentrations (>700 ppm), it loses its malodorous compound characteristics giving rise to an almost pleasant odour;

Table 2.9 Effects of hydrogen sulfide at different concentrations (Serra and Dugnani, 1988).

Effects	H_2S [ppm]
Eye irritation	10
Respiratory irritation	20
Occurrence of mild symptoms after several hours of exposure	70–150
Maximum acceptable concentration without severe symptoms after an hour of exposure	170–300
Bronchopneumonia or pulmonary oedema after prolonged exposure	250–600
Occurrence of severe symptoms after an exposure of 30 min to 1 h	400–700
Sudden loss of consciousness and coma	700–900
Immediate loss of consciousness, apnoea, and death within minutes	1000–2000

- *ammonia*: is the cause of irritation of the bronchi and lungs, while prolonged exposure to low concentrations can provoke chronic bronchitis or emphysema;
- *amines*: irritant effects have been encountered on the mucus of the primary respiratory tract, though possible irritation of the eyes with subsequent corneal damage cannot be ruled out;
- *ketones*: at low concentrations, these cause irritation of the eyes and nasal mucus, while high concentrations can cause damage to the central nervous system;
- *aldehydes*: these is general have an irritation effect on the eyes and on the mucus of the respiratory tract;
- *organic acids*: although these do not lead to any pathogenic effects at low concentrations, prolonged exposure can cause irritation of the respiratory tract.

From these basic notions, it can be seen how osmogenic compounds are potentially capable of causing serious pathogenic phenomena. However, in real life, these effects are, in the first place, alleviated by the presence of a normally very low olfactic threshold and as such immediately perceivable, prompting the person to automatically distance himself, thus avoiding exposure times and concentrations that could lead to experiencing the most serious effects reported, and in the second place by the normally low concentrations emitted by environmental protection plants. It is for this reason that odorous compounds from environmental protection plants are not so associable to toxicological risks as they are to stress factors. Odours, in fact, have a highly emotional nature, in that they possess the power to stimulate the imagination and recall even far off memories, leading to behavioural type reactions (Bertoni *et al.*, 1993; Kehoe *et al.*, 1996). The entire nervous system is influenced by odours; the effects on heartbeat, respiration and on other reflexes are well recognized. It has also been demonstrated that certain odours perform the action of potent stimulators (Bertoni *et al.*, 1993).

The unpredictability of the disturbance, its continuing presence over time and the impossibility of defending against odours also create a negative synergetic effect at a psychological level, generating tension and states of anxiety, with the resulting protests by citizens. Odours are quite often considered as warning signals: for example, the odour of smoke is associated with the idea of burning or a fire, while gas is associated with a leak.

In the collective imagination, bad odours are linked to conditions of air 'insalubrity'. In fact, on numerous occasions, more importance is attributed to bad odours than to more hazardous pollutants not directly perceivable by our senses.

2.5   Odour Annoyance and Nuisance

Health, according to the well-known definition of the World Health Organization (WHO), is not only absence of disease but a state of complete physical, mental and social wellbeing. Disturbed wellbeing in general is nonspecific, that is, without reference to a causative condition, which is typically attributed subjectively.

If, in an environmental context, adverse psychological states bear a demonstrable or at least plausible relation to an external factor, such as noise, odour or dust, the terms annoyance or nuisance are used to characterize this psychological adversity.

The term odour annoyance will be used for adverse psychological effects following odour exposure. According to Van Harreveld (2001), annoyance is the complex of human reactions that occurs as a result of an immediate exposure to an ambient stressor (odour) that, once perceived, causes negative cognitive appraisal that requires a degree of coping.

Annoyance potential is the attribute of a specific odour (or mixture of odorants) to cause a negative appraisal in humans that requires coping behaviour when perceived as an ambient odour in the living environment (Van Harreveld, 2001). It is an attribute of an odour that can cause annoyance or nuisance. Annoyance potential indicates the magnitude of the ability of a specific odorant (mixture), relative to other odorants (mixtures), to cause annoyance in humans when repeatedly exposed in the living environment to odours classified as 'weak' to 'distinct odour' on the perceived intensity scale (VDI 3882: 1997).

Nuisance is the cumulative effect on humans, caused by repeated events of annoyance over an extended period of time that leads to modified or altered behaviour (Van Harreveld, 2001). This behaviour can be active (e.g. registering complaints, closing windows, keeping 'odour diaries', avoiding use of the garden) or passive (only made visible by different behaviour in test situations, e.g. responding to questionnaires or different responses in interviews). Odour nuisance can have a detrimental effect on human sense of well-being, and hence a negative effect on health. Nuisance occurs when people are affected by an odour they can perceive in their living environment (home, work environment, recreation environment) and:

- the perception of the smell is negative;
- the appraisal is repeatedly;
- it is difficult don't perceive odours; and
- the smell is connected (often incorrectly) to a negative effect on health.

According to the definition of *Nuisance*, the Nuisance potential is the characteristic of an exposure situation, which describes the magnitude of the nuisance that can be expected in a community when exposed to an odour intermittently, but over an extended period of time, in their living environment (Van Harrevel, 2001). Nuisance potential is a function of many factors, such as the attributes of the odorant (mixture) in question, the frequency and dynamics of variation of the exposure (caused both at source and as a result of atmospheric dispersion) and attributes of the specific population that is exposed (Van Harreveld, 2001).

Nuisance sensitivity is an attribute of a specific community (or an individual) that indicates the propensity, relative to that of other individuals or populations, to experience nuisance when exposed to an odour intermittently, but over an extended period of time, in their living environment (Van Harreveld, 2001).

Studies of environmental exposure to odour at differing concentrations over differing periods of time have led to a series of conclusions about the way in which individuals perceive odour, and how this is established and then retained in memory. Surveys of communities show that where an odour nuisance is abated, the perception of odour impact is reported for prolonged periods by those living in the area, even years, after the odour is no longer present. It is evident that:

- the nuisance suffered is not caused by short-term exposure to environmental odours and similarly not reduced by short periods of mitigation or prevention;

- the association between an individual's perception and experience of nuisance from an odour is persistent and prolonged. For these individuals, exposure to the same odour at lower concentrations causes greater nuisance than for others with no history of exposure;
- the perception of annoyance/nuisance appears to be cumulative, developing over long periods of time. Memory of periods of heightened or intense exposures alongside other unwanted outcomes such as the disturbance to wellbeing or lack of influence are all important. These appear to dominate the overall perception of the odour impact and perceived history of the complaint.

All these aspects are strongly connected to factors contributing to odour impacts and are discussed in Section 7.2.

2.5.1 Odour Exposure

The term odour exposure is used here to simplify the complexity of exposure of residents in the vicinity of odour sources. If fluctuating concentrations of odour emissions in the field exceed the odour threshold, thus resulting in odour perception, this does not necessarily induce an adverse effect in terms of odour annoyance.

Individual responses to odour vary greatly and not all unpleasant odours are considered offensive at all times. Examples of this are well established where communities have become accustomed to 'wood smoke' or 'wet grass'. Equally, these same odours can trigger complaints and can impact upon people's daily lives where exposure to 'manure odours' or 'organic solid waste' is perceived as unwanted and objectionable. A feature of these differences amongst humans is the phenomenal range of choice in foods, perfumes and products linked to olfaction that are available and continue to be developed.

The formation of odour annoyance can be best understood, if odour exposure is characterized using the FIDOL factors (Dalton, 2002). When an individual exposed to odour perceives this as unwanted, it is argued that the following factors are the main determinants:

- Frequency of the odour exposure;
- Intensity of the odour;
- Duration of exposure to the odour;
- Offensiveness of the odour;
- Tolerance and expectation of the exposed subjects (location).

The importance of duration has mainly or exclusively been studied within the process of olfactory adaptation, namely the temporary decrease of olfactory sensitivity due to continued olfactory stimulation. The remaining factors frequency, intensity and offensiveness have received considerable attention in field studies. Observed odour frequency or calculated odour concentrations in the vicinity of odour sources are used as odour exposure measures and related to odour annoyance. This relationship is modified by the offensiveness degree of environmental odours (e.g. industrial, agricultural, waste-water), that is, their position on the pleasantness-unpleasantness dimension, also often termed hedonic odour-quality. All these factors will be discussed in greater detail in Section 7.2.

2.5.2   People Response

Our reactions to odour can be short-term or prolonged, as well as intense or mild in the same way as the exposure and unpleasantness of the sensation. Studies of communities exposed to unwanted odours show that exposure can lead to evidence of stress induced symptoms such as sleep disorders, headaches, respiratory problems, nausea and anxiety as well as less extreme but equally prolonged complaints, but learned responses may play a role in the impairment of mood. If exposure to odours with negative appraisal occurs repeatedly, this can affect our wellbeing and cause stress related symptoms, that is, a public health concern.

When exposed to odours that are then perceived as unwanted or unpleasant, these cause us to have a 'negative appraisal' of our local environment. This effect is regarded as an 'environmental or ambient stressor' in just the same way as other environmental stressors such as noise or unwanted lighting. When exposed to such a stressor, the individual requires some form of coping behaviour to respond and adjust.

It is becoming increasingly clear that certain communities can become sensitized to odours. Triggers for such sensitization include one-off or rare but highly adverse events that permanently change people's perceptions about the odour. When a community displays signs of sensitization, common features are a high level of complaints over the long term, and a general mistrust of those responsible for the perceived source of the odour.

To better understand the nature of an individual's response to odours, it is useful to understand two processes that occur in all sensory systems: adaptation and sensitization.

Adaptation is a reduction in responsiveness (a decrease in perceived odour intensity) during or following repeated exposure. Adaptation can occur on either a short-term or long-term basis. Short-term adaptation primarily occurs as a result of olfactory fatigue. Long-term adaptation results in a more persistent reduction in response, which can be measured in hours or even days following exposure and can account for situations where people who work in odorous environments cannot comprehend complaints from neighbours who only receive intermittent odours (Schauberger, 2001). Conversely, sensitization results in increased responsiveness during or after exposure. Individuals who may not be particularly sensitive to odours may become sensitized through acute exposure events, or as a result of repeated exposure to nuisance levels of odours. Often symptoms such as headaches, nausea, throat irritations and sleeplessness are reported at exposure levels barely exceeding the odour threshold.

2.5.3   Sensitivity of Receptors

Odour annoyance and nuisances are then strongly related to the sensitivity of the receptors. Generally, we could classify the sensitivity of receptors according to their localization in the land use plan. Naturally, many different types of land use and location can occur in immediate proximity to an odorous activity. In any case, the land uses can be grouped into three degrees of sensitivity:

- High sensitivity:
  - residential/living (high-density residential)
  - light commercial/retail/business/education/institutional
  - open space/recreational
  - tourist/conservation/cultural

- Moderate sensitivity:
  - rural residential (low-density residential)
  - light industrial
- Low sensitivity:
  - rural land
  - heavy industrial

These categories should be regarded as guidelines only, as there will always be exceptions. Particularly they should be used in accordance with amenity values defined in district plans for various land-use zones (see Part 6).

As already mentioned, the sensitivity of the receiving environment is also dependent on the offensiveness of the odour relative to the location in question, particularly if that odour is new to the area. For example, an odorous industrial chemical in a rural environment could turn that environment into a sensitive receiving environment, simply because it is unacceptable even to the very few people living in the area. People living and working in an industrial environment might not be as sensitive to that same chemical. Generally, the sensitivity of the receiving environment is dependent on the experiences and expectations of the people already in that environment, and the odours they are currently experiencing.

References

Agostinelli, S. (2005) Inquinamento odorigeno. *Regioni e Ambiente*, **6**, pp. 29–31.

Azzeri, R. (1997) La Rimozione Biologica degli Inquinanti Gassosi. *Biologia Ambientale*, **6**, pp. 24–34.

Bertoni, D., Mazzali, P. and Vignali, A. (1993) Analisi e controllo degli odori. *Quaderni di Tecniche di Protezione Ambientale*, Vol. 28, Pitagora Editrice Bologna.

Castano, P., Cocco, L., De Barbieri, A., *et al.* (1992) *Anatomia Umana*, Edi-Ermes, Milano, ISBN 9788870510249.

Centola, P., Sironi, S., Capelli, L. and Del Rosso, R. (2004) *Valutazione di Impatto Odorigeno di Una Realtà Industriale*. AIDIC Servizi Srl.

Cernuschi, S. and Torretta, V. (1996) Processi e tecnologie impiantistiche per il controllo degli odori negli impianti di trattamento delle acque di scarico, *IA Ingegneria Ambientale*, XXV (5), Milano, CIPA Editore, pp. 248–264.

Dalton, P. (2002) *Olfaction*, in S. Yantis and H. Pashler (eds), *Steven's Handbook of Experimental Psychology, Vol. 1, Sensation and Perception*, 3rd edn. John Wiley and Sons, Inc., New York, pp. 691–746.

Davoli, E., Rotilio, D. and Desiderio, M. (2001) Campionamento e speciazione degli odori. *Quaderni della Ricerca*, **74**, Regione Lombardia, Centro di Salute Ambientale 'G. Paolne', Consorzio Mario Negri Sud, pp. 37.

Frechen, F.B. (2001) Regulations and policies, in *Odours in Wastewater Treatment: Measurement, Modelling and Control*. (eds. R. Stuetz and F.B. Frechen), IWA Publishing, pp. 16–30.

Gostelow, P. and Parsons, S.A. (2000) Sewage treatment works odour measurement. *Water Sci. Technol.*, **41** (6), 33–40.

Gostelow, P., Parsons, S.A. and Stuetz, R.M. (2001) Odour measurements for sewage treatment works. *Water Research*, **35** (3), 579–597.

Hvitved-Jacobsen, T. and Vollertsen, J. (2001) Odour formation in sewer networks, in, *Odours in Wastewater Treatment: Measurement, Modelling and Control.* (eds R. Stuenz and B.F. Frechen), IWA Publishing, pp. 33–65.

Kehoe, J., Harcus, J., Smith, M., and Warren, M. (1996) Acquisition, review and correlation of odour literature for the air and waste management association. *EE-6 Odour Committee*, University of Windsor.

Koe, L.C.C. (1989) Sewage odors quantification, in. *Encyclopedia of Environmental Control Technology, Wastewater Treatment Technology*, Vol. 3, (ed P.N. Cheremisinoff), pp. 423–446. Gulf Publishing Company, Houston, TX.

Laraia, R., Centola, P., Il Grande, M., *et al.* (2003) Metodi di misura delle emissioni olfattive. *APAT Manuali e linee guida* 19/2003.

McGinley, M.C., McGinley, A. and Michael, P.E. (2002) Odor Testing Biosolids for Decision Making. *Water Environment Federation Specialty Conference: Residuals and Biosolids Management Conference*, Austin, TX.

Nimmermark, S. (2011). Influence of odour concentration and individual odour thresholds on the hedonic tone of odour from animal production. *Biosystems Engineering*, **108** (3), 211–219.

Rafson, H.J. (1998) *Odor and VOC Control Handbook*. McGraw-Hill, New York.

Ruth, J.H. (1986) Odour thresholds and irritation levels of several chemical substances: a review. *Am. Ind. Assocc. J.*, **47** (3), A142–A151.

Schauberger, G., Piringer, M. and Petz, E. (2001) Separation distance to avoid odour nuisance due to livestock calculated by the Austrian odour dispersion model (AODM). *Agriculture, Ecosystems and Environment*, **87**, 13–28.

Serra, R. and Dugnani, L. (1988) Qualità, effetti e misura degli odori nell'ambiente, *IA Ingegneria Ambientale*, vol. XVII, n.5, Milano, CIPA Editore.

Stuetz, R. and Frechen, F.B. (2001). *Odours in Wastewater Treatment: Measurement, Modelling and Control*. IWA Publishing, ISBN 1 90022246 9.

Van Harreveld, A.P. (2001) From odorant formation to odour nuisance: new definitions for discussing a complex process, *Water Science and Technology*, **44** (9), 9–15.

Vincent, A.J. (2001) Source of Odours in Wastewater Treatment, in, *Odours in Wastewater Treatment: Measurement, Modelling and Control.* (eds R. Stuenz and B.F. Frechen), IWA Publishing, pp. 69–90.

Wise, P.M., Olsson, J.M. and Cain, W.S. (2000) *Quantification of Odor Quality*. Oxford University Press.

Part 3

Instruments and Methods for Odour Sampling and Measurement

Sections 3.1–3.4
T. Zarra, V. Naddeo and V. Belgiorno
SEED – Sanitary Environmental Engineering Division, Department of Civil Engineering, University of Salerno, Italy

3.1 Introduction

When we try to characterize an odour, there are different kinds of features to consider in order to obtain reliable and representative results. The collection and analysis of odour samples can produce some difficulties which are reflected in the results of the work.

The first step of errors is related to the sampling phase. The quality of the measurement results, independently from the adopted measurement technique (dynamic olfactometry, chemical analysis or electronic nose), strongly depends on appropriate sampling. It is important to know the nature and characteristics of the odour source as well as their emission so as to identify the suitable sampling instruments, material and methods. It is also important to preliminary define a specific sampling program which represents the result of the characterization of the source/s to be to investigated, relating, for example, to their physical characteristics or the frequency and duration of their emissions. In particular, three types of source must be considered (point, area, volume sources), which can have an outward flow or not. The second criticism is related to the choice of the type of analysis, with all of them having advantages and disadvantages and there being no single recognized method for all types of analyses.

The current measurement methods are generally divided into three categories: sensorial, analytical and mixed methods. Another type of classification distinguishes them into methods for odorous substances (analytical and some mixed methods) and odour measurement (sensorial and mixed methods).

Odour Impact Assessment Handbook, First Edition. Edited by Vincenzo Belgiorno, Vincenzo Naddeo and Tiziano Zarra.
© 2013 John Wiley & Sons, Ltd. Published 2013 by John Wiley & Sons, Ltd.

The identification of the most appropriate sampling technique and quantification method of odours, therefore represents the basic element to be chosen to obtain a reliable and representative measurement, namely objective and repeatable.

In this part, the instruments and methods for odour sampling and measurements are presented and illustrated.

3.2   Sampling Techniques

The aim of sampling is to obtain representative information on the typical characteristics of an odour source by means of the collection of a suitable volume fraction of the effluent, allowing for its quantification.

Therefore, it must be analysed prior the typical characteristics of an odour source, such as:

1. geometrical configuration (point, area or volume source);
2. temporal trend of the emission;
3. transfer modalities of odorous substances from the source to the atmosphere.

The sampled operating conditions, the number and duration of samplings should be chosen so as to allow for a representative and complete evaluation of the monitored source emission and its associated odour impact.

There are two different types of sampling methods: dynamic and static sampling (Figure 3.2.1).

Dynamic sampling provides the air flow to be analysed and ducted directly from the source to the measurement device. The duct may be heated in order to minimize the occurrence of adsorption or condensation phenomena on the duct walls. In this case, the measurement device must be installed in proximity of the odour source, in situ or in a mobile laboratory in order to minimize pressure drops. If the possibility of particulate precipitation in the device exists, the sampled flow should be opportunely filtered through a glass-fibre filter before entering the device. In order to avoid condensation phenomena, the filter should be heated at the sample temperature.

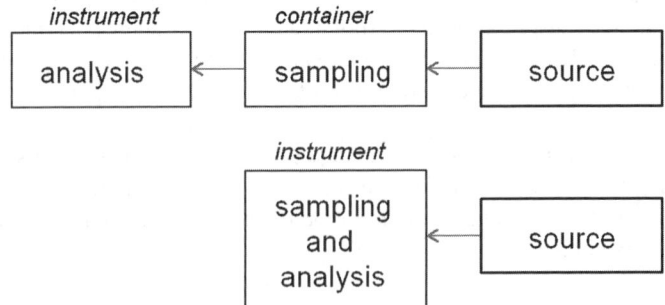

Figure 3.2.1   *Static (up) and dynamic (down) sampling methods.*

Static sampling provides the sample to be enclosed in a suitable container (canister or bag) which is connected to the measurement device in a second moment. To avoid condensation and adsorption phenomena during the sample storage, it is necessary to use specific sampling materials. To minimize any interaction between the sampled gas and the sampling container surfaces that are in contact with it, the sampling equipment should be odourless and, those surfaces, as well as the storage times, should be reduced. The EN 13725 standard in this sense defines a maximum time within which the analysis should be performed from the sampling phase.

Dynamic sampling has the advantage of minimizing the possibility of sample modifications due to adsorption on the sampling equipment or chemical reaction between the compounds contained in the sampled gas, which may occur during the time of sample storage from the moment of the collection to the analysis in laboratory.

As far as olfactometric analyses are concerned, which require a panel of selected human examiners, dynamic sampling has the drawback of being very expensive, as it entails the need to take the examiners on site. Moreover, the fact of taking the panel on site might influence their responses, due on one the hand to the consciousness about the sample provenance and on the other, to the possible presence of background odours.

For these reasons, static sampling is by far the most widely used methodology in the case of odour measurement.

3.2.1 Regulations and Guidelines

Until now, there are no specific guidelines on odour sampling. The only standard which provides information for the sampling phase is the European Standard EN 13725:2003 on olfactometry 'Air quality – Determination of odour concentration by dynamic olfactometry' (CEN, 2003). This norm includes a section dedicated to sampling, even though, already in the introduction of the document, it is specified that 'Improvements in sampling may be the subject of a future revision of this European Standard'.

In more detail, two sections of the EN 13725:2003 deal with sampling materials, whereas one section defines the possible sampling methods (dynamic versus static) and the procedures to be adopted in order to maintain the olfactory characteristics of the sample as constant as possible from the moment of the sampling to the analysis. The given indications are not exhaustive, thus leaving many degrees of freedom in the choice of sampling procedures and equipment. This represents one of the main criticisms of this standard and of odour measurement in general, as it has led to the development of a considerable number of different odour sampling methods and devices all over the world (Hudson and Ayoko, 2008b).

Given these difficulties and the importance of the problem, recently, the German VDI (Verein Deutscher Ingenieure: Association of German Engineers), which historically has always been very active in this field (the EN 13725:2003 was derived almost entirely from the German VDI 3881, 1986), has constituted a working group with the aim of defining a specific guideline on odour sampling. This guideline, the VDI 3880, which is still a draft, discusses in greater detail all the aspects of odour sampling, especially as far as the aspects that are partially overlooked in the EN are concerned, thus giving precise information and reporting examples of the procedures and equipment to be used for odour sampling on different source types.

3.2.2   General Aspects

Prior to starting a sampling, it is important to analyse and define the following aspects: the goal of the sampling, the boundary conditions in which the sampling must be done, materials to be used for sampling, the duration and number of samples to be taken, the storage and transport of sample, the type of odour substances and sources to be collected.

3.2.2.1   Objective of a Study

The objective of a survey is, in fact, the preliminary key to be defined before beginning any sampling. The goal influences both the materials, sampling times and sampling procedures. Given the absence of specific guidelines that define the sampling phase, it is therefore necessary to precede this stage by a careful study of literature, before starting to operate. Thus, with respect to sampling methods, if carrying out a characterization of emissions from a wastewater treatment plant, it is important to conduct the samplings when the plant is working at full capacity, so that the odour emissions are maximized. While in the case of plants with variable emissions, it is necessary to collect one sample for each of the conditions that cause major odour emissions.

If the efficiency of abatement devices should be evaluated, samples should be collected before and after the abatement system, with the plant working at full capacity.

If the results of the survey are to assess the odour impact from the plant, it is necessary that the sampling is conducted in order to obtain representative and exhaustive information of the plant emissions under the different operating conditions.

3.2.2.2   Working Conditions

In relation to the working conditions, showing that the sampling locations are easily practicable and conform to the legislative security requirements, secure for the sampling operator and the investigated source must be such so as to allow the eventual measurement of the effluent velocity, temperature, and so on, is recommended.

3.2.2.3   Sampling Materials

The main requirements that the sampling materials must have include:

1. *Inertia*: materials should minimize the possibility of interactions between the sampled gas. Inert materials are: Polytetrafluoroethylene (PTFE, Teflon™), copolymer of Tetrafluoroethylene and Hexafluoropropylene (FEP), Polyethyleneterephtalate (PET, Nalophan™), glass (drawback: fragility), steel (advantage: high mechanical and thermal stability; drawback: not always chemically inert, condensations or other depositions cannot be visually verified).
2. *Smooth surface*.
3. *Odourless*: they should not add odour to the sampled gas.
4. *Low permeability*: they should avoid sample losses by diffusion or incoming of external air.
5. *Non-adsorbing odours* or non-reactive with odorous samples.
6. *Robust* and without loss.
7. *Opacity*: to protect the compounds that are sensible to the light.

8. *Impervious*: to prevent significant loss of odour components between the collection and measurement time.
9. *Sufficient volumetric capacity*: to ensure adequate representation of the sample.

Recent studies have shown that some of the materials that are most widely used for the realization of odour sampling bags have non negligible diffusion coefficients with respect to specific odorous substances (Beghi *et al.*, 2008; Koziel *et al.*, 2005; Mochalski *et al.*, 2009; Zarra *et al.*, 2012), especially if soluble in water, such as ammonia (NH_3), hydrogen sulfide (H_2S) or formaldehyde (CH_2O). The diffusion of specific molecules may depend on intrinsic factors, such as the bag thickness as well as extrinsic factors, such as temperature and humidity at which the sample is stored.

The existing materials employed for the realization of odour sampling bags that meet the requirements as reported above are:

1. polyvinylfluoride (Tedlar®),
2. polyethyleneterephtalate (PET, Nalophan™),
3. polytetrafluoroethylene (PTFE, Teflon™).

3.2.2.4   Sample Duration and Number of Samples

The sample duration should be long enough to guarantee the sample to be representative of the monitored emission. The EN 13725 does not set a minimum duration, whereas the minimum sampling duration for the collection of each sample mentioned in the VDI 3880 is of 30 min.

The number of samples is related to the goal of the study. Moreover, it must be such as to obtain complete and representative information for the source and emission characterization.

3.2.2.5   Storage Time and Temperature

The time between the sample collection and its analysis should be minimized in order to reduce the possibility of sample modifications during storage. The EN 13725 sets a maximum storage time of 30 h, whereas this interval is reduced to 6 h in the recent VDI 3880. The storage time of each sample should be indicated on the analysis report.

Another parameter to be controlled during storage is the temperature. It should not exceed 25°C, but it should be kept above the sample dew point in order to avoid condensation.

3.2.2.6   Transport Conditions

Samples must be protected from mechanical damage and external contaminations should be avoided.

Samples should not be exposed to direct sun light in order to minimize (photo)chemical reactions and diffusion.

3.2.2.7   Odour Source Types

Generally, it is possible to define three odour source types (point, area and volume source) belonging to two categories (active and passive/fugitive source).

Point sources are discharges from a small opening such as a stack or vent. Area source are sources with a large surface area such as a landfill surface, a pile of solid material, or a liquid surface. Volume source are as example a building (diffuse source such as from within a building). Active source are sources with an outward flow such as a biofilter (active area source). Fugitive source are sources without an outward flow that cannot be readily quantified or defined, such as a pile of compost.

3.2.3 Sampling Program

For an exhaustive characterization of the emissions of an odorous plant, it is necessary to organize a suitable monitoring plan in order to obtain the largest amount of significant information about the emissions, thus avoiding measurement errors or useless replicates.

Sampling and analysis must be conducted with the aim of obtaining results that should be representative of the monitored plant emissivity. For this purpose, it is important to obtain sufficient information about the plant and its emission sources before sampling.

First, a detailed knowledge and analysis of the production cycle as well as all the plant activities are fundamental in order to identify its main odour sources.

It may be important to know the chemical composition of the emissions and obtain information about the possible presence of toxic compounds in the sampled effluents. This knowledge is important for security reasons, for the sampling operator as well as the examiners that perform the olfactometric analysis. For this reason, it might be useful to characterize a sample from a chemical point of view prior to the sensorial assessment.

A sampling program must clearly identify the sampling points, the methods used for the sampling (i.e. material, bag volume, etc.), the operator, the meteorological data (temperature, pressure, humidity, wind direction and velocity) and the main characteristics of the investigated sources as well as their specific process operating conditions.

3.3 Measurement of Odorous Substances

Measurements of odorous substances are usually performed through analytical and mixed or sensor-instrumental methods.

3.3.1 Gas Chromatography and Mass-Spectrometry (GC/MS)

Among the analytical methods, the most used system to measure odorous compounds is gas chromatography coupled with mass spectrometry. The principle of the gas chromatographic method is the separation of components of the mixture based on their affinity to a support present in a column through which the analyte flows, transported by the gas stream. Downstream of the separation, a chromatogram similar to the one reported in Figure 3.3.1 is obtained.

The identification of the peaks of the chromatogram, representative of the different separate substances constituting the mixture odorous investigated, is then carried out thanks to mass spectrometry.

In the analysis method, some key moments can be distinguished (Centola *et al.*, 2004):

1. The pre-concentration of the sample.
2. The transfer of the analytes trapped by the scan tool.

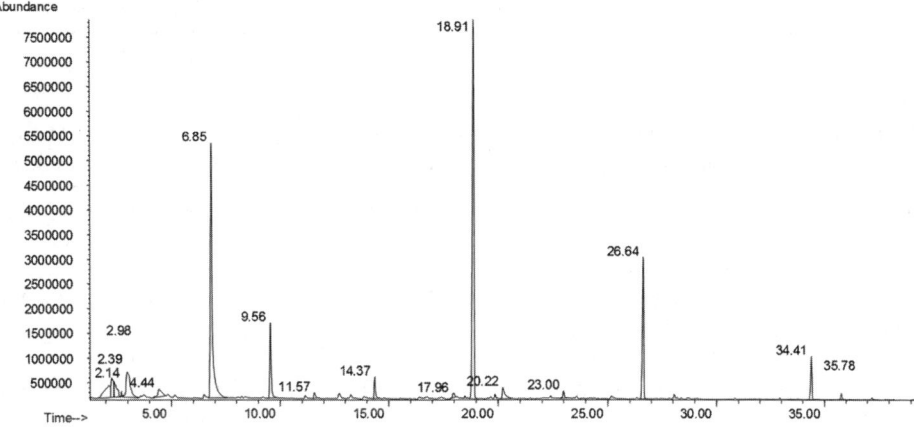

Figure 3.3.1 *Example of a chromatogram for the measurement of odorous substances.*

3. The separation of the different detected components.
4. The qualitative and quantitative analysis of the identified components.

Due to the generally low threshold of olfactory odorants perceptibility, the direct analysis is rarely practicable, but a sampling that involves a pre-concentration step is required.

The pre-concentration of the sample can be done by:

1. absorption of reactive chemicals or specific cleaning solutions;
2. adsorption on solid substrates selected according to the type of compounds to be sampled;
3. condensation at low temperatures by means of freezing techniques;
4. adsorption at low temperatures on a substrate, creating a gas-liquid interface.

The subsequent phase of desorption may be carried out thermally, through the use of solvents or by the stripping of volatile organic compounds (VOCs) in a current of steam. The first method has the advantage of a quick execution and does not require manipulation of the samples by the operator. It cannot, however, be used in the case of thermolabile substances. However, a more selective desorption is obtained when using the solvents, but the technique is more prone to errors due to the very low concentration levels (Centola *et al.*, 2004).

Several international studies have highlighted the application of this analytical technique to the quantification of odorous compounds. Table 3.3.1 shows the main references of this studies with an indication of the method used for the analysis.

GC-MS measurements relate to the physical or chemical properties of the odorous compounds, although the most common measurement made is odorant concentration. From a GC-MS analysis, it is possible to obtain indications on the numerous substances that principally form the odorous mixture. Therefore, it is possible to evaluate the efficiency of technological odour reduction systems such as scrubbers or biofilters through the presence of typical substances (Zarra, 2007; Zarra *et al.*, 2008; Zarra *et al.*, 2009).

Other types of gas chromatographs and mass spectrometers are those equipped with olfactory detection port (ODP) as well as the portable type.

The GC-MS with ODP (Figure 3.3.2) allows for the detection of the odorous substances with the identification of the quality and associated hedonic tone. The effluent is split

Table 3.3.1   *Case studies using GC-MS methods for odorous compounds characterization.*

Bibliographic reference	Investigation field	Analysis method		Mass range
		Temperature profile		
		Δt	Ramp	
Hobbs *et al.* (1995)	'Assesment of odours from Livestock Wastes ...'	27–220°C	15°C/min	35–350
Arena *et al.* (2006)	'Comparison of odours compounds ...'	10–40–240°C	6°C/min	35–280
Yasuhara (1987)	'Identification of volatile compounds ...'	10–180–250°C	15°C/min 8°C/min	20–330
Olsson *et al.* (2002)	'Detection and quantification of volatile compounds ...'	35–220°C	8°C/min	25–250

as it leaves the column so that it arrives simultaneously at the nose and the detector. Thus, the additional information is gained on compounds that are responsible for specific odours.

The portable GC-MS is a recently diffusion instrumentation category, especially applied in dangerous ambients. They make it possible to identify and measure in real-time the volatile substances present in the ambient air and quantify their concentration. The

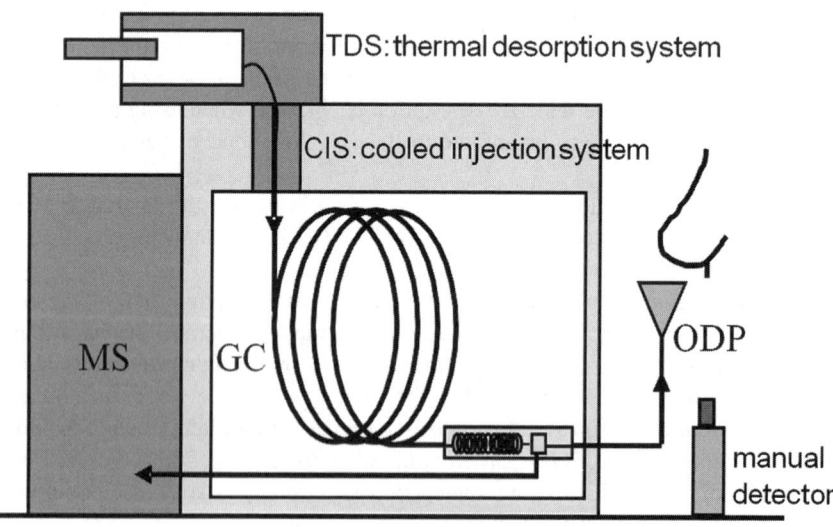

Figure 3.3.2   *Function scheme of a GC-MS with olfactory detection port.*

main disadvantage related to this instrument is the high costs as well as the difficulty of the measuring of low molecular weight compounds, usually representatives from the compounds with higher annoyance characteristics (such as example NH_3 and H_2S) (Zarra, 2007).

3.3.2 Colorimetric Tubes

The colorimetric tubes are some transparent vials filled with particular chemical compounds capable of reacting. They are used to measure the concentration of certain compounds on site and can be used only once.

The concentration is determined by applying a pump, manual or automatic, which is able to put inside the colorimetric tubes a specific air volume, which will react with the compound inside the tube. This will change the colour and make it possible to read the concentration of the compound on the basis of the length of the colour. In fact, outside the box of the tube, there is a scale that says how much of the concentration is in the air of the substance, according to the length of the colour obtained (Figure 3.3.3).

The colorimetric tubes present actually in commerce are very practical, cheap and are available for 600 different kinds of substances. However, they present a set of practical limitations:

1. some kinds of reaction with compounds which are different from the one being monitored could be present;
2. the reacting compound, inside the tube, could be damaged over time;
3. the tubes can present mistakes, but not with a constant rate, it depends on the production batch.
4. the right tube must be carefully chosen according to the supposed present concentration;
5. there are effects of particular temperature, humidity and atmospheric pressure conditions.

For these reasons, it might be useful only for occasional and screening monitoring.

These kinds of tubes are usually used in wastewater treatment plants to detect the presence of pollutant compounds, such as ammonia and sulfur compounds, which can be hardly detected with other methods.

3.3.3 Portable Multi-Gas Detectors

The portable multi-gas detectors are equipment able to detect and analyse continuously and simultaneously, more types of gaseous compound. The substances analysed are acquired by means of electrochemical sensors composed of an immersed electrolyte, a measuring electrode (anode), a counter-electrode and a reference electrode. A potentiostat keeps a constant voltage between the measuring electrode and the reference electrode. The voltage

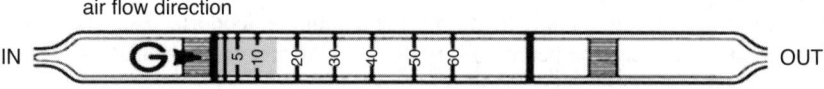

Figure 3.3.3 *Function scheme of a colorimetric tube.*

is set in such a way that a specific gas or vapour in air will oxide on the anode and detect the concentration.

The orders of detectable concentration are usually included in a range from 0–200 ppm. These devices make it possible, once identifying a key substance, to monitor its continuous development, immediately showing any possible anomalies. The main uses of such instruments are in the monitoring of emissions of H_2S and NH_3 on wastewater treatment plants and composting facilities.

Their main drawback is related to the considerable influence of the external ambient air conditions (for example temperature, humidity, etc.). For these reasons, they might be useful only for occasional and screening monitoring.

3.3.4   Gas Analysers

Gas analysers are laboratory instruments that are generally installed in static cabins or mobile laboratories (a van) that carry out the sampling and analysis phase measuring the concentrations of individual compounds. Unlike the portable multi-gas detectors, the gas analysers allow for a high precision, but are not easily movable, they need a power supply and are significantly more expensive. Their use is recommended where there is a need to monitor a single parameter with high precision for a relatively long time.

3.4   Determination of Odour Concentration by Dynamic Olfactometry

Dynamic olfactometry is currently the most used method and the only one standardized by a European Standard (EN 13725:2003) for odour measurements. It is a sensorial technique, which uses a dilution instrument (namely, the olfactometer) to present an odour, at different concentrations levels, in a controlled way to a panel of assessors. Therefore, it is possible to can record and evaluate their answers in a statistical way, in order to obtain final results of the measurements.

This technique is useful for making comparisons of odours from different sources, in that measurements should satisfy two principles, which are objectivity and reproducibility.

Dynamic olfactometry allows us to determine the odour concentration (C_{od}) of an odorous air sample relating to the sensation caused by the sample directly on a panel of opportunely selected people. C_{od} is expressed in European odour units per cubic metre (ou$_E$ m^{-3}), and it represents the number of dilutions with neutral air that are necessary to bring the odorous sample to its odour detection threshold concentration. Dilution may be static or dynamic. The first includes the mixing of fixed volumes of odour with neutral air, while the second includes the mixing of separate volumes.

The dilution factor to the perception threshold is expressed by means of some parameters, substantially equivalent:

1. Threshold Odour Number (TON). In a text approved by the 'Standard methods committee' (1985), TON is defined by the following relationship TON = (A + B)/A, where A is the amount of odorous sample in ml, B is the amount of deodorized air in ml necessary for the mixture to reach the perceptibility threshold (Serra *et al.*, 1988).
2. Effective dose at the 50% level (ED50).
3. Dilution to Threshold (D/T).

4. Odour unit (OU) = Amount of odorant(s) that, when evaporated into 1 m^3 of neutral gas at standard conditions, elicits a physiological response from a panel (detection threshold) equivalent to that elicited by one European Reference Odour Mass (EROM), evaporated in one cubic metre of neutral gas at standard conditions. EROM is an accepted reference value for the European odour unit, equal to a defined mass of a certified reference material. One EROM is equivalent to 123 mg n-butanol (CAS-nr. 71-36-3). Evaporated in 1 m^3 of neutral gas this produces a concentration of 0.040 μmol mol^{-1}.

The odour concentration (ou_E/m^3) is statistically equal to the dilution factor of the perception threshold: for example, a concentration of 200 ou_E/m^3 means that the sample has been diluted two hundred times to reach the panel threshold.

There are two standardized methods for the presentation of an odour sample to the panel: yes/no method and forced choice. In the first method, each examiner sniffs from a single port and communicates if an odour is detected or not (Figure 3.4.1). Odour samples diluted with neutral air, or neutral air alone, can exit from the sniffing port.

In the second method, two or more sniffing ports are used (Figure 3.4.2); the odour sample is presented at one port, and neutral air at the other port(s). In this case, the panel has to compare different samples and choose the port from which the odour exits.

Sampling odour mixtures at different dilutions are presented to a group of selected panellists for sniffing and their responses are recorded. Generally, the first mixture presented to the odour panellists is diluted with a very large volume of air in order to be undetectable by the human nose. In subsequent samples, the volume of diluent is decreased by a predetermined and constant factor.

Different measurement cycles are carried out and the final result is calculated as the geometric mean of the values obtained for the single series. The concentration corresponding to the odour threshold of perception is reached when 50% of the panellists are able to detect

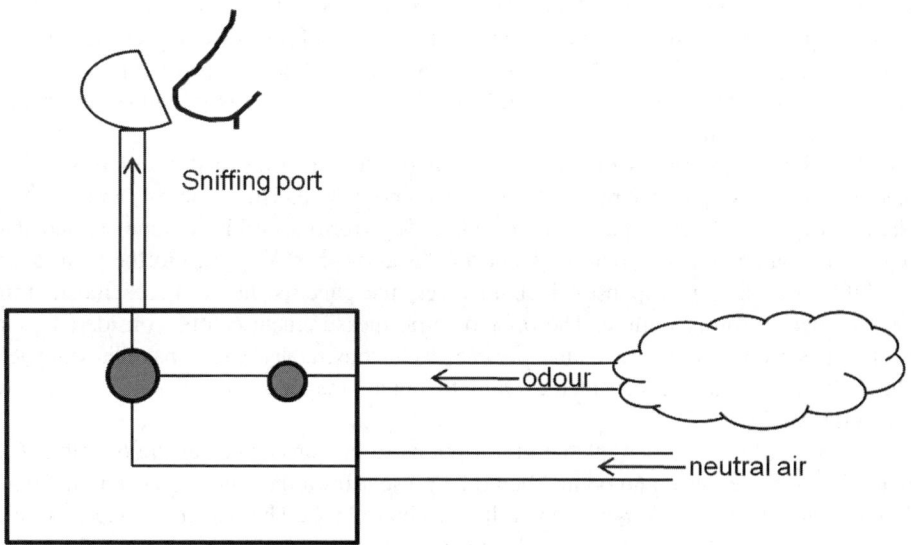

Figure 3.4.1 *Scheme of the yes/no method.*

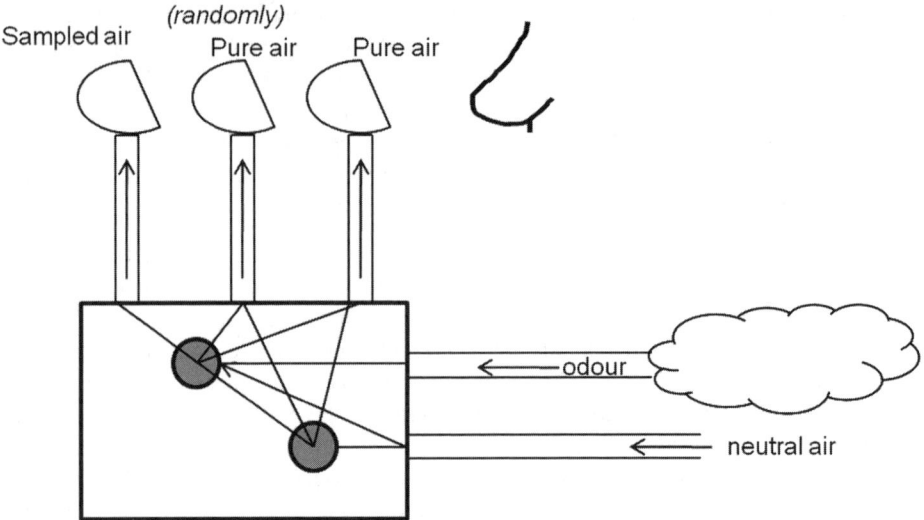

Figure 3.4.2   *Scheme of the forced choice method.*

the odour. To make the analysis as objective as possible using dynamic olfactometry, the EN 13725 defines specific requirements that must have the panellists. Panellists are qualified examiners used as sensors in olfactometric analysis and their olfactive response (odour threshold) is the measured parameter for calculating odour concentrations. The examiners are selected in order to choose individuals with average olfactive sensitivity, who constitute a representative sample of the human population. The screening is usually performed using reference gases. In particular, the most used reference gas is n-butanol and only assessors who meet predetermined repeatability and accuracy criteria for this gas are selected as panellists: the average n-butanol odour threshold is in a range of 20–80 ppb (40 ppb represents the accepted odour threshold for n-butanol) and the antilog standard deviation of individual responses is less than 2.3.

Elaboration of the results of every olfactometric test requires quality controls on the panellists' responses, as reported in the European normative. These controls aim to check whether, during a test, the panellist exceeds 20% of errors on blank samples, and that the average value of his responses is not too distant ($-5 < \Delta < 5$) from the team average value (uncommon response). In both cases, the panellist has to be excluded from the data elaboration procedure. The olfactometric measurement can be considered valid if, after these controls and eventually some exclusions, at least four panellists are left whose responses can be considered valid to obtain the odour concentration measure (Piccinini, 2002).

The EN 13725 also describes the prescriptions to be considered for the olfactometric room; for example, inside no odour should be present from the components and the temperature should be kept at a constant value of about 25°C. The panellists also require a continuous replacement of fresh and clean air as well as a comfortable surrounding environment. The number of people who compose the panel changes according to the kind of analysis that has to be carried out. To evaluate the odour threshold, at least four people are required, while to evaluate the intensity 8–16 to evaluate the hedonic tone.

With dynamic olfactometry, it is not possible to explore the substances causing the olfactory annoyance as well as their single concentrations. The principal source of uncertainty of the olfactometric method is the biological high variability of the olfactory sensibility. Even when performed according to the EN 13725: 2003, the group of panellists does not necessarily represent a statistically representative sample of the exposed population, but only a group of subjects endowed with medium olfactory sensibility. Sensorial analysis, being assigned to the 'human sensor', with its own nature being irreproducible, is the cause of a considerable uncertainty, due to the unavoidable human component that interferes in the evaluation (Sneath, 2001; Koster, 1985; Stuetz *et al.*, 2001).

3.5   Determination of Odour Concentration by the Triangular Odour Bag Method

T. Higuchi

Graduate School of Science and Engineering, Yamaguchi University, Tokiwadai, Ube, Yamaguchi, Japan

The triangular odour bag method is an air dilution method in which odour concentration and index are measured. Odour concentration is the dilution ratio when odorous air is diluted by odour-free air until the odour becomes unperceivable. The odour index is the logarithm of odour concentration, multiplied by ten. The triangular odour bag method was first developed by the Tokyo metropolitan government in 1972 (Iwasaki *et al.*, 1972; Iwasaki *et al.*, 1978). In this section, the measurement procedures of the triangular odour bag method are explained, with the quality control framework for olfactometry in Japan introduced.

3.5.1   Equipment and Apparatus

The triangular odour bag method is applied through the use of the following equipment and apparatus:

1. Pump for supplying air. A pump that has an ability to supply air at over 30 l/min is used.
2. Odour-free air distributor. This is an apparatus that can remove odours from the air to be supplied, as well as from a pump, when transferring odour-free air into an odour bag. It consists of a column packed with activated carbon and multi-way piping.
3. Syringe. A syringe made of glass is used to inject odours into an odour bag. A gastight syringe is also used if the volume is less than 1 ml. A plastic syringe that is airtight, odour-free and has low odour adsorption can also be used.
4. Odour bag. An odour bag is made of polyester film. It is odour-free, has low odour adsorption and low permeability. The odour bag is equipped with a glass tube that has an inner diameter of 10 mm and a length of 6 cm as a sample outlet port. The inner volume is 3 l.
5. Sniffing mask. A sniffing mask that is odour-free and made of plastics connects to the outlet of an odour bag and is structured to cover a nose.
6. Silicone rubber stopper. This is used as a tight stopper for the outlet port of an odour bag.

Table 3.5.1   *Concentrations of standard odour solutions used in the panel screening test.*

Compound	Concentration (w/w)	Odour description
β-Phenylethyl alcohol	$10^{-4.0}$	Floral smell Smell of rose petals
Methyl cyclopentenolone	$10^{-4.5}$	Sweet burning smell Smell of burned caramel pudding
Isovaleric acid	$10^{-5.0}$	Smell of sweat Smell of stuffy socks
γ-Undecalactone	$10^{-4.5}$	Smell of ripe fruit Smell of canned peaches
Skatole(3-Methyl indole)	$10^{-5.0}$	Musty smell Smell of faeces

3.5.2   Panel

The panel consists of six or more members. They are required to have passed the screening test, using five kinds of standard odour, that is, β-phenylethyl alcohol, methyl cyclopentenolone, isovaleric acid, γ-undecalactone and skatole (3-methyl indole). The panel screening test is carried out according to the following procedure.

1. A set of five strips of test paper (smelling strips), 14 cm long and 7 mm wide, marked with the numbers 1–5 is prepared. The top 1 cm of any two smelling strips are soaked in a standard odour solution. The remaining three smelling strips are soaked in odour-free liquid paraffin, using the same method. The concentrations of the standard odour solutions are shown in Table 3.5.1.
2. A set of five smelling strips is handed to the subjects, who choose the two smelling strips with the odour by using their olfaction.
3. Steps (1) and (2) are carried out for each of the five standard odour solutions, with a person who answers correctly for all of five having normal olfaction.

3.5.3   Timing for the Sensory Test

A sensory test (a test in which a panel uses olfaction to judge the presence of odour in an odour bag) is conducted as soon as possible, on the same day or the day after the sample is collected.

3.5.4   Procedures of the Sensory Test

Two different procedures are used depending on where the sample is collected:

- *Procedure for samples collected at environment (ambient air):*
  (a) Three odour bags numbered from 1–3 per panel member are prepared.
  (b) These odour bags are filled with odour-free air passed through the activated carbon column, and plugged up with silicone rubber stoppers.

(c) Odorous air is injected into one of three odour bags until a given dilution ratio is obtained.

(d) Each member of the panel removes the stopper and sniffs by bringing the odour bag close to their nose. After sniffing three odour bags, they should choose only one odour bag which is likely to contain odorous air out of the three bags, and write down the number of the bag chosen.

(e) The test is carried out three times per each panel member for the same dilution ratio.

(f) The responses given by the panel members are collected and compiled. The rate of correct answer 1.00 is assigned to a correct reply, 0.00 to an incorrect reply and 0.33 to a reply 'I cannot identify.'

(g) The mean rate of correct answers for all the responses is calculated. If the mean is 0.58 or more, a next session in which the sample is diluted ten times further is carried out. If the mean is less than 0.58, the test series ends.

(h) Odour concentration and odour index of the sample are calculated by Equation (3.5.1) and (3.5.2), respectively:

$$X = M \cdot 10^{(r_1 - 0.58)/(r_1 - r_0)} \qquad (3.5.1)$$

$$Y = 10 \log X \qquad (3.5.2)$$

where:

$X =$ odour concentration,
$M =$ the highest odour dilution ratio when the mean rate of correct answers is 0.58 or more,
$r_1 =$ the mean rate of correct answers when odour dilution ratio is M,
$r_0 =$ the mean rate of correct answers when odour dilution ratio is $10\,M$,
$Y =$ odour index.

The starting dilution ratio of this method is ordinarily fixed at 10. When the rate of correct answers at the dilution ratio 10 is less than 0.58, the odour concentration and odour index should be 'below ten.' In this method, odour concentration and odour index values less than 10 cannot be measured, since it is difficult to get sample dilution ratio less than 10.

- *Procedure for samples collected at source* (Figure 3.5.1):

Steps (a)–(d) are identical to those of the procedure for ambient air.

(e) The responses given by the panel members are collected and compiled. If a panellist gives a correct response, he participates in the next session in which the sample is diluted three times further, and repeats the above steps (a)–(d). If a panellist gives an incorrect response or could not identify the odour, the series of tests ends. The test should be continued until all the panel members give incorrect responses.

(f) The odour concentration and odour index of the sample is derived from the following steps:

First, the logarithm of the odour threshold value of each panel member is calculated by Equation (3.5.3):

$$Z_i = (\log M_{1i} + \log M_{0i})/2 \qquad (3.5.3)$$

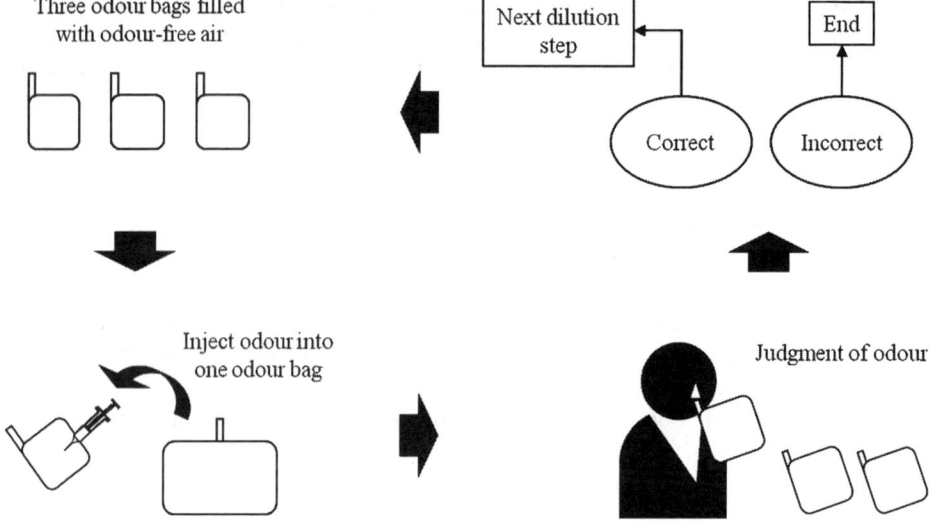

Figure 3.5.1   *Measurement procedure for source air.*

where:

Z_i = logarithm of odour threshold value of a panel member,

M_{1i} = the highest odour dilution ratio when the panel member gave a correct response,

M_{0i} = odour dilution ratio when the panel member gave an incorrect response or could not identify the odour.

Then, among these odour threshold values of the panel members, the maximum and minimum are removed and the mean of the rest (Z) is calculated. Finally, the odour concentration and odour index are obtained by Equation (3.5.4) and (3.5.5), respectively;

$$X = 10^Z \tag{3.5.4}$$

$$Y = 10\ Z \tag{3.5.5}$$

An example of the measurement is shown in Table 3.5.2.

3.5.5   Quality Control Framework

In recent years, the need to develop quality control systems for olfactometry and standardization of measurement procedures for the promotion of nationwide diffusion of olfactometry in municipalities has been recognized. In this part, the quality control framework for the triangular odour bag method in Japan is introduced.

3.5.5.1   *Reference Odours*

Reference odours are necessary in order to conduct interlaboratory comparisons of olfactometry as well as routine verification of measurement results in each olfactometry

Table 3.5.2 *Example of measurement for source air sample.*

Panel	Dilution step	1	2	3	4	Logarithm of odour threshold: Z_i
	Amount of odour injected (ml)	30	10	3	1	
	Dilution ratio: D	100	300	1000	3000	
	log D	2.00	2.48	3.00	3.48	
A	Number of bag with odour	2	1	1	2	3.24 (max.)
	Answer	2	1	1	3	
	Judgment	CA	CA	CA	IA	
B	Number of bag with odour	3	1	1	3	3.24
	Answer	3	1	1	2	
	Judgment	CA	CA	CA	IA	
C	Number of bag with odour	2	3	1		2.74
	Answer	2	3	3		
	Judgment	CA	CA	IA		
D	Number of bag with odour	3	1			2.24 (min.)
	Answer	3	3			
	Judgment	CA	IA			
E	Number of bag with odour	2	3	1		2.74
	Answer	2	3	2		
	Judgment	CA	CA	IA		
F	Number of bag with odour	2	1	2	2	3.24
	Answer	2	1	2	3	
	Judgment	CA	CA	CA	IA	

Calculation:
$Z = (3.24 + 2.74 + 2.74 + 3.24)/4 = 2.99$
Odour concentration: $X = 10^{2.99} = 977 = 980$
Odour index: $Y = 10 \cdot 2.99 = 29.9 = 30$
CA: Correct answer; IA: Incorrect answer.

laboratory. Four odorous compounds (i.e. 1-butanol, ethyl acetate, *m*-xylene, and dimethyl sulfide) have been proposed as reference odours. In Europe, 1-butanol is defined to be a reference odour in CEN standard EN 13725: 2003. Ethyl acetate is considered to be 'specific offensive odour substances' as designated by the Offensive Odour Control Law in Japan, and *m*-xylene and dimethyl sulfide are compounds that have been used as reference odours in previous interlaboratory comparisons in Japan. Reference odours for olfactometry should fulfil the following requirements:

- Odour sample should be prepared easily and accurately.
- Odour sample should remain stable for the measurement period.
- Odour threshold values of the panellists should not vary widely.
- Odour quality should be easily recognized.
- Low health and psychological effect on operators and panel members should be ensured.

Considering all these aspects, ethyl acetate was selected as a reference odour for the triangular odour bag method in Japan. Although 1-butanol is designated as a reference odour in CEN EN 13725, it was not selected because there is less measurement data for 1-butanol in Japan and ethyl acetate has the advantages in sample preparation and data accumulation.

3.5.5.2   *Preparation of Reference Odours*

An easy-to-operate and cost-effective technique for reference odour preparation is necessary in order for it to be adopted nationwide in municipalities and olfactometry laboratories. On the assumption that reference odour sample with odour concentration of two to three thousand is appropriate in the quality control process, the concentration of ethyl acetate is calculated to be around 2000 ppm in consideration of an odour threshold of 0.87 ppm (Nagata and Takeuchi, 1990).

Four preparation methods for reference odour (i.e. steel cylinder method, standard gas generator method, odour bag/vacuum bottle method, and handy gas cylinder method) were proposed. A steel cylinder containing ethyl acetate of 2010 ppm was specially ordered. In the odour bag/vacuum bottle method, an odour bag or a glass vacuum bottle is used to vaporize the ethyl acetate reagent. These four preparation methods were verified at three olfactometry laboratories and confirmed to be applicable to quality control processes.

3.5.5.3   *Interlaboratory Comparison of Olfactometry*

In 2000 and 2001, an interlaboratory comparison of olfactometry was carried out in order to collect basic data for the establishment of a quality control procedure and the determination of quality control criteria for the triangular odour bag method.

1. *Methods*

In 2000, the interlaboratory comparison was conducted by using a measurement method for samples taken at the sources. A total of seven olfactometry laboratories in Japan participated in the test. A three-litre-capacity sampling bag filled with ethyl acetate of around 2000 ppm was delivered to each laboratory four times. The odour index of each sample was measured according to the official procedure of the triangular odour bag method. The tests were conducted six times over four days (i.e. three times for the second sample and only once for the other three samples). The steel cylinder method was used to prepare the reference odour, that is, ethyl acetate of 2010 ppm. The gas concentration of each sample was analysed with GC-FID just before the delivery.

In 2001, the interlaboratory comparison was carried out by using a measurement method for samples taken at the environment. A sampling bag with the capacity of 20 l filled with ethyl acetate of around 50 ppm was delivered to each laboratory four times. The steel cylinder method was used to prepare the reference odour, that is, ethyl acetate of 50.9 ppm. The other measurement conditions were identical to those in 2000.

2. *Results*

The mean values, repeatability standard deviations, and reproducibility standard deviations of detection thresholds were calculated from the results according to JIS Z 8402-2 (1999), which is the Japanese version of ISO 5725-2: 1994 as shown in Table 3.5.3.

Table 3.5.3   *Mean values (m), repeatability standard deviations (sᵣ), and reproducibility standard deviations (sᵣ) of logarithms of detection thresholds obtained from interlaboratory comparison in 2000 and 2001.*

Measurement method (Year)	m	s_r	s_R
For ambient air (2001)	−0.10	0.13	0.24
For source air (2000)	−0.26	0.17	0.22

In practice, the detection threshold logarithms were used for the calculation of these values. The statistical data reported in Table 3.5.3 can be used to determine the quality control criteria for the triangular odour bag method.

3.5.5.4   Quality Control Manual

On the basis of ongoing discussions about reference odour and interlaboratory comparisons, a quality control manual for laboratory use was published in 2002. Figure 3.5.2 shows the quality control framework for the triangular odour bag method in a laboratory.

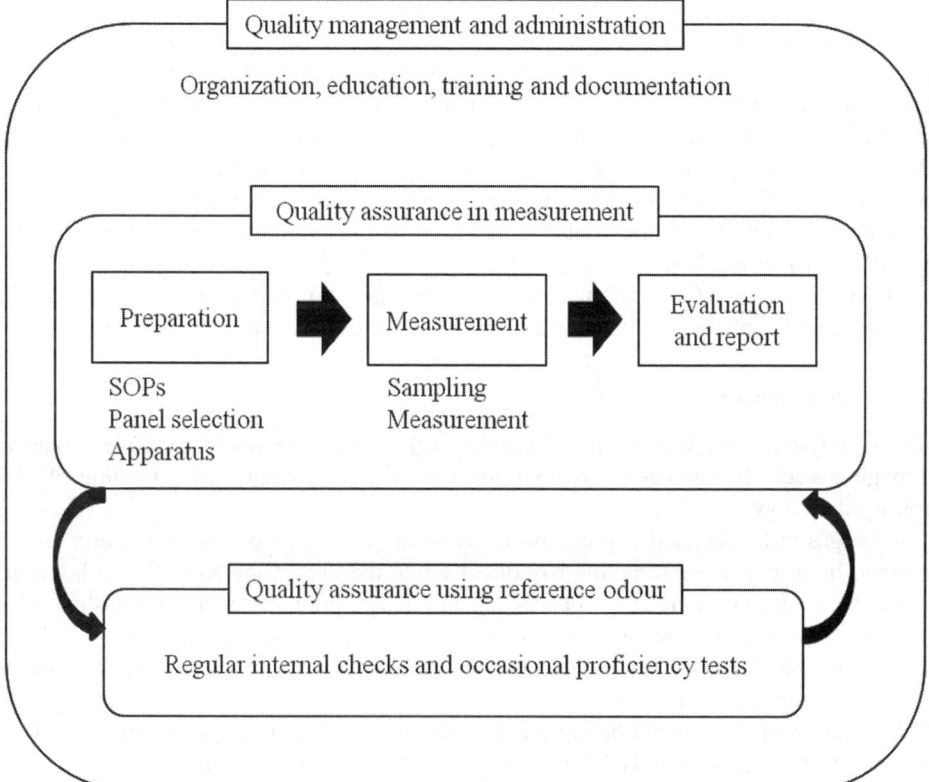

Figure 3.5.2   *Quality control framework for the triangular odour bag method.*

The fundamental topics in the manual are as follows:

- Establishment of quality control system and organization in a laboratory
- Education and training of staff concerned
- Documentation of measurement processes
- Preparation of standard operating procedures (SOPs)
- Evaluation and report of measurement results
- Regular internal quality checks using reference odour
- Occasional proficiency tests using certified reference odour.

On the basis of the collaborative assessment experiment, accepted reference values, repeatability, and reproducibility of reference odour have been obtained. Therefore, individual olfactometry laboratories are able to carry out regular quality checks and compare the results with these values.

3.6   Estimation of Emission Rate

T. Zarra, V. Naddeo and V. Belgiorno

SEED – Sanitary Environmental Engineering Division, Department of Civil Engineering, University of Salerno, Italy

When measuring odours, especially when the goal of the activity is impact evaluation, it is not sufficient to measure odour concentration in isolation, but it is necessary to account for the air flow associated with the monitored odour source, as, in most cases, these parameters are related to each other. The fundamental parameter to be considered is the Odour Emission Rate (OER), which is expressed in odour units per second (ou_E/s), and is obtained as the product of the odour concentration and the air flow associated with the source. The volumetric air flow should be evaluated under normal conditions for olfactometry: 20°C and 101.3 kPa on wet basis.

The technique used for sampling depends on the source typology (Gostelow *et al.*, 2003; Bockreis and Steinberg, 2005) and is as important as the chosen measurement method.

3.6.1   Point Sources

In a point source, an odour is emitted from a single point, generally in controlled manner through a stack. In this case, sampling consists of the withdrawal of a fraction of the conveyed air flow.

If the gas to be sampled is pressurized, the withdrawal may be conducted in a direct manner, by inserting the sampling bag directly into the duct. Otherwise, the withdrawal should be conducted by creating a depression. For this purpose, the sampling bag should be inserted into a suitable container. The air inside the container is sucked out by a pump. Due to the depression realized inside the container, the gas to be sampled is indirectly sucked inside the sampling bag (Figure 3.6.1).

The used container should be airtight, in order to avoid the intake of external air. The advantage of this procedure is that the gas to be sampled does not come into contact with the pump. The sampling point should be chosen on a section where the air velocity is as uniform as possible.

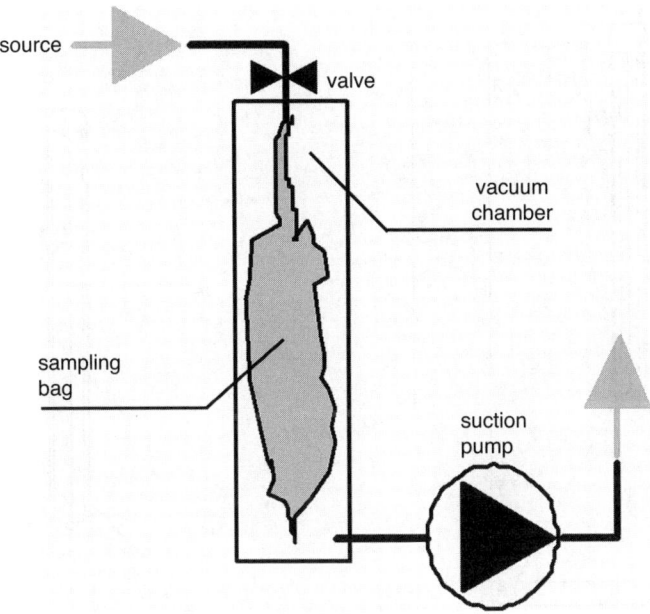

Figure 3.6.1   *Scheme of sampling by means of depression pump.*

In the case of point sources, it is possible to calculate the emitted air flow by measuring the air velocity as well as the duct transversal section. The OER can be obtained as follows:

$$OER = Q_{air} \cdot c_{od} \tag{3.6.1}$$

where:

OER = Odour Emission Rate (ou_E/s)
Q_{air} = effluent volumetric air flow (m^3/s)
c_{od} = measured odour concentration (ou_E/m^3)

3.6.2   Area Sources

In the case of area sources, emissions typically come from extended solid or liquid surfaces. Two different kinds of area sources may be distinguished:

1. *Active sources*: sources having an outgoing air flow (e.g. biofilters or aerated heaps).
2. *Passive sources*: sources without outward air flow. The mass flow from the solid or liquid surface to the air (volatilization) is due to phenomena such as equilibrium or convection (natural or forced). Examples include landfill surfaces and wastewater treatment tanks.

In some cases, the distinction between the two kinds of sources may not be clear. An example may be represented by the biological oxidation tanks of a wastewater treatment plant, which have an outward flow, even if very low. For such cases, it is necessary to establish a volumetric air flow limit to distinguish between active and passive sources. The VDI 3880 sets a specific flux limit of 50 $m^3/h/m^2$.

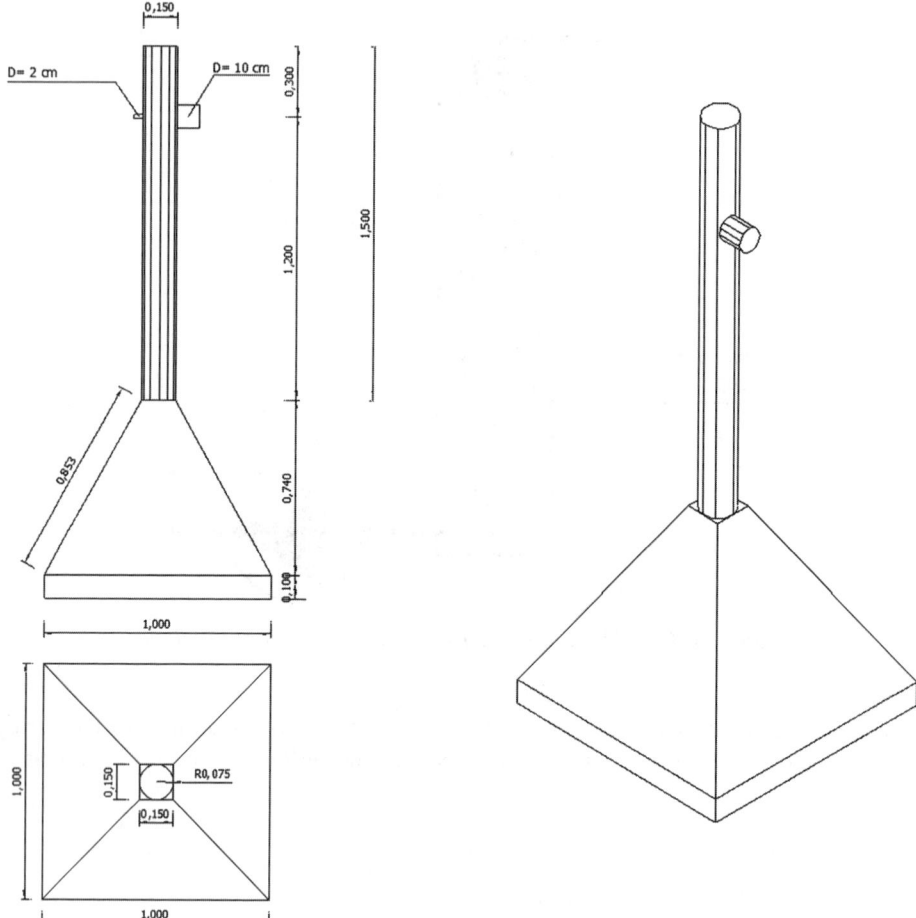

Figure 3.6.2   *Static hood for active area source sampling.*

In the case of active area sources, sampling is performed by means of a 'static' hood that isolates a part of the emitting surface thus channelling the outward air flow into the hood outlet duct, where the sample is finally collected with the same modality used for sampling from point sources.

The static hood consists of two main parts. The first is a cone or pyramid frustum with known base area (e.g. 1 m²). The second one, surmounting the first, is a stack, generally cylindrical, with a diameter of 10–20 cm (Figure 3.6.2).

One or more openings are realized on the stack in order to allow the sample collection and the measurement of the emission physical parameters (e.g. temperature, relative humidity, velocity). The sampling hood should be realized with allowed materials, which must be odourless and inert (e.g. steel or aluminium with an internal PTFE coating). For sampling, the hood must be positioned on the emitting surface with the aim of isolating the sampling point from the external atmosphere and avoiding that the wind may dilute the emitted gas before it is sucked into the sampling bag.

In order to obtain representative data of the entire source, it is necessary to carry out more samplings in different points, which should be uniformly distributed over the emitting surface. Indicatively, the surface portion sampled by means of the static hood should be about 1% of the total emitting surface (for instance, on a biofilter with a surface of 500 m², five samples in five different points uniformly distributed over the biofilter surface should be sampled).

Each sample is collected by inserting the PTFE tube of the sampling bag into the proper sampling hole in the hood stack, after a sufficient time for the odorous flux to have filled internally the whole body of the hood. The opening used for the sample collection should also be used for the insertion of the instrumentation for the determination of the emission physical parameters. The determination of the effluent velocity allows for the evaluation of the flow distribution throughout the emitting surface. It is important to stress that the effluent velocities measured at the outlet of an active area source should not be used for the determination of the emitted flow.

Nonetheless, the verification of the flow uniformity throughout the emitting surface is important in order to define the average emitted odour concentration, i.e. the average value that, multiplied by the effluent flow, gives the OER. Two different cases may be distinguished:

1. active area sources with homogeneous flow distribution; and
2. active area sources with non-homogeneous flow distribution.

Active area sources should be considered to have an homogeneous flow distribution if the differences between the measured effluent velocities on the monitored surface portions are below a defined factor (for this purpose, the VDI suggests a factor of 2). In such cases, the average odour concentration is obtained as a geometric mean of the odour concentration values of the collected samples, according to the following equation:

$$\bar{c}_{od} = \sqrt[n]{\prod_{i=1}^{n} c_i} \qquad (3.6.2)$$

where:

$\bar{c}_{od}$ = average odour concentration (ou$_E$/m³)
c_i = odour concentration measured on the i-th surface portion (ou$_E$/m³).

In the case of active area sources with non-homogeneous flow distribution (the differences between the velocities measured on the different surface portion are higher than the fixed factor, e.g. 2), the average odour concentration is calculated as a weighted geometric mean; according to the following equation:

$$\bar{c}_{od} = \frac{\sqrt[n]{\prod_{i=1}^{n} (c_i \cdot v_i)}}{\sum_{i=1}^{n} v_i} \qquad (3.6.3)$$

where:

$\bar{c}_{od}$ = average odour concentration (ou_E/m^3)
c_i = odour concentration measured on the i-th surface portion (ou_E/m^3)
v_i = effluent velocity measured on the i-th surface portion (m/s).

In the case of passive area sources, the estimation of the OER is a rather complicated process, as it is difficult to measure a representative odour concentration, and, most of all, to determine a well-defined air flow rate.

In general, the estimation of emission rate values from passive area sources may be performed by adopting two different approaches (Gostelow *et al.*, 2003; Hudson and Ayoko, 2008a):

1. indirect measurements using micrometeorological methods, where emission rates are derived from the simultaneous measurements of wind velocities and concentrations across the plume profile downwind the source; or
2. direct measurements using an enclosure of some sort, i.e. so called 'hood methods'. In this case, emission rates are derived from the data regarding the concentration of the compounds of interest measured in the samples collected at the outlet of the sampling device combined with the dimensions of the device and the operating conditions.

Currently, there are two types of hood: the static flux hood and the wind tunnel. The main differences between the two hoods are the ventilation flow rate and the airflow dynamics that are established over the liquid surface. SOER measurements collected with a wind tunnel are not equivalent to static flux hood results, nor is there a consistent ratio between the two types of measurement. Overall, wind tunnels are currently the favoured type of sampling hood for a passive area source.

Indirect techniques such as micrometeorology do not perturb the emission process because a sampling device is not used. However, a large number of samples are required to characterize the considered emission, thus making such techniques impractical for odour assessments. For this reason, hood methods are by far the most widely used techniques for the evaluation of emission rates from passive area sources, and will be therefore treated more extensively in this section.

Various sampling devices have been designed and tested for the collection of samples from a range of area sources (Frechen *et al.*, 2004; Hudson and Ayoko, 2008b; Capelli *et al.*, 2009). All these devices are based on the same principle: to isolate a portion of the emitting surface by means of a hood, to insufflate a neutral (i.e. odourless) air stream inside the hood and finally measure the odour concentration at the hood outlet (see Figure 3.6.3).

Figure 3.6.4 reports an example of a wind tunnel tool used for the passive area source sampling. In particular, the plant and a tri-dimensional view with its dimensional characteristics are represented (Capelli *et al.*, 2009).

Based on this principle, the estimation of the OER requires the calculation of another significant parameter, that is the Specific Odour Emission Rate (SOER), expressed in odour units emitted per surface and time unit ($ou_E/m^2/s$), according to the following equation:

$$SOER = \frac{Q_{air} \cdot c_{od}}{A_{base}} \tag{3.6.4}$$

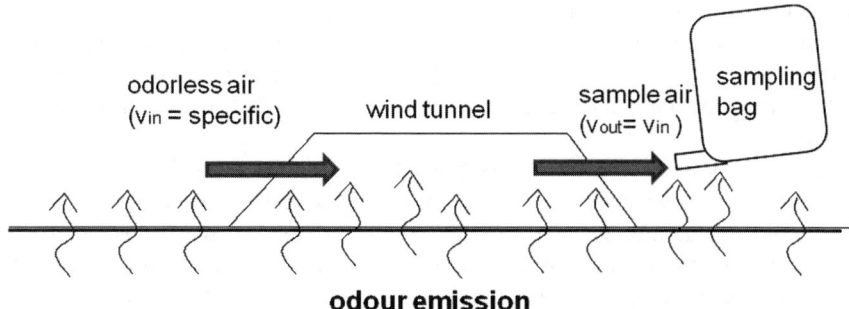

Figure 3.6.3   *Functioning principle of a passive area sources sampling tool.*

where:

$SOER$ = Specific odour Emission Rate ($ou_E/m^2/s$)
Q_{air} = air flow rate inside the hood (m^3/s)
c_{od} = measured odour concentration (ou_E/m^3)
A_{base} = base area of the hood (m^2).

Finally, to calculate the OER, it is sufficient to multiply the SOER by the emitting surface of the considered source:

$$OER = SOER \cdot A_{em} \qquad (3.6.5)$$

where:

OER = Odour Emission Rate (ou_E/s)
$SOER$ = Specific odour Emission Rate ($ou_E/m^2/s$)
A_{em} = emitting surface of the considered source (m^2).

In order to obtain representative results of the real emissive scenario, the sampling should be carried out with particular care so as to prevent the sampling hood, which covers and isolates a portion of the emitting surface, altering the emissivity of that surface portion. For

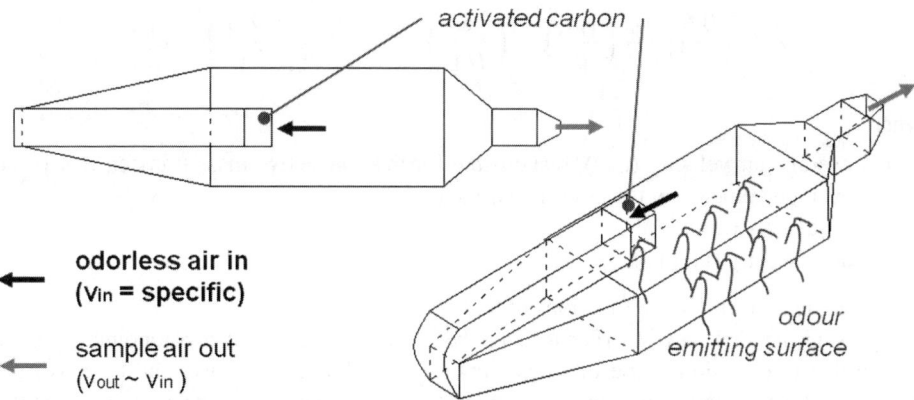

Figure 3.6.4   *Plant and tri-dimensional view of a wind tunnel.*

instance, a pressure variation inside the hood might suppress or promote the volatilization of odorous substances. For this reason, one important aspect during sampling is to leave sufficient time between the hood positioning on the sampling surface and the sample collection. The time interval depends on the characteristics of the hood (e.g. geometry and dimensions).

Sampling on passive area sources is generally performed using wind tunnels. These kinds of hood are designed with the aim of simulating the atmospheric condition of parallel flux without vertical mixing: a horizontal neutral air flow causes the volatilization of odorous compounds from the sampled surface, thus originating an odour emission. Wind-tunnel hoods are best suited to sampling from area sources that have a medium to high SOER, where the collected sample odour concentration will be more than 100–200 OU/m³, thereby reducing errors that can occur when operating at the lower limit of performance of olfactometers.

In the case of liquid surfaces, this phenomenon of mass transfer to a gas phase, called forced convection, may successfully be described by means of the Prandtl boundary layer theory (Thibodeaux and Scott, 1985). This theory affirms that the mass transfer coefficient relevant to a compound can be expressed with the following equation:

$$K_C = \frac{0,664 D_i}{l} \mathrm{Re}^{1/2} \mathrm{Sc}^{1/3} \tag{3.6.6}$$

where:

K_C = mass transfer coefficient (m/s)
D_i = molecular diffusivity of i in the monitored liquid (m²/s)
l = surface length in the flow direction, which generally corresponds to the length of the base of the wind tunnel (m)
Re = Reynolds number
Sc = Schmidt number.

The Reynolds and Schmidt numbers are adimensional groups that have the function of characterizing the convective mass transfer:

$$K_C = \frac{0,664 D_i}{l} \left(\frac{v \rho l}{\mu} \right)^{1/2} \left(\frac{\mu}{D_i \rho} \right)^{1/3} = 0,664 \left(\frac{D_i^4 \rho}{l^3 \mu} \right)^{1/6} v^{1/2} \tag{3.6.7}$$

where:

v = average air velocity (m/s), responsible for the convective mass transfer, that is the average velocity on the emitting surface.
ρ = air density (kg m⁻³)
μ = air viscosity (kg/m/s).

Density and viscosity are functions of temperature.

By using the Fick law in order to describe the mass transfer of a compound in a gas stream due to diffusion at the interface between liquid and gas phase, and by making a material balance of the compound i, assuming that the molar flow of the compound i at the wind tunnel outlet is equal to the molar flow of i emitted at the liquid-gas interface, it is possible to obtain an equation that relates the concentration of the generic compound i to the transfer coefficient as well as the gas stream velocity:

$$c_i \propto \frac{K_C}{v} \tag{3.6.8}$$

According to the above described Prandtl boundary layer theory, K_C is proportional to $v^{1/2}$. For this reason, in the case of convective mass transfer from a liquid to a gas phase it may be proven that:

$$c_i \propto v^{-1/2} \tag{3.6.9}$$

And, analogously, if odour is considered:

$$c_{od} \propto v^{-1/2}$$

Given the relationship between odour concentration, SOER and OER, it may be derived that (Bliss *et al.*, 1995; Sohn *et al.*, 2005):

$$SOER \propto v^{1/2} \tag{3.6.10}$$

$$OER \propto v^{1/2} \tag{3.6.11}$$

Based on this model, it is evident that the odour concentration, as well as the SOER and OER, are a function of the air velocity on the sampled surface. In other words, environmental odour emissions from area sources depend on the wind speed on the emitting surface. Analogously, during sampling, odour emissions are a function of the air velocity inside the sampling hood.

The previously described model has, up to now, been validated for liquid area sources, showing a good correspondence between theoretical and experimental results (Capelli *et al.*, 2009). On the contrary, such good correspondences have not been found for solid area sources. This may be due to the fact that mass transfer from a solid to a gas phase is a more complex process, involving other phenomena such as diffusion within the solid and effects of turbulence on the monitored surface. For this reason, other variables should be taken into account, such as the solid porosity and composition, the surface roughness, and so on. Nonetheless, the volatilization of organic compounds from the surface to the atmosphere may be described by similar models as those used for the case of liquid area sources (Zhang *et al.*, 2002). In more detail, the SOER and OER may be expressed as an exponential function of the air velocity:

$$SOER, OER \propto v^{n} \tag{3.6.12}$$

where n is an experimental exponent that depends on the considered conditions, which is in general not equal to 0.5, as it is for liquids.

According to the Equations (3.6.7–3.6.12), it is possible to observe that the odour concentration measured at the wind tunnel outlet decreases with the air velocity. In general, odour concentration values below 50–100 ou_E/m^3 are not easily measured by dynamic olfactometry, as they tend to be similar to the typical values of nonodorous ambient air. For this reason, sampling on not highly emissive surfaces (e.g., wastewater oxidation tanks, exhausted landfill surfaces) may entail some difficulties. In such cases, it may be useful to conduct the sampling operations using low air flows in order to have low air velocities on the surface to be sampled (about 1–10 cm s^{-1}), with the aim of preventing that the odour concentration values at the hood outlet be too low to be measured (Capelli *et al.*, 2009; Frechen *et al.*, 2004). Given the direct dependence of the measured odour concentration from the air velocity on the sampled surface, it is important to indicate the velocity adopted for odour sampling on the olfactometric report.

3.6.3   Volume Sources

Volume sources are typically buildings from which odours come out, intentionally, through naturally ventilated ducts, as well as unintentionally, through doors, windows or other openings. The OER estimation in such cases is complicated, as it is difficult to measure a representative odour concentration and, in general, it is not possible to define a precise air flow. In order to evaluate the OER, the air velocity in correspondence of the openings should be determined, or the air flow coming out of the building should be estimated using a tracing gas.

The OER is then calculated as follows:

$$OER = Q_{air} \cdot c_{od} \tag{3.6.13}$$

where:

> OER = Odour Emission Rate (ou_E/s)
> Q_{air} = volumetric air flow coming out from the building (m^3/s)
> c_{od} = measured odour concentration (ou_E/m^3).

3.6.4   Odour Emission Capacity

The odour emission capacity (OEC) is used to measure the potential odour emissions in liquids. The method, not universally accredited, was firstly presented by Frechen and Köster in 1998. The method defines the Odour Emission Capacity of a liquid as the total amount of odorants, expressed in ou_E/m^3 Liquid, which can be stripped from 1 m^3 of the liquid under given standardized conditions, and was determined by the following equation (Zarra *et al.*, 2011):

$$OEC = \int (C_{0d} - C_{100}) \ / \ V_L \ dV \tag{3.6.14}$$

where:

> C_{0d} = concentration of odour emission from air samples collected at suitable volume of fluxed air after beginning ($V_{to} = 0$) to V_{to} (volume at time t), detected by dynamic olfactometry in accordance with EN 13725:2003, expressed in terms of OU/m^3;
> C_{100} = concentration of odour emission established as the limit which defines the end of the test, fixed at 100 OU/m^3 (Frechen and Köster, 1998);
> V_L = volume of liquid sample used for the analyses.

Figure 3.6.5 illustrates the scheme of a test reactor implemented for OEC measurement.

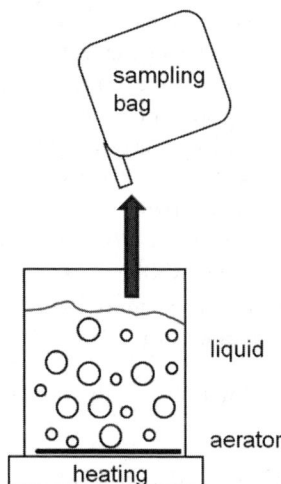

Figure 3.6.5 *Test reactor used for OEC measurement.*

3.7 Measurement of Odour Exposure by Field Assessment
T. Zarra, V. Naddeo and V. Belgiorno

Odour exposure can be performed using different sensorial, analytical and mixed methods.

3.7.1 Field Inspection

Field inspections are usually applied according to the German guideline VDI 3940 *Part 1 – Measurement of odour impact by field inspection: Measurement of impact frequency of recognizable odours. Grid measurement.* Grid measurements in Germany are mainly employed in approval and monitoring proceedings as well as urban development planning as means of determining the prior or total odour load in the assessment area. Grid measurement is a statistical survey method in which over a certain period, assessors record detected odours at grid intersection points in the assessment area in accordance with a certain methodology. The results are used for calculating the odour impact characteristic (number of hours per year of odour impact) for the assessment squares in the analysed area. The assessment area is considered as a circle whose centre coincides with the location of the source and whose radius is equal to 30 times the stack height or, for lower heights (<20 m), is at least 600 m. These distances can be changed for particular cases. In urban development planning, the assessment area is identical to the planning area.

The assessment square is the unit when considering the result of a single measurement, because an area value is better than a point one. At the beginning, a square with a side of 250 m is considered, which can be increased if there is no observed significant odour in the zone, or decreased if there are particular conditions or recorded nuisances. The position of the grids can be optimized according with the local features: wind direction, characteristic land use, accessibility of measurement points. Point inspections can also be implemented.

The survey period usually lasts six months or a year, during which 52 (13 × 4 measurement points per assessment square) or 104 measurement days (26 × 4 measurement points) must be obtained, respectively. However, it should be representative of several periods of the year and it must consider the cycles inside the factory. However, these factors affect the choice of a suited measurement plane: with a survey on the scale of 52 field inspections in six months, two to three field inspections are necessary for week; 104 field inspections in six months, four to five field inspections are necessary for week; and 104 field inspections in a full year, two to three field inspections are necessary per week.

To evaluate the odour impact characteristic, all the positive measurements (when the percentage odour time exceeds the value of 10%) must be referred to. To identify the emitter, the quality of the recorded positive measurement can also be considered. In this case, further data have to be recorded. The sum of the positive assessments (odour hours) is calculated for each assessment square as following:

$$n_A = n_{MP1} + n_{MP2} + n_{MP3} + n_{MP4} \qquad (3.7.1)$$

where:

n_A = number of odour hours per assessment square;
A = running index of assessment squares;
= n_{MP1}, n_{MP2}, n_{MP3}, n_{MP4} = number of positive single measurements (odour hours) at the measurement points ($_{MP1}$ to $_{MP4}$) of an assessment square (grid square).

The square-related odour impact characteristic is calculated from the number of odour hours per assessment square and the total number of samples with the following equation:

$$H_{rel,A,i} = \frac{n_{A,i}}{N} \qquad (3.7.2)$$

where:

$H_{rel,A,\ i}$ = square-related odour impact characteristic as the relative frequency of hours with odour, specified according to odour quality i;
$nA,\ i$ = number of odour hours per assessment square, specified according to odour quality i;
i = running index of the surveyed odour quality per square;
N = total number of samples ($N = 52$ or 104).

The square-related odour impact characteristics may have to be calculated for every odour quality surveyed. This differentiation is used for surveying the odour impact from individual companies, facilities or processes.

3.7.2   Community Surveys

Community surveys are usually implemented through the use of questionnaires to investigate about the annoyances caused by the odour releases in the neighbouring population. They can be aimed at different purposes, thus the type of question used can change. The most common purposes are to:

- provide a daily on site record that will demonstrate adequate running of the site;
- recording odour events to quantify the degree of annoyance;
- evaluate the extension of the involved area and the eventual presence of interference features due to other installations.

People are usually questioned about: presence of odour events in the area, their characterization (expressing quality, durability and frequency of the odours), individuation of the source, recording the related complaints, evaluation of probable negative effects on the health and behaviour linked to them. It is important to choose a representative receptors sample, it has been observed that differences on social status and age can give different results, so the sample should be as heterogeneous as possible. It is also important to choose people who are located in an optimal way around the investigated source. It has sometimes been observed that 'about half of the people had not/would not complain in the event of experiencing odours' (McKendry *et al.*, 2002). At the same time, they do not know what the right institutional body to which refer for this scope is and only half of them are aware of the existence of already existing institutional groups or local forums for the population dealing with this problem (McKendry *et al.*, 2002). Therefore, deciding to start this kind of investigations can be useful to make the population aware about the processes present in the area and their harmful effects. In fact, it has been observed that an odour nuisance is often directly linked to harmful effects, but it is usually not true. It becomes important to be clear when involving people in the analysis so as to obtain their collaboration. Example of odour surveying questionnaires are described in several guidelines, for example in the 'Good Practice Guide for Assessing and Managing Odour in New Zealand' (2003) and the VDI Guidelines (VDI 3883 Part I, VDI 3883 Part II).

3.7.3   Odour Diaries

Odour diaries, usually sought from a sub-group of the community rather than being passively received by a regulatory agency or industry, quantify the zone of influence from a specific odour source and the associated characteristics of the odour exposure pattern. This method measures the extent of adverse effects caused by the odour exposure pattern and provide a method for obtaining information from the community about odour impacts.

The most common purposes to carry out an odour diary programme are to:

- collate exposure pattern information over a defined period of time (comprehensive diary programme, CDP). Collected data can be used to define the percentage of time (hours per year) that people are exposed to a specific odour source, as well as the typical strength and character of the impacts.
- confirm whether a odorous plant is causing occasional odour impacts (basic diary programme, BDP).

Obviously, the CDP requires more detailed information and analysis than the BDP. The CDP is appropriate when it is necessary to establish the likelihood that adverse odour effects are occurring. On the contrary, the BDP is used more as a diagnostic tool, which can be implemented with minimum effort to confirm whether or not odour emissions from a specific source can be detected by a neighbouring community. This information is not sufficient to establish the extent of adverse effects being caused, so the main programme used is the CDP.

Generally, to have a complete representation, the information that collected in a CDP should include the date and time of the investigated day, the duration of the event, the main meteorological conditions as well as the character and strength of the odour. A common criticism of odour diaries is that the information received is often not filled in correctly, and that volunteer diarists can quickly lose enthusiasm for the programme. For this reason, this technique is not very widespread.

3.7.4   Plume Measurement

Plume measurements are usually applied according to the German guideline VDI 3940 *Part II – Measurement of the Impact Frequency of Recognizable Odours: Plume measurement*. The main field of application of plume measurement is for the reverse calculation of emission source strengths by using suitable dispersion models. Furthermore, it can also be used in the calibration and validation of dispersion models, for the determination of plume extent and in the pinpointing of causers of emissions. The methods used to reach these results are widely explained in this guideline. Further instructions about the creation of a dispersion model (main required variables to determine the plume extension) are contained in VDI 3788. Plume measurements are used for gathering information on odour impact within an odorant plume. The investigation takes place in specific (measured) meteorological conditions (VDI 3940 Part II). The most important are: wind direction, wind speed, stability categories (or dispersion categories which depend on the daytime hours and wind speed). As for grid field inspections, the percentage time with odour detection (percentage odour time) during a single measurement is used as the measured variable for odour impact. The criterion of the positive single measurement/odour hour is employed for plume measurements only in order to define the plume boundary. To calibrate or validate dispersion models, the actual percentage time of odour detection during the measurement cycle is required.

The odour plume of an emission source is the extension of the area in which the odours are clearly recognizable and which is dependent on the operating condition and current dispersion situation of the source. Plume measurement is always tied to the facility in question. The plume boundary is reached by definition when the percentage odour time reaches a predetermined percentage. A percentage odour time of 10% of the measurement cycle has been conventionally agreed. The plume axis is the line in the direction of dispersion on which close to ground level in relation to the plume's transverse extension the respective peak percentage odour times are located. It usually coincides with the wind direction. The dispersion direction is determined by ascertaining the current wind direction 2 m above the ground (potential impact range). For this, estimation methods using a compass, flag, balloon or smoke cartridge are sufficient.

A plume measurement consists of measurements each lasting ten minutes at several intersection lines at right angles to the current wind direction. An intersection line measurement consists of at least five measurement points and five assessors. The distance between the intersection lines and measurement points depends on the anticipated extension of the odour plume, which is affected by the structural height of the emission source, the odorant flow rate, the current meteorological conditions as well as buildings/vegetation (orography) and topography. They do not necessarily have the same size. The intersection lines and measurement points are entered on a map. The exhaust air plume of a facility emitting odorants is directly affected by wind direction, wind velocity and dispersion category (stability of the atmosphere) There are six dispersion categories; from extremely stable to extremely unstable, they are determined by the wind speed and time of day. The results of the single measurements of a plume measurement are compiled in a table. For a good description of the plume, a sufficiently large number of intersection line measurements are necessary. According to convention, a total of at least 30 intersection line measurements are required on at least five measurement days (total data set). The main differences to be considered in plume measurements for different purposes are described in the guidelines. The main considered purposes are: estimating odour emissions (by reverse calculation of source strength), validation of dispersion models and determining plume extent.

3.8 Measurement of Odour by Sensor Arrays

N. Dutta and M. Bhuyan
Department of Electronics and Communication Engineering, Tezpur Central University, Tezpur,
Assam, INDIA

3.8.1 Odour Sensors

Odour sensors have been in practical use for many years. Odour sensors incorporate a sensor element that reacts to odours. The odour recognition process in an artificial nose begins in the senor system which is responsible for the measurement of the odorant stimulus through the sensitivity of its sensors. The gas sensor area consists of several different types of sensing materials that contribute to the gas sensor. Some of the most commonly used gas sensors include, metal oxide semiconductor, conducting polymer sensors, acoustic wave sensors, field-effect gas sensors, pellistors, and fibre-optic sensors. The basic principle of the E-nose is that each odour leaves a characteristic pattern or fingerprint on each sensor array. The degree of selectivity and type of odours that can be detected largely depend on the choice and number of sensors in the sensor array. The sensors are often mounted in an airtight chamber containing gas inlets and outlets in order to control the gas flow. The signals from each sensor are measured and processed, usually by an analogue to digital conversion that is performed by a computer. After the signal processing, the data is transformed by a variety of pre-processing techniques designed to reduce the complexity of the multi-sensor response. As shown in Figure 3.8.1, the headspace generation ensures the concentration of volatiles of organic solvents before and during sampling. During the collection time, the sensors are exposed to a constant flow of gases through pipelines inside the electronic nose, with the response beginning. During the purging operation, the sensor heads are cleared with a blow of fresh air so that the sensors go back to their baseline values.

3.8.1.1 *Types of Sensors*

Among the different types of sensors, conducting polymers, bulk acoustic devices and metal oxide semiconductor (MOS) are most commonly used.

Conducting Polymer Sensors.   An active material of conducting polymer such as polypyrolles, thiophenes, indoles, or furans are used in this type of sensor. The polymer material is electropolymerized between two electrodes with a gap of about 10–20 μm. On application of odorants, the bonding of the molecules changes and affects the transfer of electrons along the polymer chain, causing a change in the conductivity. Conducting polymer sensors can work at ambient temperature and do not need heaters like MOS sensors. However, their

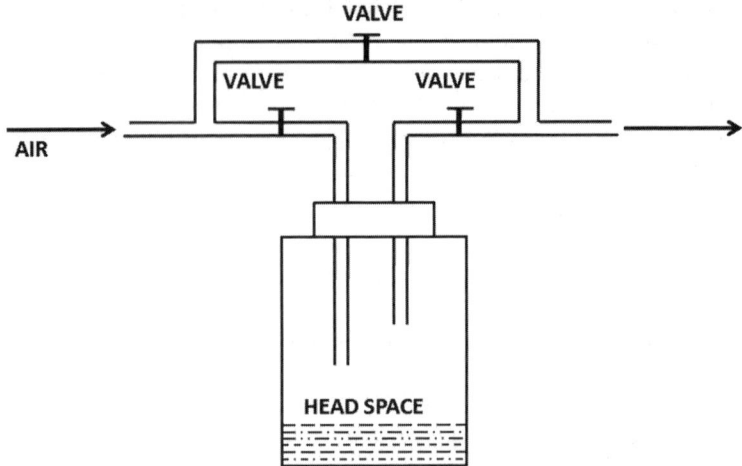

Figure 3.8.1 *A headspace sampling system.*

major disadvantage is that the manufacturing of the sensors by electro polymerization is difficult, time consuming and prone to be affected by humidity.

Acoustic Sensors. The most common acoustic sensors are quartz crystal microbalance (QCM) and surface acoustic wave (SAW) devices as shown in Figures 3.8.2(a,b). These sensors are operated by a mass changing principle. When exposed to the odorant, the surface of these sensors absorbs the gas molecules decreasing the resonance frequency of the resonating disc. Upon exposure to a reference gas, the resonance frequency of the sensor goes back to its original value. The selectivity and sensitivity of the QCM sensor depends on the coating of the polymer material. The size and mass of the quartz crystal can control the response and recovery time of the sensor. The SAW sensor operates at a much higher frequency than the QCM sensor, hence the change in resonance frequency is also much higher in this sensor. A typical SAW sensor produces a resonant frequency of several hundred megahertz. Due to the higher operating frequency, the signal-to-noise ratio is found to be less in SAW sensors. The only drawback of these sensors is that they require more complex electronic processing circuits compared to conductivity sensors.

Metal Oxide Semiconductor (MOS) Sensors. These sensors are based on change of resistance when exposed to volatile organic compounds. The cause of the sensor resistance change is due to an ionosorption process and explained in terms of electron transfer from the semiconductor to adsorbed surface species. The adsorption process that is responsible for the sensor signal is strongly influenced by the presence of the pre-adsorbed species (such as ionosorbed oxygen, hydroxyl groups, carbonates, etc.) and by only measuring the change of resistance upon exposure to the target gas, is the electrical effect of quite complex surface reactions measured. MOS sensors are the most popular in E-nose applications because of their low-cost and flexibility associated to their production, simplicity of use and higher discriminating power. Upon exposure to an oxidizing substance, the surface of the MOS sensor undergoes a chemical reaction, which translates into a measured change in

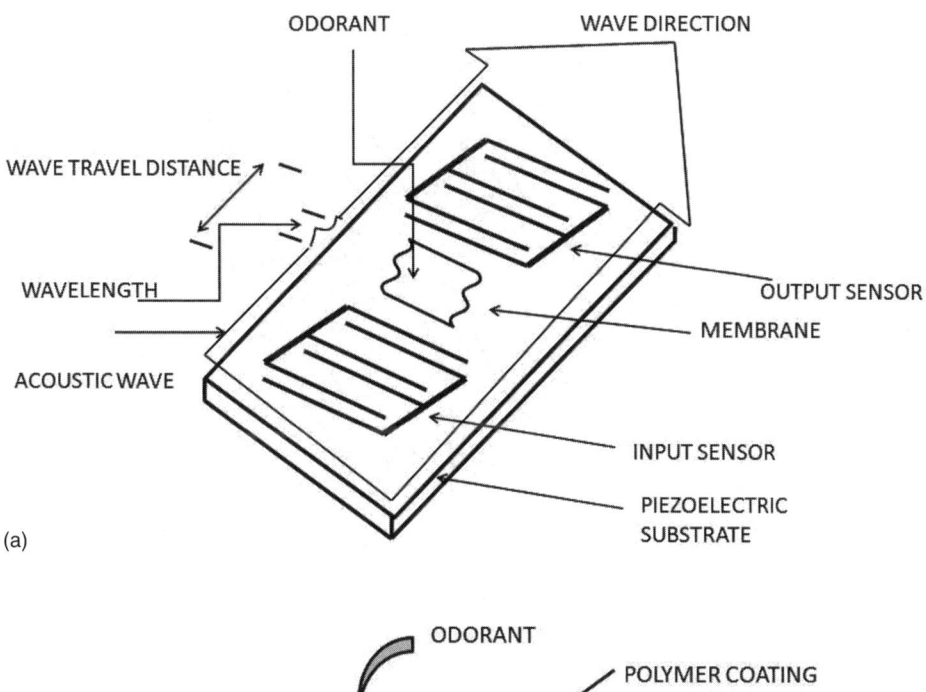

(a)

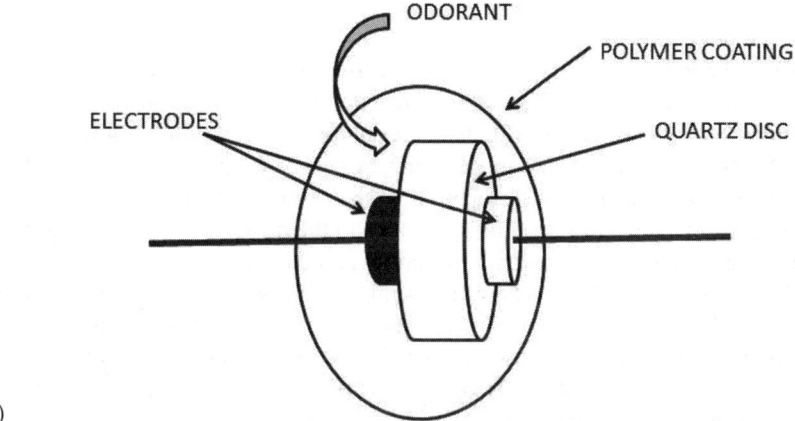

(b)

Figure 3.8.2 *(a) SAW sensor. (b) QCM sensor.*

conductance across the sensor. The interest value during the data collection from the nose is the change in voltage over a period of time. Figure 3.8.3 shows the constructional feature of the MOS sensor.

The basic principle of the device is that it allows gaseous compounds to react with the catalytic metal and produce species that are able to diffuse through the metal film and absorb onto a metal insulator. A reversible reaction occurs where the analyte binds to the surface of the sensing material. The binding is determined by the intermolecular forces between the analyte and the sensing material but is usually characterized by a hydrogen bonding. When the odour concentration is removed, the analyte does not change but will

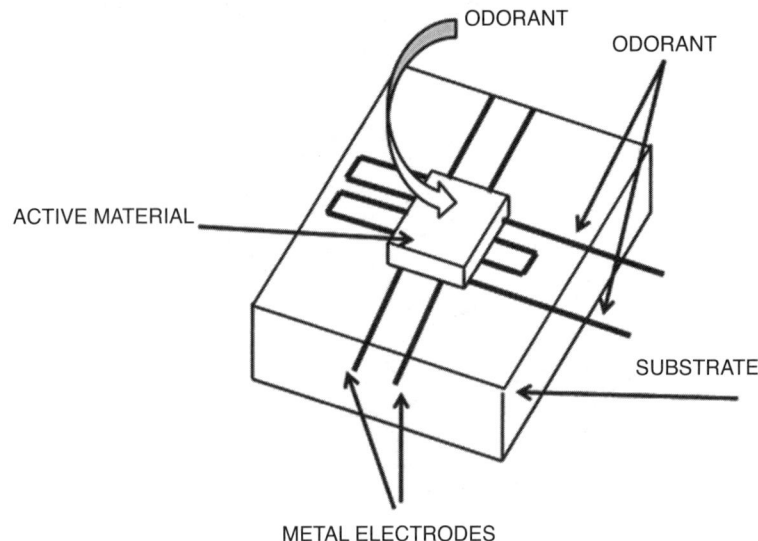

Figure 3.8.3   *MOS sensor.*

dissociate from the sensing material. This type of reaction is similar to the interaction between odours and receptor proteins in biological systems. When an analyte undergoes a chemical change at the sensor surface, that is, catalysis, an irreversible reaction occurs. Sensors exhibiting such reactions are mostly SnO_2 sensors, which have a high sensitivity to specific odours. The change in the properties of the sensors causes the change in conductivity. Gas sensors operate by binding molecules to the device surface through one or more mechanisms including adsorption, absorption, and chemisorption and coordination chemistry. The choice of binding mechanism has important implications for the selectivity and reversibility of the sensing system. Chemisorption is preferred when a highly selective system is required. The high binding strength will, however, result in poor reversibility. Chemisorption of oxygen molecules in the air environment takes place in the forms of O^{2-}, O^- and O^{2-}. An equilibrium state exists among the chemisorbed species on the surface of SnO_2 at a constant temperature. During the interaction of the sensing material with gas molecules to be detected, the conductance of the gas sensors varies. The surface reactions occur at low temperatures while at higher temperatures bulk reactions between point defects in the SnO_2 lattice and gaseous oxygen molecules take place. Therefore the sensor material is heated up to a temperature of the order of 400°C to 600°C with the help of a heater deposited in the sensor module. In both cases, adsorption at active sites occurs first, followed by some surface catalytic reactions. Similar reactions also occur at grain boundaries or at three-phase boundaries (i.e. at metallic contacts on surface metallic clusters). The differences in response patterns between the sensors are subjected to multivariate analysis and other statistical data processing, and classification of odour types are performed.

Gas sensors have attracted the attention of many researchers interested in gas sensing under atmospheric conditions due to the: low-cost and flexibility; simplicity of use; large number of detectable gases/possible application fields (Williams, 1999; Barsan *et al.*, 1999; Korotcenkov, 2005; Ihokura and Watson, 1994).The initial research on metal oxide-gas reaction effects was conducted by Heiland (1954), Bielanski *et al.* (1957) and Seiyama *et al.* (1962) and the decisive step was taken when Taguchi brought semiconductor sensors based on metal oxides to an industrial product (Taguchi-type sensors, 1971). Nowadays, there are many manufacturers producing this type of sensors, such as Figaro, Fast Ion Sensor (FIS), Magnetospheric Ion Composition Sensor (MICS), Underground Storage Tank (UST), CityTech, Applied-Sensors, NewCosmos, and so on. Over the last decades, the gas sensors deploying metal oxide semiconductor (MOS) sensors have been manufactured using TiO_2, WO_3, SnO_2 and Ga_2O_3 powders and thick-films in particular. The sensors known as Taguchi-sensors or TGS (Taguchi Gas Sensor) are the most commonly used gas sensors. The MOS gas sensors are typically used in ventilation control, combustible gas leak warning, combustion control and breathe alcohol detection applications and so on.

Recently, MOS gas sensors have been based on the micro hot plates fabricated using advanced silicon technology while conventional Taguchi gas sensors are based on ceramic fabrication processes. The advantages of micro hotplate MOS sensors are faster startup, faster response, lower power consumption, more accurate temperature control as well as more accurate detection and identification characteristics (Utrianinen *et al.*, 2008).

Field Effect (FE) transistors operating in the diode coupled mode have also recently been used as gas sensors (Utrianinen *et al.*, 2008). The gas sensitive properties are achieved by depositing a catalytic metal stack on the device and the gate attains a gas sensing sensitivity. The sensitive layer can be designed to be specific and sensitive to hydrogen containing toxic gases, such as NH_3, HCN and H_2S. Combinations of the dual device used are:

(a) One FE sensor and one MOS sensor on the same header.
(b) Two different MOS sensors.

The ideal gas sensor would exhibit reliability, robustness, sensitivity, selectivity and reversibility. As high selectivity with high reversibility in the sensors is difficult to attain, either a compromise is necessary or the sensor detection layer must be regenerated. Today, the use of chemical sensors to measure and analyse odours is a growing field that attracts interests from the sensor and pattern recognition communities. A variety of sensing technologies are available and presently there are several different kinds of commercial electronic noses which use these sensors for different applications.

3.8.2   Electronic Noses

The odour sensor if used alone can generate a signal proportional to the concentration of the odour. However, they are not able to obtain statistical metrics, signal features and patterns as any human or animal sentience can (Bhuyan, 2010).

The five human senses: sight, hearing, smell, touch and taste; all adopt an intelligent sensing mechanism which is not possible in a single sensor system. The human senses use an array of hundreds of biological sensing cells that works in parallel to generate a signal pattern, which is transmitted to the brain. The brain recognizes the pattern of the signal rather than its magnitude. The pattern can be obtained because the biological receptor cells have different sensitivity to different classes of signal. This technique of human sensing by array based sensors is mimicked by the E-nose.

In an artificial olfactory system such as the E-nose, the olfactory receptor cells are replaced by a chemical sensor that is able to characterize different gas mixtures, just like the human nose. It is an intelligent instrumentation system, which consists of three functional components that work in tandem; an integrated chemical sensor array, an interfacing electronic circuitry and a pattern recognition (PARC) software paradigm. Sensor electronics convert the chemical signal into an electrical signal as well as amplify and condition it. Interfacing electronics circuitry is used to digitize and store the response signal for processing. Patterns from known odours are used to build a database and train a pattern recognition system so that unknown odours can subsequently be classified and identified. The degree of selectivity and type of odours that can be detected largely depend on the choice and number of sensors in the sensor array. Hence the E-nose is a combined chemical sensing and data analysis system.

An E-nose, as the human olfactory system, discriminates new patterns and associates them to specific odours by training the system, using a set of known data samples. Sample delivery systems are used to transfer the odour from the source (typically by a miniature pump) to a sensor chamber in which an array of selected gas sensors is installed. The process of odour delivery can be summarized as follows: At the beginning of a sampling process, the odour delivery system drives each sensor to a known reference state by applying a reference gas (for example, fresh air) to the sensor chamber. The readings of the sensors in this state are known as the *baseline level*. Then, the delivery system exposes the sensors to a given odour, producing first a transient response as the compounds starts interacting with the surface and bulk of the sensor's active material. After a few seconds to a few minutes, the sensors reaches a steady state and, finally, the odour is removed from the system by pumping the reference gas to prepare the system for a new measurement cycle. The above mentioned steps are known in e-nose literature as a '*three phase sampling process*' and they are usually carried out in chambers where humidity, temperature and exposure to the analyte are controlled (Trincavelli *et al.*, 2009). Figure 3.8.4 shows a typical response of

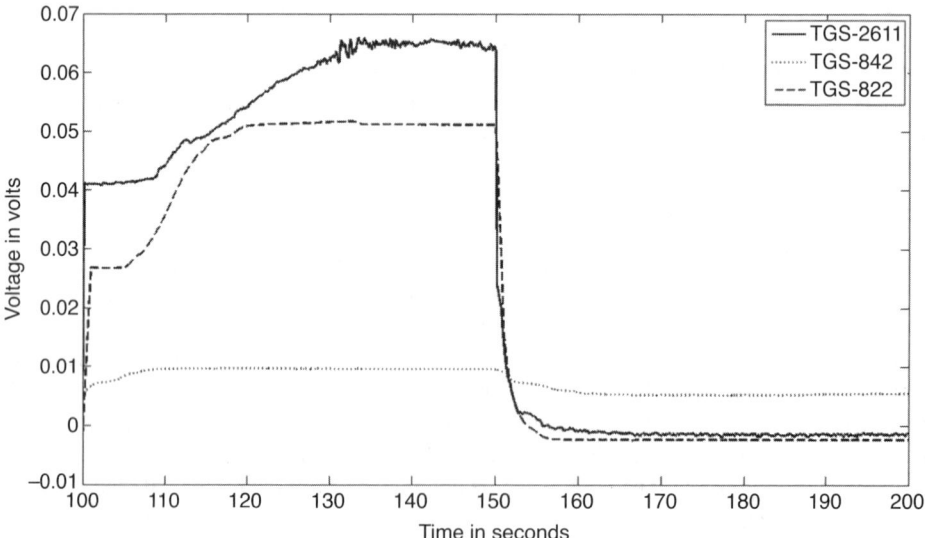

Figure 3.8.4   *Response of an array of E-Nose sensors.*

a given array of gas sensors to a three-phase sampling process. The three phase sampling process has been widely used in laboratory-based applications with a significant amount of success. Many articles on this subject have been published over the last years, mainly in relation to the food and beverage industry (Olafsoon *et al.*, 1992; Winquist *et al.*, 1993; Kleperis *et al.*, 1999; Hermle *et al.*, 1999). Commercial E-nose systems also invariably employ these basic stages of operation.

In general, the sensor array comprises of a selected group of non-specific gas sensors, which means that the sensors in the array have different responses or selectivity to certain compounds. The different response rates and intensity levels of the sensors in the array will produce a characteristic response pattern (i.e. *'finger print'*) when exposed to volatiles with similar chemical content, whereas a different response pattern will be produced when the array is exposed to a volatile with different chemical characteristics. When a sensor array composed of *n* sensors (say), where each sensor will produce a response x_{ij} during a given experiment *j* then the array response can be represented by a vector given by:

$$x_j = (x_{1j}, x_{2j}, \ldots . x_{nj})^T \tag{3.8.1}$$

When *N* experiments are repeated with the same sensor array, the response can be represented as a response matrix *X* given by:

$$X = \begin{pmatrix} x_{11} & x_{12}\ldots & a_{1N} \\ x_{21} & x_{22} & x_{2N} \\ a_{n1} & a_{n2} & a_{nN} \end{pmatrix} \tag{3.8.2}$$

Where each column represents a response vector associated with a particular experiment, whereas the rows are the responses of an individual sensor.

3.8.3 Signal Processing and Pattern Recognition

It has already been mentioned that odour classification and discrimination is performed by an E-nose with the help of algorithms supported by learning, training and decision making. These algorithms are basically data and signal processing that work sequentially. The four sequential stages of E-nose signal processing and pattern recognition are as follows:

1. Pre-processing
2. Feature extraction
3. Classification
4. Decision making

Figure 3.8.5 shows the different stages of the E-nose signal processing stages.

3.8.3.1 Pre-Processing

The sensor response in terms of variation of resistance in the case of chemoresistive type and polymer sensors, or variation of mass and resonance frequency in the case of piezoelectric sensors is processed to obtain a variation in voltage or current. Sensor drift is one of the factors that have to be compensated for in this stage. The most important function of this stage is the sensor output normalization. Since the sensors in the array have different sensitivities, the voltage levels of the output signals from the sensors will be different.

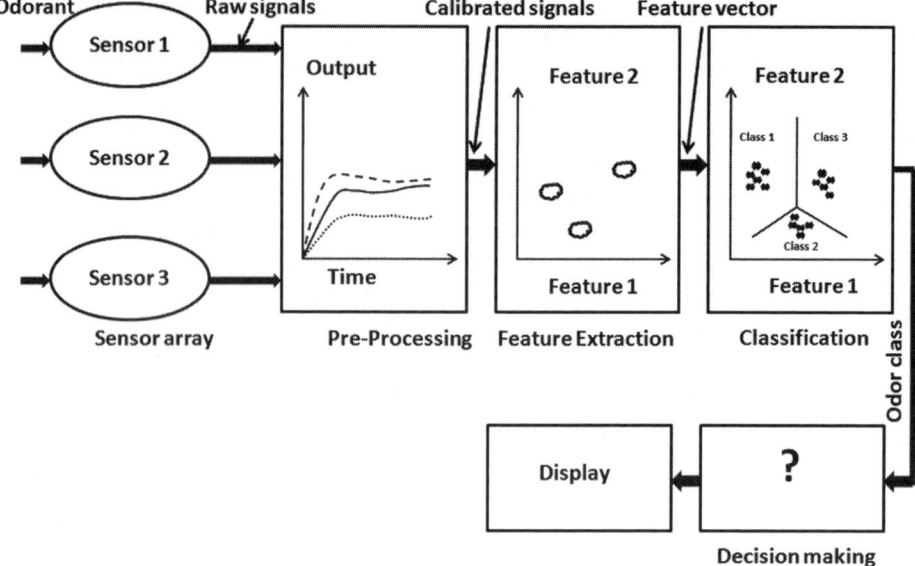

Figure 3.8.5   *Block diagram of various stages of the E-nose system.*

Hence, the signal levels need a standardization or normalization. Various normalization methods are available such as:

1. *Liberalization:*

$$X_{ij} = \log \left| (Y_{ij}^{\max} - Y_{ij}^{\min}) \right| \tag{3.8.3}$$

$$X_{ij} = \sqrt{\left(\left| (Y_{ij}^{\max} - Y_{ij}^{\min}) \right| \right)} \tag{3.8.4}$$

2. *Vector Normalization:*

$$R_{ij} = X_{ij} \Big/ \sum_{ij} (X_{ij})^2 \tag{3.8.5}$$

3. *Sensor Normalization:*

$$R_{ij} = (X_{ij} - X_{ij}^{\min}) / (X_i^{\max} - X_i^{\min}) \tag{3.8.6}$$

4. *Sensor Auto Scaling:*

$$R_{ij} = \left| (X_{ij} - X_{ij}^{mean}) \right| \sigma_i \tag{3.8.7}$$

Where σ is the standard deviation. X's are the calculated or pre-processed values and Y's are the observed or measured values for sample odour and $Y_{ij}^{\max}$ and $Y_{ij}^{\min}$ are its maximum and minimum values. The normalized data are stored in the memory for pattern recognition.

3.8.3.2   Feature Extraction

This process uses mathematical techniques to find relevant but hidden information from the data. The techniques adopted for feature extraction are generally supervised linear transformations such as principal component analysis (PCA), where the data is expressed in terms of a linear combination of orthogonal vectors accounting for a certain amount of variance in the data. PCA orthogonalizes the components of the input vectors, orders the resulting orthogonal components so that those with the largest variation come first and eliminate those components that contribute the least to the variation in the data set. Each principal component in PCA is a linear combination of the original variables. The first principal component is a single axis in space. When each observation is projected onto that axis, the resulting values form a new variable, while the variance of this variable is the maximum among all possible choices of the first axis. The second principal component is another axis in space, perpendicular to the first. A new variable is generated when the observation on this axis is projected, the variance of which is the maximum among all possible choices of this second axis. The full set of principal components is as large as the original set of variables. However, the sum of the variances of the first few principal components exceeds 80% of the total variance of the original data (Haykin, 1999). The principal components of the raw data can be computed when all the variables are in the same units. Standardizing the data is important when the variables are in different units or when the variance of the different columns is substantial. The first three principal components are generally considered in most of the applications.

3.8.3.3   *Classification*

The process of separating individual samples from universal samples space is called classification. The classification helps to sort out samples to the nearest matching class or even to the absolute class of the sample as per the flexibility of algorithm rules. When the E-nose response data are projected on an appropriate low-dimensional space, the classification stage is used to identify the patterns that are representative of each odour. The classification stage is able to assign to the data a class label to identify the odorant by comparing its patterns with those compiled during training. The different tools used for performing classification work are shown in Figure 3.8.6.

The decision making stage makes judgment to the classification and even determines that the unknown sample 'does not belong' or 'it belongs' or 'it nearly belongs to' any one of the databases.

3.8.4   Artificial Neural Network

An artificial neural network (ANN) is man's attempt to computationally mimic the brain. The most commonly used pattern recognition technique with electronic noses is the ANN. They are highly parallel mathematical constructs that have been inspired by the biological nervous system. ANN consists of a number of processing elements called 'neurons' which represent the biological (olfactory) neurons, and their interconnections, the synaptic links. The strengths of these connections are called 'weights' and are determined either during a training phase (or learning phase) for supervised ANN, or by an algorithm for unsupervised ANN. There are many different types of ANN structures that have been applied to solve odour identification problems such as in artificial neurons. This signal transfer is simulated by multiplication of the input signal, x, with the synaptic weight, w, to derive the output signal y. The artificial neuron is the heart of every neural network. A single neuron receives input signals, x_i, from n neurons, aggregates them by using the synaptic weights, w_j, and

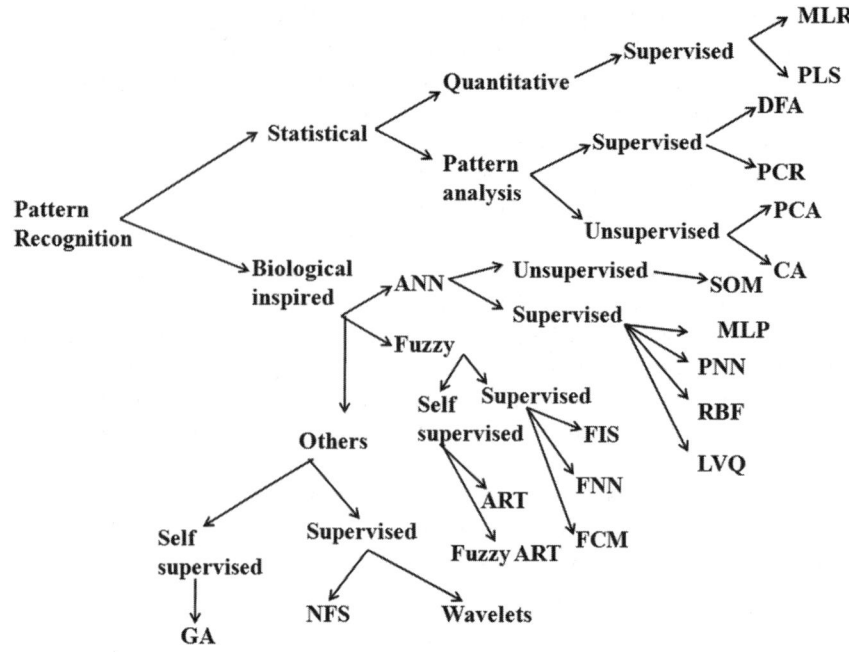

Abbreviations

ANN-Artificial Neural Network	NFS-Neuro Fuzzy System
ART-Adaptive Resonance Theory	PCA-Principal Component Analysis
DFA-Discriminate Function Analysis	PCR-Principal Component Regression
GA-Genetic Algorithm	PLS-Partial Least Square
LDA-Linear Discriminant Analysis	RBF-Radial Basis Function
MLR-Multiple Linear Regression	SOM-Self Organizing Map

Figure 3.8.6   *Tree diagram showing various classification schemes.*

passes the result after conversion by a transfer function (activation function) as the output signal Y_i (Figure 3.8.7). Some important activation functions are given in Table 3.8.1 A general neural network consists of an input layer, one or more hidden layer(s), and an output layer and all these layers are fully connected. Figure 3.8.8 shows that the output signals (y_i) of the neurons of one layer act as the input signals (x_i) for the neurons of the following layer.

The input layer of the ANN interfaces the signals from the external world to the ANN. Hidden layers are the real classifiers that work for the classification algorithm of the ANN. The output layer is considered a collector of the features detected and producer of the target or result.

The main advantages of neural networks are its adaptability in terms of learning, self-organization, training, and noise-tolerance. ANN is the most commonly used pattern recognition techniques with electronic noses. ANN makes it possible to detect a greater number of compounds than the number of unique sensor types. Furthermore, the less selective sensors can be rendered much more selectivity when used in conjunction with an ANN. This is because, once a network is trained for a particular odour, the consecutive recognition

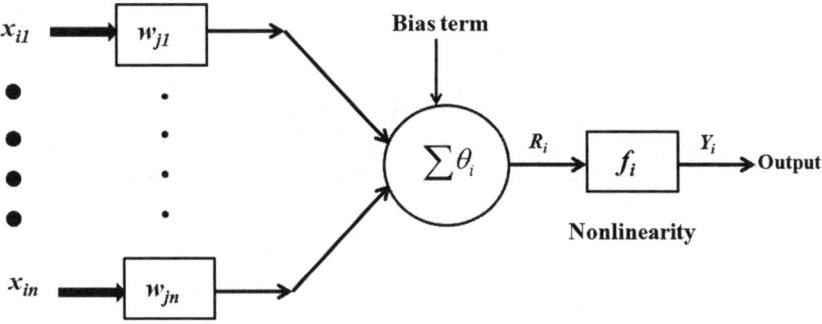

Figure 3.8.7   *Operation of a single neuron.*

process can occur rapidly and effectively. Learning in a neural network is accomplished through an adaptive procedure, known as 'learning rule'. The weights of the network are incrementally adjusted so as to improve the performance over time. The basic learning rules are the supervised learning, unsupervised learning and reinforcement learning.

3.8.5   Temperature Modulation

To improve the selectivity of the sensors, the operating temperature of the MOS sensors has been widely investigated in many works. Researchers have also explored different ways to achieve selectivity either by enhancing gas adsorption or promoting specific chemical

Table 3.8.1   *Important transfer functions for neural networks.*

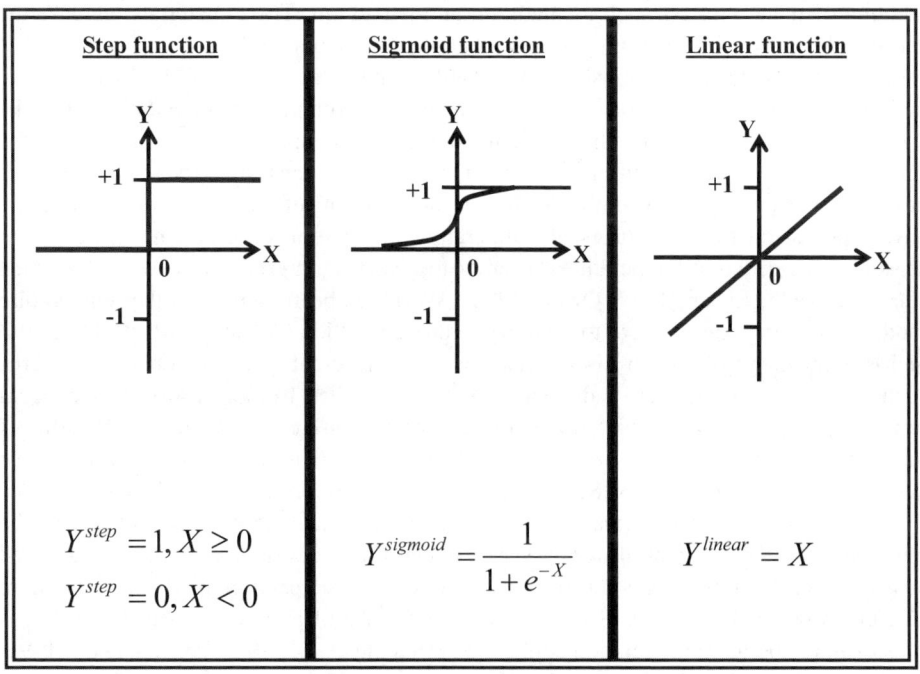

Step function	Sigmoid function	Linear function
$Y^{step} = 1, X \geq 0$ $Y^{step} = 0, X < 0$	$Y^{sigmoid} = \dfrac{1}{1 + e^{-X}}$	$Y^{linear} = X$

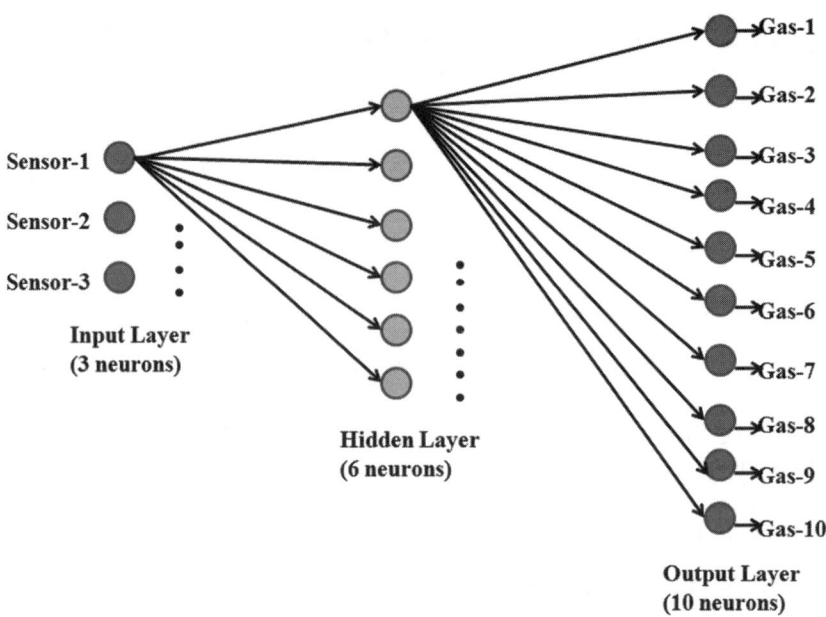

Sensor-1

Sensor-2

Sensor-3

**Input Layer
(3 neurons)**

**Hidden Layer
(6 neurons)**

Gas-1
Gas-2
Gas-3
Gas-4
Gas-5
Gas-6
Gas-7
Gas-8
Gas-9
Gas-10

**Output Layer
(10 neurons)**

Figure 3.8.8 *Architecture of an ANN for an E-Nose.*

reactions via catalytic or electronic effects using bulk dopants, surface modification methods and by the addition of metallic clusters or oxide catalysts (Göpel, 1994; Demarne and Sanjines, 1992). The selectivity of chemical sensors can also be strongly influenced by the addition of metal clusters such as platinum and palladium. These materials increase the sensor selectivity for reducing gases, for example, CO (Hohl, 1990). Apart from material selection, one of the most established ways of enhancing the selectivity of MOS sensors is by periodically varying the sensors' operating temperature. Temperature modulation alters the kinetics of the sensor through changes in the operational temperature of the device. As the sensor response changes at different working temperatures, measuring the sensor response at different temperatures is similar to having an array of different sensors. Researchers have reported on the advantages of temperature modulation on a ceramic metal oxide sensor at two different temperatures in order to detect the presence of carbon monoxide (Le Vine, 1975; Eicker, 1977; Owen, 1980). Work has been carried on the temperature modulation using square wave to quantify hydrogen sulfide (Advani *et al.*, 1983; Lantto and Romppainen, 1988). To discriminate between different gases, modulating patterns such as sawtooth, triangular and square were also applied to the sensors (Bulkowiecki *et al.*, 1986). The sinusoidal variation in the temperature also enhanced the classification of different gases. A number of works on the cyclic variations of the sensor heater voltage have been reported by many authors (Sears *et al.*, 1989a; 1989b). The response of the gas sensors to modulating temperature primarily depends on the analytical model which is based on the physical and chemical properties of the sensor material. The development of micromachined substrates for metal oxide gas sensors ensured operating temperature modulated in a more efficient way. Cavicchi *et al.* (1995) introduced the use of micromachined tin oxide gas sensors in temperature modulation applications (Suehle *et al.*, 1993; Cavicchi *et al.*, 1995; Cavacchi *et al.*, 1996). In many works (Heilig *et al.*, 1997), quantitative analyses of gases

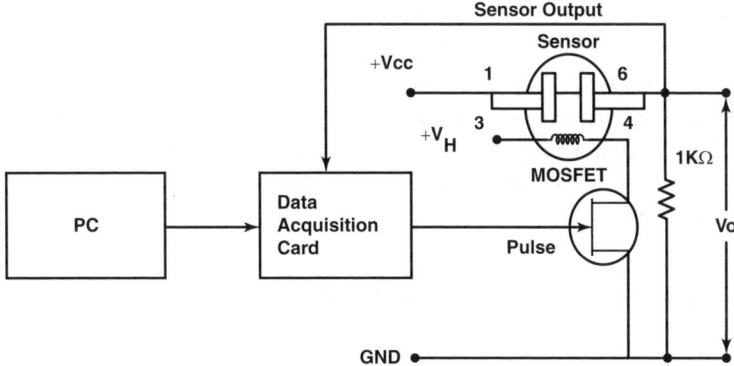

Figure 3.8.9 *Heater modulating circuit.*

with temperature modulated gas sensors have been reported. Temperature modulation of the MOS sensors can be done by applying a pulsating voltage to the heater of the sensor with the help of a heater driver circuit as shown in Figure 3.8.9 The voltage is pulsed by an oscillator circuit generating variable frequency and variable duty cycle signals. In PC based E-nose systems, the frequency and duty cycle is controlled by the PC.

3.8.6 Feature Selection

Feature is an optimal subset of measurements based on an evaluating criterion. Over recent years, many works related to feature extraction, feature construction, space dimensionality reduction, sparse representations have been performed by a large number of the researchers, with the resulting applications including bio-informatics, chemistry, text processing, pattern recognition, speech processing, vision perception, flavour discrimination and quality prediction. Feature selection is a pre-processing step to machine learning and is useful in reducing dimensionality, removing irrelevant data, increasing learning accuracy, and improving result comprehensibility. Feature selection based on static performance data, which are used as the training set, may not always provide the optimal classification results in dynamic environments. Hence, to enable the feature selection to adapt to the dynamic operation, it requires the system to be able to dynamically update the training set data. Selecting the most relevant variables is usually suboptimal for building a predictor, particularly if the variables are redundant. Conversely, a subset of useful variables may exclude many redundant, but relevant, variables.

3.8.7 Embedded E-Noses

In a conventional electronic-nose system, the data pre-processing, feature extraction and feature classification operation are performed by a personal computer either in offline or online mode. In an exclusively on-line technique, the recent trends in portable computing, the use of embedded systems at a reasonably low cost and size is the most effective solution.

The use of embedded technology has several advantages: availability of an abstraction layer for signal acquisition and control via an operating system, high level programming of the signal processing algorithms, large data storage in solid state disks, commercial-off-the-shelf hardware for Ethernet connectivity, serial ports, hardware for interfacing various types of displays, and so on. Development of a hardware based field type E-Nose system has the following features and advantages:

- Increased acceptability of the E-Nose by the user
- Shifting the PC based system to an 'electronic gadget' and field type system
- Portable and stand-alone type
- The inbuilt ANN and fuzzy logic can be trained locally
- Minimum user functions
- Low cost

The embedded E-Nose can be implemented onto three types of platforms; Microcontroller, FPGA and VLSI.

3.8.7.1 *Microcontroller-Based E-Nose*

The data acquisition, signal processing, sampling control and classification can be implemented in microcontroller based hardware with the following tasks:

- sensor array and sampling controller interface
- multiple channel A/D and D/A conversion
- controlling of the gas sampling valves and pumps
- programming of EEPROM to upload and download the sensor data and parameters
- programming to calculate ANN parameters
- programming to calculate ANN outputs

The block diagram of the embedded microcontroller based E-Nose is shown in Figure 3.8.10.

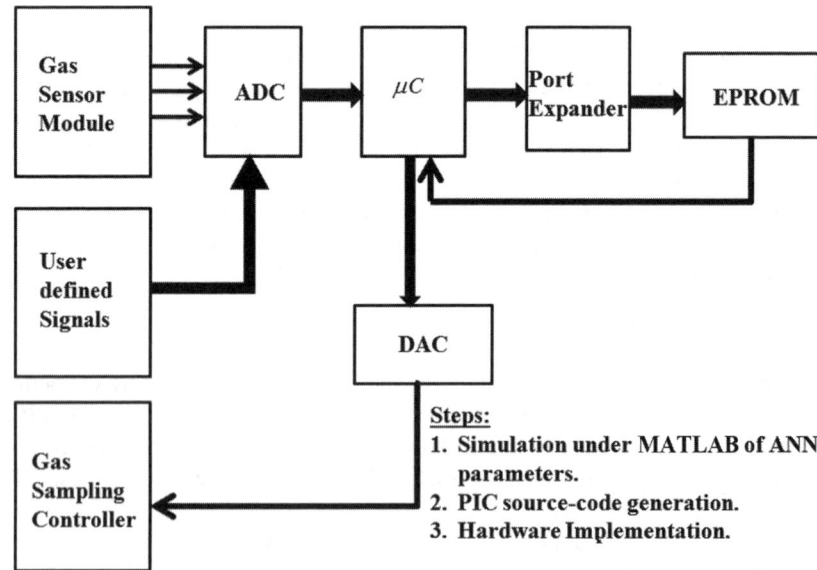

Figure 3.8.10 *Block diagram of the embedded microcontroller based E-Nose.*

The microcontroller based E-Nose is operated in two modes: Train mode and Test mode. The Train mode operations are:

- Initialization (software variables, constants and specific registers and I/O ports)
- The microcontroller performs the ADC and DAC operations
- The representative portion of the pattern signals are processed for storing
- The user defined samples are graded and stored in the memory
- The sensor data is uploaded to EEPROM which are used for training of the ANN
- The data are downloaded from the EEPROM to train the system by ANN Neuron and sigmoid function implementation
- The trained parameters (weights) are stored in EEPROM and the data are cleared

The test mode operations are:

- Initialization (software variables, constants and specific registers and I/O ports)
- The microcontroller performs the ADC and DAC operations
- The representative portion of the pattern signals are processed
- The trained data are downloaded from the EEPROM to activate the ANN
- The ANN outputs are calculated by microcontroller on-line for classification
- The classification output is displayed in display board

The design steps of an embedded E-Nose are:

1. Before implementing the system algorithm in actual hardware, the ANN model is simulated in MATLAB to have an idea of the odour output level so as to scale appropriately the PIC computed classification to be output via ports.
2. The discrete time model is implemented using the corresponding difference equation which will be coded in software to run the hardware.
3. The algorithm is first turned into code that runs on the microcontroller using a compiler to get the source code downloaded to the microcontroller.

3.8.7.2 FPGA-Based E-Nose

The functions of the FPGA based E-Nose are:

- Neuron and ANN network implementation
- Timing and control
- Display and
- ADC /DAC /EPROM interfacing

The main components of the FPGA based E-Nose are:

- Programmable Logic Design
- Configurable Logic Blocks (CLB)
- Neuron Implementation
- Input: 8 MOS Sensor signal (8-bit integer or 32-bit FP)
- Hidden Layer: with sigmoid neurons
- Output Layer: Single Linear Function (classification)

A basic Integer mode Multiply-Accumulate module is shown in Figure 3.8.11(a,b). There are two mathematical functions: addition and parallel multiplication. IEEE 754 format-32

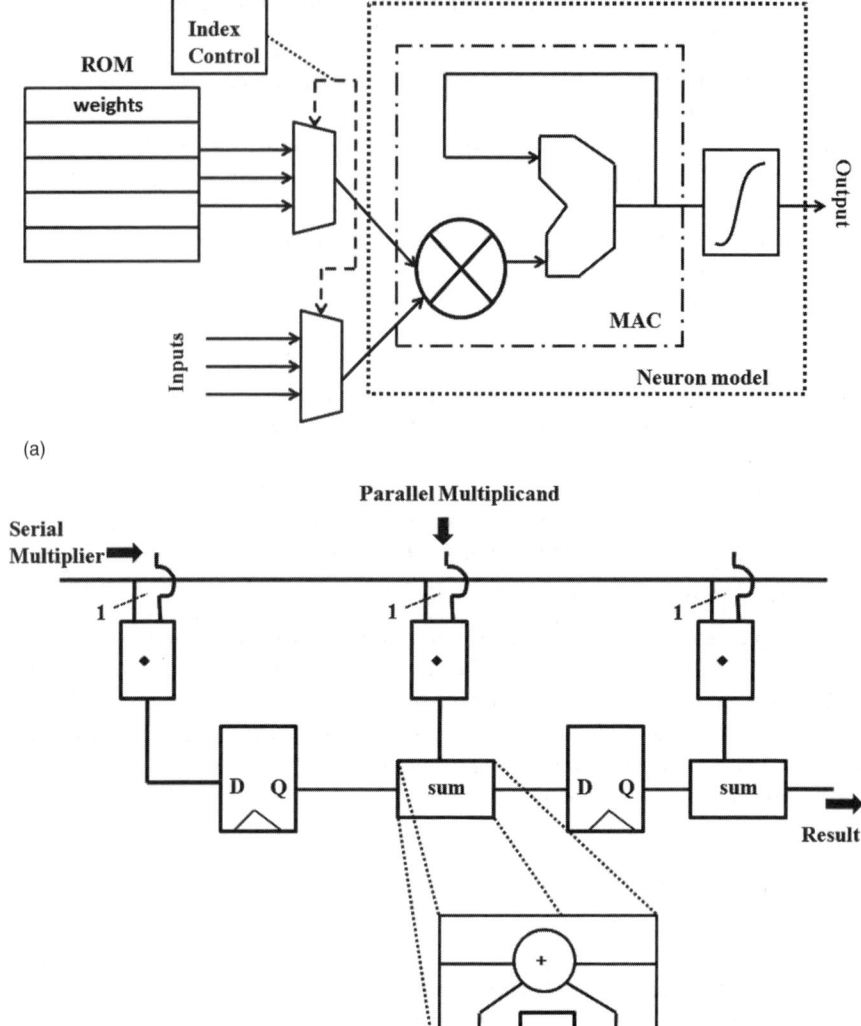

(a)

(b)

Figure 3.8.11 *FPGA-based multipliers: (a) Serial parallel multiplier (b) Integer mode Multiply-Accumulate module.*

bits: FP number one bit for the sign, eight bits for the exponent, and 23 bits for the fraction.

3.8.7.3   VLSI-Based E-Nose

The basic features of VLSI based E-Noses are:

- Advanced microsystems that include, sensors, interface-circuits, and pattern-recognition integrated monolithically or in a hybrid module.

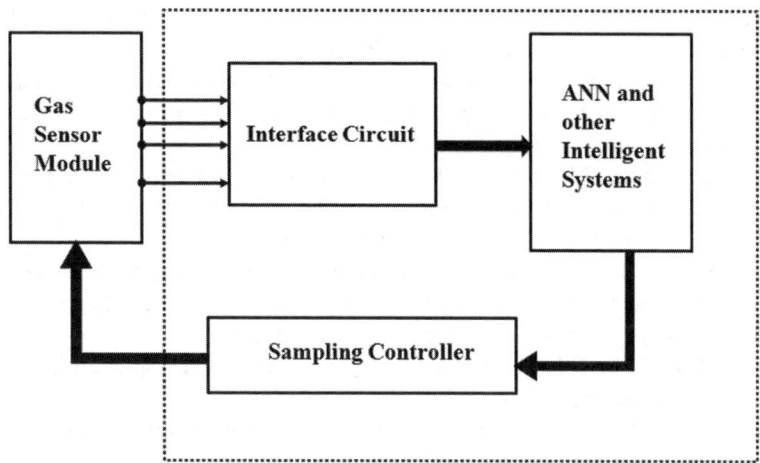

Figure 3.8.12   *– VLSI-based E-Nose configurations.*

- Artificial Neural Networks (ANN)-based ASIC can be designed.
- They have the advantage of working fast (after the training phase) even with large amounts of data.

A VLSI-based E-Nose is shown in Figure 3.8.12.

References

Advani, G.N., Beard, R. and Nanis, L. (1983) Gas measurement method, U.S. Patent 4399684, August 23, 1983.

Arena, E., Guarrera, N., Campisi, S. and Nicolosi Asmundo, C. (2006) Comparison of odour active compounds detected by gas-chromatography–olfactometry between hand-squeezed juices from different orange varieties. *Food Chemistry*, **98** (1), 59–63.

Barsan, N., Schweizer-Berberich, M. and Gopel, W. (1999) Fundamental and practical aspects in the design of nanoscaled SnO2 gas sensors. A status report, *Fresen. J. Anal. Chem.*, **365**, 287–304.

Beghi, S. and Guillot, J.M. (2008) Use of poly(ethylene terephtalate) film bag to sample and remove humidity from atmosphere containing volatile organic compounds. *Journal of Chromatography A*, **1183**, 1–5.

Bhuyan, M. (2010) *Intelligent Instrumentation, Principles and Applications*, pp. 16–17, CRC Press.

Bielanski, A., Deren, J. and Haber, J. (1957) Electric conductivity and catalytic activity of semiconducting oxide catalysts, *Nature*, **179**, 668–669.

Bliss, P.J., Jiang, J.K. and Schulz, T.J. (1995) The Development of a Sampling System for Determining Odor Emission Rates from Area Surfaces: Part II. Mathematical Model. *Journal of the Air & Waste Management Association*, **45**, 989–994.

Bockreis, A. and Steinberg, I. (2005) Measurement of odour with focus on sampling techniques. *Waste Management*, **25**, 859–863.

Bulkowiecki, S., Pfister, G., Reis, A. *et al.* (1986) Gas or vapor alarm system including scanning gas sensors, U.S. Patent 4567475, January 28, 1986.

Capelli, L., Sironi, S., Del Rosso, R. and Céntola, P. (2009) Design and validation of a wind tunnel system for odour sampling on liquid area sources. *Water Science and Technology*, **59**, 1611–1620.

Cavicchi, R.E., Suehle, J.S., Kreider, K.G. *et al.* (1995) Fast temperature programmed sensing for micro-hotplate gas sensors, *IEEE Electron Device Lett.*, **16**, 286–288.

Cavicchi, R.E., Suehle, J.S., Kreider, K.G. *et al.* (1996) Optimized temperature-pulse sequences for the enhancement of chemically specific response patterns from micro-hotplate gas sensors', *Sensors Actuators B*, **33**, 142–146.

CEN (2003) EN 13725:2003. Air quality – Determination of odour concentration by dynamic olfactometry, Brussels.

Centola, P., Sironi, S., Capelli, L. and Del Rosso, R. (2004) Valutazione di impatto odorigeno di una realtà industriale. AIDIC Servizi Srl.

Demarne, V. and Sanjines, R. (1992) Gas sensors principles, operation and development, (ed. G. Sberveglieri) Kluwer Academic, Dordrecht, the Netherlands, pp. 89–116.

Eicker, H. (1977) Method and apparatus for determining the concentration of one gaseous component in a mixture of gases, U.S. Patent 4012692, March 15, 1977.

Frechen, F.B., Frey, M., Wett, M. and Löser, C. (2004) Aerodynamic performance of a low-speed wind tunnel. *Wat. Sci. Tech.*, **50**, 57197764.

Frechen, F.B. and Köster, W. (1998) Odour emission capacity of wastewaters – Standardization of measurement methods and application. *Wat. Sci. Tech.*, **38** (3), 61–69.

Good Practice Guide for Assessing and Managing Odour in New Zealand (2003) Ministry for the Environment, Wellington, New Zealand.

Göpel, W. (1994) New materials and transducers for chemical sensors, *Sensors and Actuators B*, **18–19**, 1–21.

Gostelow, P., Longhurst, P., Parsons, S. and Stuetz R. (2003) *Sampling for measurement of odours*. IWA Scientific and Technical report No. 17. IWA Publishing, London.

Haykin, S. (1999) *Neural Networks: A Comprehensive Foundation*, Pearson Education.

Heiland, G. (1954) Zum Einfluss von Wasserstoff auf die elektrische Leitf-ahigkeit von ZnO Kristallen, *Zeit. Phys.*, **138**, 459–464.

Heilig, A., Bârsan, N., Weimar, U. *et al.* (1997) 'Gas identification by modulating temperatures of SnO_2-based thick-film sensors', *Sensors Actuators B*, **43**, 45–51.

Hermle, T., Weimar, U., Rosenstiel, W. and Göpel, W. (1999) *Performance of Selected Evaluation Methods for a Hybrid Sensor System*, ISOEN.

Hobbs, P.J., Misselbrook, T.H. and Pain B.F. (1995) Assessment of odours from livestock wastes by a photoionization detector, an electronic nose, olfactometry and gas chromatography-mass spectrometry. *J. Agric. Engng. Res.*, **60**, 137–144.

Hohl, D. (1990) The role of noble metals in the chemistry of solid-state gas sensors, *Sensors and Actuators B*, **1**, 158–165.

Hudson, N. and Ayoko, G.A. (2008a) Odour Sampling 1: Physical chemistry considerations. *Bioresource Technology*, **99**, 3982–3992.

Hudson, N. and Ayoko, G.A. (2008b) Odour Sampling. 2. Comparison of physical and aerodynamic characteristics of sampling devices: A review. *Bioresource Technology*, **99**, 3993–4007.

Ihokura, K. and Watson, J. (1994) *The Stannic Oxide Gas Sensor: Principle and Application*, CRC Press Inc.

Iwasaki, Y., Fukushima, H., Nakaura, H. *et al.* (1978) A new method for measuring odors by triangle odor bag method (I) – Measurement at the source, *Journal of Japan Society of Air Pollution*, **13** (6), 246–251 (in Japanese).

Iwasaki, Y., Ishiguro, T., Koyama, I. *et al.* (1972) On the new method of determination of odor unit, in *Proceedings of the 13th Annual Meeting of the Japan Society of Air Pollution*, Japan, p. 168 (in Japanese).

Kleperis, J., Lusis, A., Zubkans, J. and Veidemanis, M. (1999) *Two Years' Experience with Nordic E-Nose*, ISOEN, pp. 11–14.

JIS Z 8402-2: (1999) Accuracy (trueness and precision) of measurement methods and results – Part 2: Basic method for the determination of repeatability and reproducibility of a standard measurement method (Japanese version of ISO 5725-2: 1994).

Koster, E.P. (1985) *Limitations Imposed on Olfactometry Measurement by the Human Factor*, Elsevier Applied Science.

Koziel, J.A., Spinhirne, J.P., Lloyd, J.D. *et al.* (2005) Evaluation of sample recovery of malodorous livestock gases from air sampling bags, solid-phase microextraction fibers, Tenax TA sorbent tubes, and sampling canisters. *Journal of the Air & Waste Management Association*, **55**, 1147–1157.

Korotcenkov, G. (2005) Gas response control through structural and chemical modifications of metal oxide films: state of the art and approaches, *Sens. Actuators B*, **107**, 209–232.

Lantto, V. and Romppainen, P. (1988) Response of some SnO_2 gas sensors to H_2S after quick cooling, *J. Electrochem. Soc.*, **135**, 2550–2556.

Le Vine, H.D. (1975) Method and apparatus for operating a gas sensor, U.S. Patent 3906473, Sep. 16, 1975.

McKendry, P., Looney, J.H. and McKenzie, A. (2002) *Managing Odour Risk at Landfill Sites*. MSE Ltd & Viridis. Sita Environmental Trust.

Mochalski, P., Wzorek, B., Sliwka, I. and Amann, A. (2009) Suitability of different polymer bags for storage of volatile sulphur compounds relevant to breath analysis. *Journal of Chromatography B*, **877** (3), 189–196.

Nagata, Y. and Takeuchi, N. (1990) Determination of odor threshold value by triangle odor bag method, *Bulletin of Japan Environmental Sanitation Center*, **17**, 77–89 (in Japanese).

Olafsoon, R., Martinsdottir, E., Olafsdottir, G. *et al.* (1992) *Monitoring of Fish Freshness using Tin Oxide Sensors*, in, J.W. Gardner and P.N. Bartlett (eds), *Sensory Systems for an Electronic Nose*, Kluwer Academic Publishers. pp. 257–272.

Olsson, J., Börjesson, T., Lundstedt, T. and Schnürer. J. (2002) Detection and quantification of ochratoxin A and deoxynivalenol in barley grains by GC-MS and electronic nose. *International Journal of Food Microbiology*. **72** (3), 203–214.

Owen, L.J. (1980) *Gas monitors*, U.S. Patent 4185491, January 29, 1980.

Piccinini, S. (2002) *Il compostaggio in Italia*, Maggioli Editore, ISBN 88 387 2438 5.

Sears, W.M., Colbow, K. and Consadori, F. (1989a) General characteristics of thermally cycled tin oxide gas sensors, *Semicond. Sci. Technol.*, **4**, 351–359.

Sears, W.M., Colbow, K. and Consadori, F. (1989b) Algorithms to improve the selectivity of thermally cycled in oxide gas sensors, *Sens. Actuators B*, **19**, 333–349.

Seiyama, T., Kato, A., Fujiishi, K. and Nagatani, M. (1962) Anewdetector for gaseous components using semiconductive thin films, *Anal. Chem.*, **34**, 1502f.

Serra, R. and Dugnani, L. (1988) Qualità, effetti e misura degli odori nell'ambiente, *IA Ingegneria Ambientale*, **XVII** (5), Maggio, Milano, CIPA Editore.

Sneath, R.W. (2001) Olfactometry and the CEN Standard EN13725, in, *Odours in Wastewater Treatment: Measurement, Modelling and Control*. (eds R. Stuetz and B.F. Frechen), IWA Publishing.

Sohn, J.H., Smith, R.J., Hudson, N.A. and Choi, H.L. (2005) Gas Sampling Efficiencies and Aerodynamic Characteristics of a Laboratory Wind Tunnel for Odour Measurement. *Biosystems Engineering*, **92**, 37–46.

Stuetz, R. and Frechen, F.B. (2001) *Odours in wastewater treatment. Measurement, Modelling and Control*. IWA Publishing.

Suehle, J.S., Cavicchi, R.E., Gaitan, M. and Semancik, S. (1993) Tin oxide gas sensor fabricated using CMOS micro-hotplates and in situ processing, *IEEE Electr. Dev. Lett.*, **14**, 118–120.

Taguchi, N. (1971) U.S. Patent 3,631,436.

Thibodeaux, L.J. and Scott, H.D. (1985) Air/Soil Exchange Coefficients, in, *Environmental Exposure from Chemicals, Volume I* (eds B. Neely and G.E. Blau), CRC Press, Inc., Boca Raton, FL.

Trincavelli, M., Coradeschi, S. and Loutfi, A. (2009) Odour classification system for continuous monitoring applications, *Sensors and Actuators B, Chemical*, **139** (2), 265–273.

Utriainen, M. (ed.) (2008) Varpula, A., Niskanen, A.J., Juha Sinkkonen, J. *et al. Novel Hand-Held Chemical Detector with Micro Gas Sensors*, Environics Oy, Helsinki University of Technology, Applied, Sensor Sweden AB and Swedish Defence Research Agency.

VDI Guideline 3883 Part I (July 1997) *Effects and Assessment of Odours. Psychometric Assessment of Odour Annoyance*. Questionnaires.

VDI Guideline 3883 Part II (March (1993) *Effects and Assessment of Odours. Psychometric Assessment of Odour Annoyance. Parameters by Questioning*. Repeated brief questioning of neighbour panellists.

VDI Guideline 3940 Part I (February 2006) *Measurement of Odour Impact by Field Inspection. Measurement of the Impact Frequency of Recognizable Odours*. Grid measurement.

VDI Guideline 3940 Part II (February 2006) *Measurement of Odour Impact by Field Inspection. Measurement of the Impact Frequency of Recognizable Odours*. Plume measurement.

Williams, D.E. (1999) Semiconducting oxides as gas-sensitive resistors, *Sens. Actuators B: Chem.*, **57**, 1–16.

Winquist, F., Hornsten, E., Sundgren, H. and Lundström, I. (1993) Performance of an electronic nose for quality estimation of ground meat, *Measur.,Sci., Technol.*, **4**, 1493–1500.

Yasuhara, A. (1987) Identification of volatile compounds in poultry manure by gas chromatography – mass spectrometry. *Journal of Chromatography A*, **387**, 371–378.

Zarra, T. (2007) Procedures for detection and modelling of odours impact from sanitary environmental engineering plants. PhD Thesis, University of Salerno, Italy.

Zarra, T., Naddeo, V., Belgiorno, V. *et al.* (2008) Odour monitoring of small wastewater treatment plant located in sensitive environment. *Water Science and Technology*, **58** (1), 89–94. doi:10.2166/wst.2008.330.

Zarra, T., Naddeo, V., Belgiorno, V. *et al.* (2009) Instrumental characterization of odour: a combination of olfactory and analytical methods. *Water Science and Technology*, **59** (8), 1603–1609.

Zarra, T., Naddeo, V., Giuliani, S. and Belgiorno, V. (2011) Control of odour emission in wastewater treatment plants by direct and undirected measurement of odour emission capacity. *Proceedings of 4th IWA Conference on Odours and VOCs*. 17–21 October 2011, Vitoria, Brazil.

Zarra, T., Reiser, M., Naddeo, V. *et al.* (2012) A comparative and critical evaluation of different sampling materials and methods in the measurement of odour concentration by dynamic olfactometry. *Proceedings of NOSE2012*. 23–26 September 2012, Florence, Italy.

Zhang, H., Lindberg, S.E., Barnett, M.O. *et al.* (2002) *Atmospheric Environment*, **36**, 835–846.

Part 4

Strategies for Odour Control

J. M. Estrada[1], R. Lebrero[1], G. Quijano[1], N. J. R. Kraakman[2] and R. Muñoz[1]
[1]*Department of Chemical Engineering and Environmental Technology,*
Valladolid University, Spain
[2]*CH2M Hill, Level 7, 9 Help Street, Chatswood NSW 2067, Australia* and *Laboratory of*
Biotechnology, Faculty of Applied Sciences, Delft University of Technology,
Netherlands

4.1   Introduction

Odour pollution is often linked to industrial activities such as waste treatment (wastewater treatment plants, compost facilities, landfills), intensive animal farming, food processing, pulp and paper production, and so on. (Lebrero *et al.*, 2011; Zhu *et al.*, 2002). Today, the stricter environmental regulations imposed worldwide, together with the encroachment of residential areas on industrial facilities in the last decades, have resulted in an increase in the number of public odour complaints (Easter *et al.*, 2008; Stuetz and Frenchen, 2001). In fact, more than half the complaints received by the environmental regulatory agencies worldwide concern malodours (Kaye and Jiang, 2000). For instance, odour annoyance affects approximately 20% of the population in Europe, with malodours from wastewater treatment plants (WWTP) being ranked amongst the most unpleasant ones (INE, 2009). Despite not being a direct cause of disease, long-term exposure to high-strength malodorous emissions actually does negatively affect human health, causing nausea, headaches, insomnia, loss of appetite, respiratory problems, irrational behaviour, and so on. (Sucker *et al.*, 2008; Zarra *et al.*, 2008; Jehlickova *et al.*, 2008). In addition, malodorous emissions can pose a severe occupational risk within confined spaces in WWTPs or pulp and paper industries, due to the accumulation of lethal H_2S concentrations (Vincent, 2001). Odour pollution also entails a significant economic cost, and for instance, housing less than one

Odour Impact Assessment Handbook, First Edition. Edited by Vincenzo Belgiorno, Vincenzo Naddeo and Tiziano Zarra.
© 2013 John Wiley & Sons, Ltd. Published 2013 by John Wiley & Sons, Ltd.

mile from odour sources can be up to 15% cheaper (Van Broeck *et al.*, 2008). Therefore, the minimization and abatement of unpleasant odour emissions is nowadays ranked among one of the main challenges of major industrial sectors of the economy worldwide.

The minimization of odour impact (in terms of annoyance to the nearby population) involves first a detailed characterization of the key odour sources and composition of the odorous emissions, and secondly, a correct design, operation and maintenance of both the industrial facilities and if necessary the systems implemented for the abatement of the odorous emissions (Zhu *et al.*, 2002; Estrada *et al.*, 2011; Muñoz *et al.*, 2010). Most of the odour control strategies implemented nowadays in industry involve four types of measures oriented to prevent, disperse, minimize the nuisance or remove the odorants from the emission (Lebrero *et al.*, 2011):

- Prevention of odorant formation at source. This involves, for instance, process design and operation devised to maintain aerobic conditions in waste treatment facilities (since anaerobic conditions promote the generation of odorous volatile compounds like fatty acids and reduced sulfur and nitrogen compounds), good operational practices and animal nutrition in animal farms or operation at low sulfidity and anthraquinone addition during kraft pulping in pulp and paper facilities (Lebrero *et al.*, 2011; Hainong *et al.*, 2006; Hobbs *et al.*, 1996).
- Control of dispersion of the odorous emissions to guarantee an exposure of the nearby population below odour nuisance levels. The implementation of buffer zones (separation between the odour source and the potentially affected population), turbulence-inducing structures such as trees or high barrier fences or chimneys can mediate a dilution of the emission and reduce the odour concentration, with the subsequent reduction in odour annoyance (Tchobanoglous *et al.*, 2003). In this context, the installation of odour covers or enclosures (low and high level covers) has significantly improved the management of odour pollution by minimizing the odorous emissions.
- Minimization of the effects by, for example, spraying masking or inhibitory agents or neutralizers in scenarios of intermittent emissions or where the implementation of other measures is difficult. The rationale underlying this technology is the spraying of additives that mask, inhibit or neutralize the inherent unpleasant hedonic tone of the emission (Decottignies *et al.*, 2007; Bruchet *et al.*, 2008).
- Implementation of treatment technologies to reduce the odour concentration in the industrial off-gas emission before it reaches the atmosphere. Odour treatment systems are employed and implemented when prevention and control of dispersion are not sufficient to avoid odour nuisance in the surroundings of the plant/factory (Lebrero *et al.*, 2011).

The cost of these measures often increases with the degree of development of the odour pollution process. Thus, any measure oriented to preventing odour formation at source will likely entail lower costs than implementing costly additive spraying or odour treatment technologies. Figure 4.1 shows the process of odour pollution in sewer systems together with the evolution of odour control cost and degree of effectiveness. In this particular case, the potential for odour nuisance reduction is larger and its cost lower if industrial discharges are properly managed and anaerobic conditions in the sewer network are prevented (Frenchen, 2008).

This book chapter will focus on the recent advances on odour dispersion measures, impact minimization and odour abatement methods from an economic, technical and sustainable

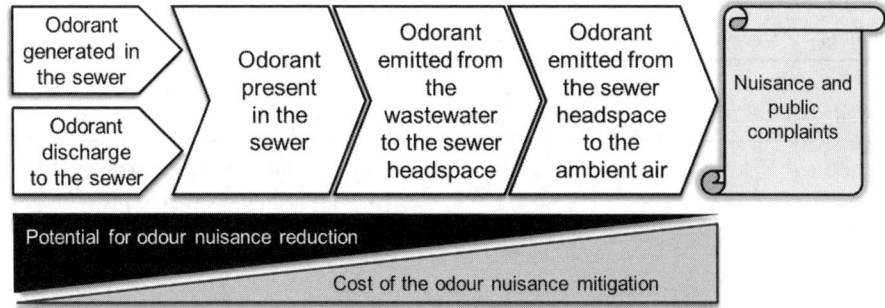

Figure 4.1 *Scheme of the odour pollution process in sewer networks (Adapted from (Frenchen, 2008) Copyright (2008) AIDIC Servizi Last accessed 25/07/2012).*

viewpoint. The sensitivity of the economics of the five most commonly applied end-of-the-pipe technologies nowadays towards design and operational parameters and commodity prices, along with their robustness, will be evaluated. Finally, a critical discussion on the different odour monitoring strategies in odour abatement facilities will be presented as a key approach to odour assessment and management.

4.2   Control of Odour Dispersion

Once an odorous emission is released from waste management or industrial facilities, the mitigation of odour nuisance to the nearby population should be a priority. In this context, odour dispersion constitutes the first tool available to reduce odour annoyance. The use of covers, turbulence-inducing structures and the establishment of buffer zones are examples of cost-efficient strategies for odour dispersion control.

The use of covers for the containment of odorous emissions is a common practice nowadays in WWTPs. When such facilities are close to residential areas, it is frequent to cover the treatment units with the highest odour potential such as headworks, primary clarifiers, sludge thickeners, sludge processing facilities and sludge load-out facilities (Tchobanoglous *et al.*, 2003). Up to three cover configurations can be used depending on the type of odour management strategy applied (Koe, 2001): (i) low-level covers, (ii) high-level covers and (iii) dual covers. Low-level covers are extensions of the treatment unit walls and their surface is very close to the odour emission source (Figure 4.2a). As a result, a small headspace

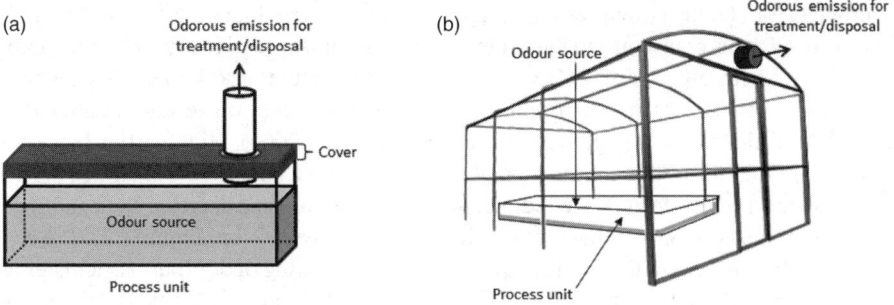

Figure 4.2 *Low-level (a) and high-level (b) cover configurations for odour containment.*

Table 4.1   *Materials used for high- and low-level covers (Reprinted from Stuetz and Frenchen, 2001, Copyright (2001) IWA Publishing).*

Material	Cover configuration	Observations
Concrete	Low-level	Concrete slabs are commonly coated with plastic materials to prevent corrosion. High lifetime.
Wood	Low-level	Wood slabs normally require plastic coating to prevent corrosion. Low lifetime.
Composites of isophthalic polyester resin and glass fibre	Low-level/ high-level	High resistance/weight ratio. Useful for low and medium corrosion conditions. Lifetime of ~20–25 years.
Composites of vinyl ester resin and glass fibre	Low-level/ high-level	High resistance/weight ratio. Useful for high corrosion conditions. More expensive than composites of isophthalic polyester resins. Lifetime of ~20–25 years.
Sailcloth-type material made of polyester and PVC	High-level	Low resistance/weight ratio. Lifetime of ~10–15 years.
Aluminium and aluminium/copper alloys	High-level	High resistance/weight ratio. Useful for low and high corrosion conditions. Lifetime of ~25–50 years.

is obtained and the odorous compounds are concentrated. This configuration is a good alternative when the minimization of the emission volume is required. However, sampling and the visual inspection of the process is significantly hindered. High-level covers require the installation of a supporting structure surrounding the process unit (mainly made of aluminium and stainless steel), which entails large headspace volumes (Figure 4.2b). The odorous compounds in this configuration are therefore diluted allowing for tasks such as sampling, maintenance and machinery installation. Dual covers combine both low and high level covers and offer a higher degree of protection versus fugitive emissions. They are used only in sensitive areas when the odour emission is very high. The materials used for covers are summarized in Table 4.1.

Turbulence-inducing structures are designed to promote the dilution effect of the wind to disperse an odorous emission and therefore to reduce its nuisance potential. This technology is based on the installation of a turbulence-inducing structure, also known as windbreak, and is commonly used as an odour nuisance reducing method in wastewater treatment and livestock facilities (Tchobanoglous *et al.*, 2003; Lin *et al.*, 2006). High barrier fences and tree belts are the most common windbreaks for this type of odour control strategy. These structures are basically designed to redirect the wind trajectory and decrease the wind speed, which induce the necessary turbulence to dilute the odorous emission (Figure 4.3) (Lin *et al.*, 2006; Cleugh, 1998). Windbreaks may also improve the settling of odorous particulates and odour carriers such as dust by reducing the wind speed (Mukhtar *et al.*, 2004). Windbreaks

(a)

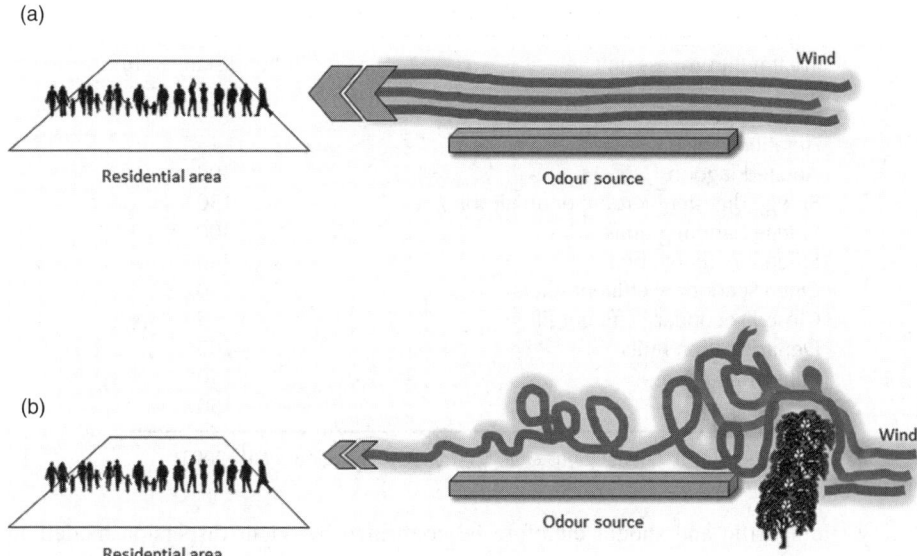

(b)

Figure 4.3   *Action mechanisms of windbreaks installed to promote odour dispersion.*

based on tree belts offer an additional advantage over fences since adsorption and absorption of odorous compounds on the foliage, followed by microbial uptake, may significantly improve the odour removal potential (Tyndall and Colletti, 2007). However, the location of the odorous emission as well as the space availability and meteorological conditions must be carefully considered for the optimal selection of the turbulence-inducing structure.

The use of buffer zones is another strategy to reduce the impact of malodours in residential zones. The concept of a buffer zone is based on establishing a minimum distance from the odour emission to the exposed population in order to avoid odour nuisances. This odour nuisance control strategy is based on the fact that the perception of odour decreases exponentially with distance as a result of its dilution Mukhtar *et al.*, 2004), although this is only true under favourable climatic conditions (e.g. low humidity, no rainfall and low wind speed). A recent odour assessment study confirmed that rainfall, moisture conditions, wind direction and wind speed are key parameters that should be considered to set buffer zones (Chemel *et al.*, 2012). Thus, the meteorological conditions play a key role in the establishment of a buffer zone since rainfall is one of the most important parameters affecting the dispersion of an odour emission. As a matter of fact, odour emission increases in open sources have been reported following a heavy rainfall (Mukhtar *et al.*, 2004). This increase, commonly observed in open livestock farms, was hypothesized to be due to a flooding out effect of the odorous material, which finally promotes the stripping of malodorous compounds (Wright *et al.*, 2005).

Therefore, odour characterization studies must be conducted in order to determine the type and magnitude of the odour source as well as the annoyance impact (Tchobanoglous *et al.*, 2003). This buffer zone distance must be established based on the current legislation of the local environmental agency where the odour emission is located. Table 4.2 shows typical buffer zones defined for odour control in several process units in WWTPs, but can

Table 4.2   *Buffer zones defined for several process units in WWTPs.* *

Treatment process unit	Buffer distance (m)
Sedimentation tank	125
Aeration tank	125
Aerated lagoon	300
Sludge digester (aerobic or anaerobic)	150
Sludge handling units	300
Effluent recharge bed	250
Open secondary effluent filters	150
Closed secondary effluent filters	75
Denitrification units	100
Polishing lagoon	150
Land disposal	150

*Adapted from Tchobanoglous *et al.* (2003) (Tchobanoglous *et al.*, 2003).

be very site specific and should therefore be confirmed by odour dispersion modelling (Lebrero *et al.*, 2011).

4.3   Control of Odour Effects on an Exposed Community

Despite there being several treatment technologies up to date available to control odour emissions, their implementation might be limited by their high investment and operational cost under specific scenarios such as diffusive emission sources or highly variable odour flowrates (Decottignies *et al.*, 2007). An alternative to avoiding odour nuisance in the nearby population is the use of chemical additives designed to mask, neutralize or minimize the perception of malodorous emissions. As a matter of fact, these additives are nowadays used for odour nuisance abatement in waste management and industrial facilities (Bruchet *et al.*, 2008; Bilsen and De Fré, 2008, Sutton *et al.*, 1999). These additives can be classified as: (i) masking agents, (ii) surfactants or (iii) neutralizers based on their action mode and can be dosed by air spraying or surface application (Table 4.3). Commercial products may combine two or more types of additives to improve the performance of the odour nuisance abatement (Decottignies *et al.*, 2007).

Masking agents are constituted by terpenic compounds and other oxygenated molecules commonly used in the perfume industry. Therefore, masking agents play an olfactory role that dominates over the nuisance of odorous emissions. In addition, some masking agents are able to block specific malodorous receptors in the nose and are sometimes named as inhibitors. For instance, eucalyptol blocks the receptors for ethyl mercaptan, while coumarin blocks the receptors for skatole. Common terpenes used in commercial products are limonene, α-pinene, β-pinene, camphene and eugenol (Decottignies *et al.*, 2007).

Surfactants, such as amphipathic molecules, are added as emulsifiers to increase the absorption of odorous compounds in water. It is well known that the addition of small amounts of surfactant affects the air-water equilibrium of volatile organic compounds leading to a decrease in their Henry's law constant (Vane and Giroux, 2000). Therefore, the application of a surfactant solution is expected to increase the apparent solubility of

Table 4.3 *Characteristics and action mode of the controlling agents for odorous emissions.*

Controlling agent	Characteristic compounds	Action mode
Masking agents	Terpenic compounds and some oxygenated molecules like coumarin.	Mask odorous emissions nuisance and block some specific malodorous receptors.
Surfactants	Amphipathic molecules such as alcohols, glycols and esters.	Increase the apparent solubility of odorous compounds in aqueous media, thus reducing the odour emission.
Neutralizers	Aliphatic and aromatic aldehydes (for direct application to odorous emissions). Fibre-degrading enzymes and plant extracts (to increase the nutrient utilization of livestock).	The aldehydes react with odorous compounds decreasing the odour annoyance. The enzymes and plant extracts decrease the nutrient excretion of monogastric livestock and therefore the odour emission.

odorants, thus resulting in a decrease of the odour concentration of the emission. Typical surfactants used in commercial products are alcohols, glycols and esters like isoamyl acetate (Decottignies *et al.*, 2007).

Neutralizers, also known as odour counteractants, can be divided into two groups: those applied directly to an odorous emission and those used to prevent the odour emission. The first group of neutralizers is composed of aliphatic and aromatic aldehydes that react with a wide variety of odorants such as hydrogen sulfide, mercaptans, ammonia and amines. Since these aldehydes are very reactive, the malodorous molecules are chemically modified and their odour annoyance is therefore abated. The second group of neutralizers is constituted by feed additives such as fibre-degrading enzymes (e.g. β-glucanases, xylanases and cellulases) and plant extracts (e.g. from *Yucca schidigera*), which are devised to increase the digestibility and nutrient utilization in animal feedstock. In this regard, it has been reported that nitrogen excretion and its further decomposition is one of the main sources of malodours in swine manure, the presence of fibres being one of the main factors reducing nitrogen utilization in monogastric animals (Varel and Yen, 1997). Thus, an increase in the feed digestibility is reflected in a reduction of nutrient excretion and consequently in a less unpleasant hedonic tone from the livestock manure emission (Sutton *et al.*, 1999).

Recent studies on annoyance reduction from odorous emissions of solid waste management facilities and a citric acid production plant by addition of commercial additives originally labelled as neutralizers revealed that these products were actually masking agents rather than neutralizers (Decottignies *et al.*, 2007; Bilsen and De Fré, 2008). Some of these commercial products were efficient for nuisance abatement only at concentrations up to 100 times higher than those recommended by the supplier. Likewise, a recent study conducted by Suez Environment alerted about the limited efficiency of commercial additives, which in some cases even contributed to increasing the odour concentration over the regulated discharge limits (Decottignies *et al.*, 2007; Bruchet *et al.*, 2008). The use of chemical additives is therefore recommended only under scenarios where the costs of other measures is prohibitive or as a temporary measure for annoyance abatement, and always preceded by empirical optimization studies. In this regard, research on the nuisance abatement efficiency of commercial additives (for each type and mixtures of them) is still necessary to optimize their formulation and the use of this technology. Systematic studies considering odour emissions with different molecules profile, odour concentration and air flow will clarify the real potential and limitation of these controlling agents.

4.4   Control of Odour Emission

Among all the measures for the control of odour emissions, end-of-pipe treatment technologies have been consistently proven to be the most effective despite usually implying the highest investment and operating costs. Odour abatement technologies have been traditionally classified into physical/chemical and biological. In this section, the main odour abatement technologies of each group will be briefly introduced and discussed using a comparative approach focusing on sustainability and robustness. This will provide up-to-date state of the art guidelines for technology selection, which will be valuable in dealing with the challenges in the field of odour management.

4.4.1   Physical/Chemical Technologies

A wide range of technologies are included under this classification, chemical scrubbers and activated carbon filters being the most applied technologies nowadays in the field of odour treatment. The two main mechanisms underlying these odour abatement techniques are chemical oxidation and solid phase adsorption, respectively, which will be discussed later on. Some other examples of physical/chemical technologies in the field of odour treatment are incineration (based on thermal oxidation), dry oxidation, photolysis, ozonization and advanced oxidation processes (Burgess *et al.*, 2001; Delhoménie and Heitz, 2005; Water Environmental Federation, 2004). However, only the most commonly applied technologies nowadays will be discussed here. The main advantages of physical/chemical technologies are the large experience gained over the years in their design and operation, their low empty bed residence times (EBRTs) and their rapid start-up (Lebrero *et al.*, 2011). These facts make them robust technologies from an operational point of view and economic alternatives in terms of investment costs (Figure 4.4). However, their main drawbacks are their high operating costs and environmental impacts, which render activated carbon adsorption and chemical scrubbing often in poorly sustainable and expensive technologies (see Figure 4.5).

4.4.1.1   Chemical Scrubbing

Chemical scrubbers are a physical/chemical technique based on the transfer in a packed tower of the odorous compounds from the gas emission to a chemical oxidant-containing aqueous phase, where they chemically react to be destroyed (Figure 4.6) (Lebrero *et al.*, 2011). Due to the nature of this technology, malodorous compounds will face a mass transfer resistance to be dissolved in the aqueous phase (and consequently to be oxidized). This resistance will increase when increasing the hydrophobicity of the compound to be treated, often becoming the limiting step of chemical scrubber performance. Thus, the mass transfer rate will be determined by the odorant concentration in the gas phase and in the aqueous

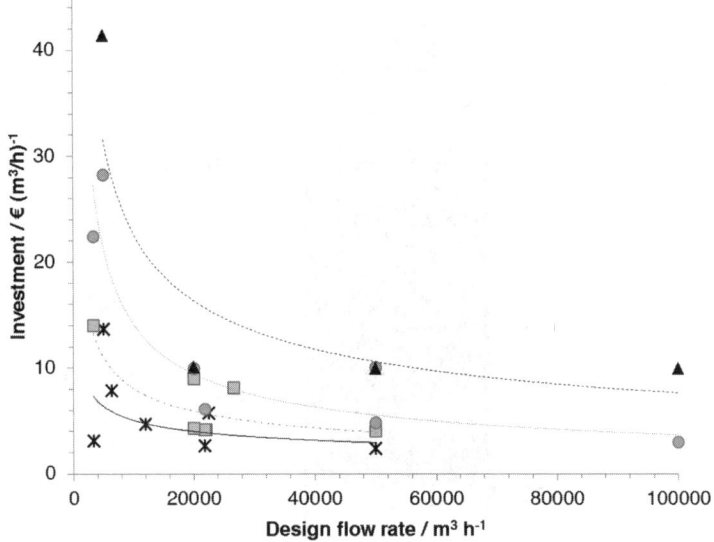

Figure 4.4 *Influence of design flowrate on the investment costs for activated carbon (black stars/solid line), chemical scrubber (squares/dash-dotted line), biofilter (circles/dotted line) and biotrickling filters (triangles/dash line) according to Estrada et al. (2011). (Reprinted with permission from Estrada et al., 2011, Copyright (2011)).*

solution, the air/water partition coefficient for that odorant (Henry's Law constant), the mass transfer resistance in the system and the interfacial area available for transfer (Revah and Morgan-Sagastume, 2005). Chemical scrubbers present removal efficiencies (RE) of up to 99.0% for water soluble odorants (Henry's Law constant lower than 0.07 M atm^{-1}) such as H_2S, while the removal of highly hydrophobic compounds, such as terpenes or hydrocarbons (Henry's Law constant higher than 20 M atm^{-1}), can be as low as 50%, this limitation being one of the main drawbacks for these systems in the field of odour abatement. The REs for odorants with moderate hydrophobicity are found within the 50–90% range.

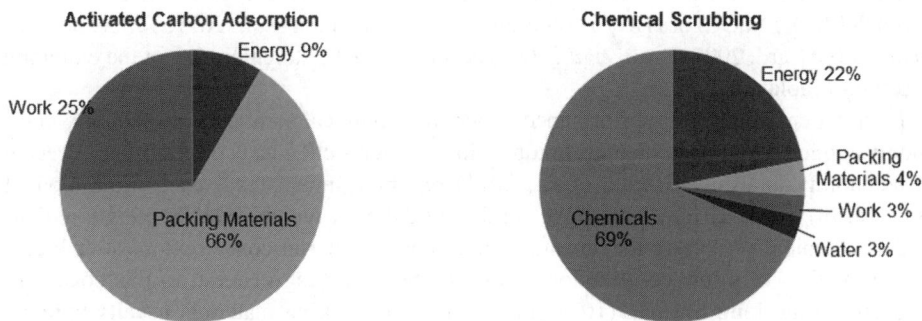

Figure 4.5 *Contribution of different operating and design parameters to the operating cost of the most commonly applied physical/chemical technologies in a WWTP treating 50 000 m^3 h^{-1} containing 21.9 mg m^{-3}. (Reprinted with permission from Estrada et al., 2012, Copyright (2012)).*

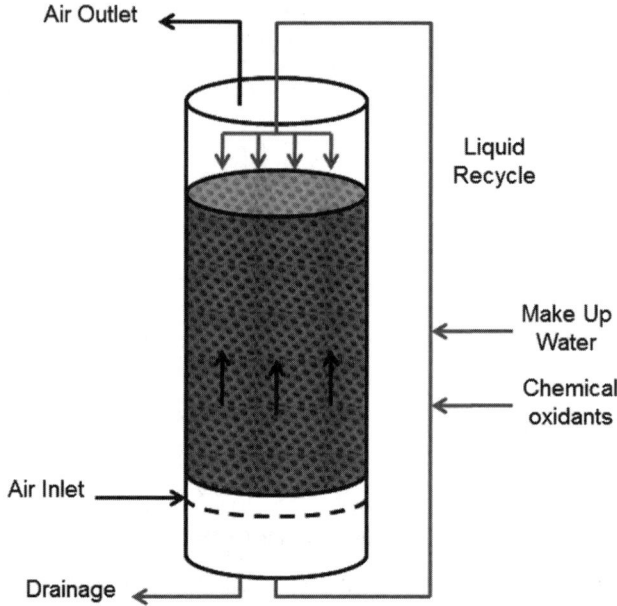

Figure 4.6   *Schematic design of a chemical scrubber.*

Although the fundamentals of this technology remain the same, chemical scrubbers can be arranged into different configurations. Counter-current packed towers (upwards flow for the gas, downwards for the liquid) are the most commonly applied, but co-flow or cross flow scrubbers can also be found (Tchobanoglous *et al.*, 2003). One or two-stage configurations are also implemented. Another type of chemical scrubbers is mist chambers, where the oxidizing liquid solution is atomized (droplets size around 10 μm) in the odorous emission, which horizontally flows across the chamber. Although this configuration achieves higher REs due to the higher mass transfer area available, it is less commonly implemented since it requires a higher maintenance and space for installation (Card, 2001).

The chemical oxidants used are often chosen depending on the target odorants to be oxidized. $KMnO_4$, $NaOCl$ or H_2O_2 are commonly added (Estrada *et al.*, 2011; Tchobanoglous *et al.*, 2003, Card, 2001; Prado *et al.*, 2009) according to the stoichiometry of the oxidation reactions (Table 4.4).

From an environmental performance viewpoint, chemical scrubbers are the only commonly applied odour abatement technique which presents chemical consumption. A recent sustainability analysis (Estrada *et al.*, 2011) reported annual reagents consumption of 2 kg $(m^3/h)^{-1}_{air\ treated}$ for a two-stage scrubber treating a typical WWTP emission. This makes chemical scrubbing the second technology in material consumption, after biofiltration, with a very low contribution of the scrubber packing material to this consumption due to the long lifespan (10 years) of the inert packing materials usually applied. This high chemical consumption also has an impact on the operating costs, accounting for a 69% of the overall annual cost (see Figure 4.5). Another important environmental impact caused by chemical scrubbers derives from their annual water consumption ($3.3 \cdot 10^{-2}$ l $(m^3/h)^{-1}_{air\ treated}$ according to the same study). The cost of water

Table 4.4 Oxidation reactions and chemicals involved in H_2S removal in chemical scrubbers (Tchobanoglous et al., 2003; Card, 2001).

Oxidant	Reaction	mg l^{-1} oxidant/ mg l^{-1} H_2S
NaOCl	$H_2S + 4NaOCl + 2NaOH \rightarrow Na_2SO_4 + 2H_2O + 4NaCl$ $H_2S + NaOCl \rightarrow S^o + NaCl + H_2O$	8–10
$KMnO_4$	$3H_2S + 2KMnO_4 \rightarrow 3S + 2KOH + 2MnO_2 + 2H_2O$ (acid pH) $3H_2S + 8KMnO_4 \rightarrow 3K_2SO4 + 2KOH + 8MnO_2 + 2H_2O$ (basic pH)	6–7
H_2O_2	$H_2S + H_2O_2 \rightarrow S^o\downarrow + 2H_2O$ (pH<8.5)	1–4

consumption does not represent a big share of the total operating costs and accounts only for a 3%. However, the quality of the water needed (at least similar to tap water) significantly increases the environmental impact of chemical scrubbers when compared to biological technologies, where water quality requirements are much lower and recycled or secondary effluent can be used. On the other hand, a key advantage of chemical scrubbers is provided by their low land occupation due to their low gas residence times (1–2.5 s) (Gabriel and Deshusses, 2004), which usually results in more compact systems than biological techniques. However, the space needed for chemical storage can significantly reduce this advantage. For instance, full scale chemical scrubbers present land needs approximately three times lower than biotrickling filters and 8–21 times lower than conventional biofilters (Estrada et al., 2011).

Chemical scrubbers are one of the most economic alternatives in terms of investment costs. This is mainly due, as above mentioned, to the small required gas contact time the accumulated experience and standardization of these systems as a result of their widespread application. As a matter of fact, chemical scrubbers have been one of the most common odour abatement technologies (Tchobanoglous et al., 2003; Edwards and Nirmalakhandan, 1996). A recent cost compilation of full-scale chemical scrubbers revealed investment costs of €3–12 $(m^3/h)^{-1}$ air treated depending on the design flowrate (Estrada et al., 2011). This compilation reported the following relationship (Equation 4.1) between investment cost and design flowrates treated:

$$Ic = 509 \times F^{-0.4} \qquad (4.1)$$

Where **Ic** is the investment cost (€ $(m^3/h)^{-1}$ air treated) and **F** the gas flowrate (m^3 h^{-1})

However, this technology ranks as the second most expensive in terms of operating costs (€3.6 $(m^3/h)^{-1}$ air treated) only behind activated carbon adsorption (Estrada et al., 2012). Another important issue when evaluating the economics during technology selection is how process economics will be affected by the operation or design variables. When evaluating the sensitivity of the Net Present Value 20 (NPV20, total spent money in 20 years of operation, including investment and operating costs (Estrada et al., 2011)), chemical scrubbers are highly sensitive towards variation in H_2S concentration (as chemical consumption depends on the amount of odorant treated) and chemicals prices, which constitute the main

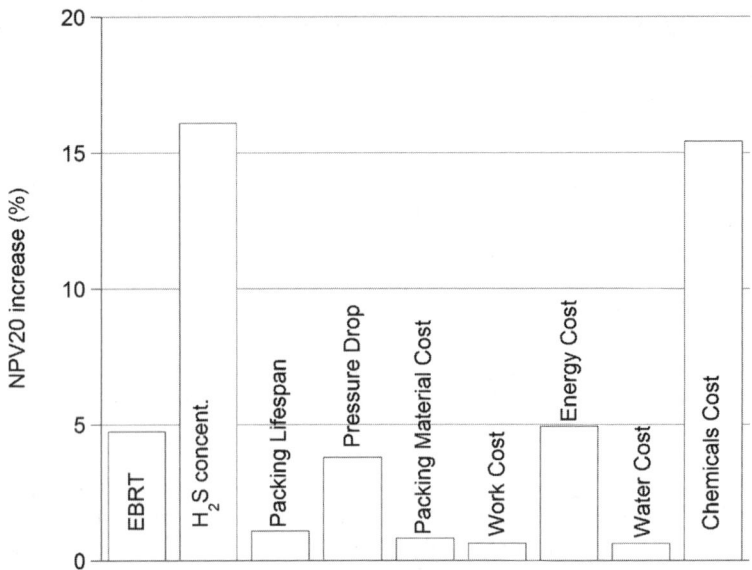

Figure 4.7   *Sensitivity of the NPV20 of a two-stages chemical scrubber applied at a WWTP (gas flowrate 50 000 m³ h⁻¹, 21 mg m⁻³ H₂S concentration, 1 m bed height, pressure drop 500 Pa, packing material cost €1200 m⁻³, EBRT of 4 s, 2 s per stage) for a 25% increase of the selected parameters.*

contributors to the operating costs. Figure 4.7 presents the sensitivity of chemical scrubber NPV20 towards variations in EBRT, H_2S concentration, packing material lifespan, pressure drop, packing material cost, work cost, energy cost, water cost and chemicals cost in order to identify the most important contributors to process economics. All the parameters were increased by a 25% to allow for a fair comparison. Hence, for a 25% increase of those two parameters, the NPV20 would increase by more than 15%. Other parameters such as the EBRT, the pressure drop or the energy price (due to its influence on the water recycling rate) would cause an increase of approx. 5% when increased by 25%. The effect of the rest of the design and operation parameters or commodity prices can be considered negligible. Thus, chemical scrubbers will not be significantly affected by the price of water, work (which includes labour, transport, handling disposal and maintenance) or packing material, or by the lifespan of the packing material (the cost of the packing material accounts for a very small amount of the total yearly operating costs; Figure 4.5).

The social impact of chemical scrubbers is related to the transport, storage and handling of the hazardous chemicals needed in the process, this technology being less safe than its biological counterparts (Estrada *et al.*, 2011). Benefits to the community and surrounding population are directly related to their odour abatement efficiency, which depends at the same time on the correct operation of the systems and on the hydrophobicity of the odorants. When compared to activated carbon adsorption, biotrickling filters, biofilters or hybrid systems (a biotrickling filter back-up by an activated carbon filter), chemical scrubbers support similar social benefits to those obtained with biotrickling filters, although ranking at the lowest benefit levels (Estrada *et al.*, 2011).

The robustness of odour abatement technologies can be analysed by accounting for the probability of an operational disorder or upset to happen and its effect on the odour abatement efficiency (Kraakman, 2003). Based on both our previous experience and surveys to odour abatement facility operators, chemical scrubbers present a moderate robustness, similar to that of biological technologies. Critical issues that impact on a proper operation and recovery are failures in the water supply and recirculation pumps and chemical dosing disorders. On the other hand, this technology efficiently withstands fluctuations of the inlet emission temperature or air supply interruptions, which are probable issues on full scale odour abatement applications (Estrada *et al.*, 2012).

Overall, chemical scrubbing is a technology which use is steadily decreasing mainly due to its usually high operating costs and environmental drawbacks such as water and chemical consumptions. Moreover, chemical scrubbers do not provide any additional benefits compared to biological techniques which are safer, more sustainable and offer similar REs and robustness at a lower cost.

4.4.1.2   Activated Carbon Adsorption

This technology is based on the physical adsorption of odorants on an activated carbon bed by intermolecular forces (Revah and Morgan-Sagastume, 2005). If the adsorbed odorants are stable and poorly reactive, they will remain trapped in the solid adsorbent. However, if the odorants are reactive, they may chemically react with other compounds adsorbed. For instance, reduced sulfur compounds are easily oxidized in the presence of atmospheric oxygen when adsorbed in activated carbon. Adsorption processes usually take place in 0.5–1.5 m height towers (higher towers would lead to high pressure drops) at gas residence times ranging from 1.5–10 s. In order to improve odour abatement effectiveness and allow for a continuous operation, adsorption systems usually consist of at least two towers that work alternatively: while odorant adsorption takes place in one of them, the regeneration of the saturated packing material (desorption of the adsorbed odorants by temperature increase, pressure decrease, or carbon washing with water) is carried out in the other one (Tchobanoglous *et al.*, 2003; Revah and Morgan-Sagastume, 2005). At the end of the carbon packing lifespan or when no regeneration of the activated carbon is possible, one of the towers will be in operation while the packing material of its counterpart is substituted.

Activated carbons are usually obtained by activation at high temperature of organic materials such as wood or anthracite coal. High performance materials such as impregnated activated carbons have been recently developed for adsorption, which are available in the market at higher prices and provide increased removal efficiencies for specific applications. In these materials, a chemical reagent (NaOH, KOH, urea, $KMnO_4$) or a catalyst (generally heavy metals) is impregnated on the surface of the activated carbon to remove a certain family of compounds or accelerate chemical reactions, respectively (Turk and Bandosz, 2001). Although impregnation usually improves the RE for specific compounds, it does not always represent an advantage due to the notorious reduction of the available surface for adsorption and the reduction of the auto ignition temperature of the material, which increases the risk of ignition during its storage in the presence of air.

Adsorption does not require the transfer of odorants to an aqueous phase, and the high affinity of the adsorbent for hydrophobic compounds supports the highest abatement efficiencies for these odorants (up to 99.9%). However, the removal efficiencies for hydrophilic VOCs are lower, ranging from 80 to 90%. Overall, activated carbon adsorption typically presents higher removal efficiencies compared to biofiltration, biotrickling filtration and chemical scrubbing. The adsorption capacity of a packing bed depends on a number of factors such as the nature of the material, the odorant concentration in the gaseous stream, the operation temperature and humidity, and the mixture of odorants present in the emission. In some specific scenarios, a high humidity in the malodorous stream or its fluctuations can hinder the design and operation of adsorption systems, as, for example, water molecules compete with odorants for the active sites of the carbon. A typical capacity value of 0.1 g $_{\text{adsorbed compounds}}$/g $_{\text{activated carbon}}$ is often considered for design purposes in some applications. The lifespan of a certain adsorption bed can be calculated by Equation (4.2):

$$t = \frac{6.7 \times 10^6 S \cdot W}{E \cdot Q \cdot M \cdot C} \tag{4.2}$$

Where S is the adsorption capacity of the packed bed [dimensionless] ≈ 0.1; W is the total mass of adsorbent (kg), E is the elimination efficiency [dimensionless] ≈ 1, Q is the flowrate to be treated (l s^{-1}), M is the molecular weight of the odorants (g mol^{-1}) and C is the odorant concentrations (ppm$_v$), which a priori is the most difficult variable to be predicted in odour abatement applications (Turk and Bandosz, 2001).

Activated carbon adsorption presents a low environmental impact when the sustainability analysis is only performed to the odour removal process (excluding activated carbon production and regeneration). Thus, a moderate annual packing material consumption ($\approx$ 0.8 kg (m^3/h)$^{-1}$ $_{\text{air treated}}$ for the treatment of an odorous emission containing 21 g m^{-3} H$_2$S in a WWTP), annual energy requirements of 19 MJ (m^3/h)$^{-1}$ $_{\text{air treated}}$ and the absence of water requirements convert this technology into a low resource usage method when applied at relatively low odorous compound concentrations. In addition, the low residence times applied in this technique reduce the land needs, supporting high REs in compact systems (Estrada *et al.*, 2011). However, it is important to remark that the previously presented data did not include the environmental impacts derived from the activation, transport, regeneration (if applied) or disposal of the activated carbon. Those processes can be highly energy demanding and must be taken into account if a lifecycle analysis is performed. Activated carbon disposal as a hazardous waste can introduce an extra energy consumption and economic cost in countries where legislation requires special treatment for hazardous waste (e.g. Europe or Japan). In terms of investment costs, adsorption systems also benefit from the small required gas contact time and the wide application and accumulated experience on this technology, being one of the most economic alternatives to install (Figure 4.4) (€4–14 (m^3/h)$^{-1}$ $_{\text{air treated}}$) (Estrada *et al.*, 2011). A recent compilation of investment costs for full-scale activated carbon filters resulted in Equation (4.3):

$$Ic = 113.2 \times F^{-0.33} \tag{4.3}$$

On the other hand, the low packing material lifespan and its high price increase the yearly operating costs to $\approx$ €7.2 (m^3/h)$^{-1}$ (Estrada *et al.*, 2012), with this technology being the most expensive in terms of operating costs (despite the annual packing material requirements being moderate). According to the European Commission, the yearly operating costs can reach up to €200 (m^3/h)$^{-1}$, which shows the high variability of the operating costs associated to this technology (European Comission, 2003). An increase of 25% in all the parameters

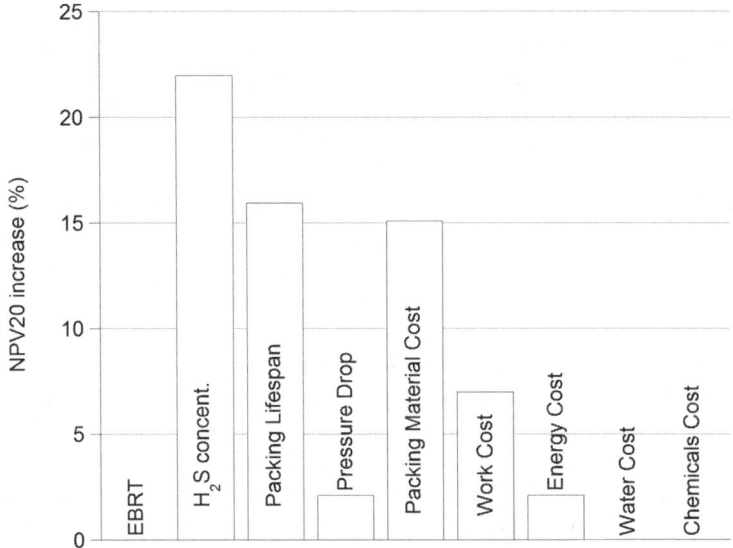

Figure 4.8 *Influence of an increase of 25% in the selected parameters on the NPV20 of an activated adsorption system applied at a WWTP (gas flowrate 50 000 m^3 h^{-1}, H_2S concentration 21 mg m^{-3}, bed height 0.5 m, pressure drop 1750 Pa, packing material cost €4.8 kg^{-1}, EBRT 2.5 s).*

susceptible to affect the operation costs shows how the odour concentration (in this case H_2S concentration) is the key variable in determining the operating costs (Figure 4.8). Other variables exhibiting an important influence are the packing material lifespan and the packing material cost, both of them causing an increase higher than 15% in the NPV20. Work cost (which includes labour, transport, handling disposal and maintenance) moderately affects the NPV20, mediating a 7% increase when increased by 25%. This work cost is also a parameter associated to the packing material and is determined by the regular maintenance and replacement of the activated carbon beds. However, activated carbon adsorption presents a low sensitivity towards the pressure drop or the energy cost and is not affected by parameters such as the EBRT, the water price (as no water consumption is needed if water-based regeneration is not considered) and chemical costs. The operational efforts in this type of systems must be always focused on maximizing the packing lifespan in order to achieve important economic and environmental savings.

The robustness analysis of activated carbon filtration ranks this technology as the most robust odour abatement method nowadays applied. The relative simplicity of the technology (no water or chemicals needed and no process control involved) implies that common issues in odour abatement scenarios such as fluctuations of inlet odorant concentrations, foul air interruptions or air temperature fluctuations will result in minor or marginal effects on the odour abatement efficiency (Estrada *et al.*, 2012). However, it must be noted that duplicate or backup systems are often needed to guarantee a consistent odour removal during packing material replacement or regeneration.

The high odour removal efficiencies supported by activated carbon adsorption systems provide the highest benefits to the nearby population and to the health and welfare of employees in the site. In addition, a high odorant removal efficiency also reduces the potential atmospheric impacts of the odorants (photochemical smog, global warming). Unfortunately, activated carbon adsorption is considered a more hazardous technology than biological techniques, although safer than chemical scrubbers. The main risk of this technology is associated to the possibility of smouldering in loaded, non-ventilated units (Estrada *et al.*, 2011).

Although activated carbon adsorption is the most expensive technology especially at moderate and high odour concentrations and not as sustainable as biological techniques, it presents some important benefits that make it suitable for some specific applications. Firstly, it can be applied in scenarios with limited space availability due to the compact designs derived from their low residence time. Secondly, activated carbon filtration provides higher removal efficiencies and robustness than any of the technologies here evaluated, which is valuable in highly sensitive scenarios (e.g. population or recreational areas very close to a remote odour source). In addition, carbon adsorption can be used as a polishing step coupled to another primary odour treatment (which reduces the odorant load to the activated carbon filter). These hybrid systems, which exhibit a superior removal efficiency and process robustness, are already on the market and will be discussed later in this chapter.

4.4.1.3   Incineration

Incineration, based on the oxidation of the gaseous pollutants at high temperature (also in the presence of catalyst), is still a commonly applied technique for industrial gaseous effluents (where VOCs are present at high concentrations and the process could be even self-maintained), but its use in odour abatement has been significantly reduced due to its high operating costs (Estrada *et al.*, 2011). Its application to malodorous emissions presents a number of drawbacks originating from the large flowrates to be treated and the trace level concentrations of the odorants in these emissions. First, additional fuels (usually natural gas) must be added to the process as a source of energy to attain the high temperatures needed for the complete and spontaneous oxidation of odorants, since the reaction at such low concentrations is not self-maintained. This additional input of fuel involves resources and energy consumption, additional costs and atmospheric impacts derived from the combustion of fossil fuels. On the other hand, since this technology is based on the combustion of the odorants in the gas phase, no mass transfer limitations are found. Thus, incineration is the technology which supports the highest REs regardless of the hydrophobicity of the target compounds: efficiencies of 99.9% are often reported for odour removal (Estrada *et al.*, 2011).

In order to minimize the energy input to the incineration process, some alternatives have been developed. Thus, in the regenerative incineration, a part of the heat transferred to the odorous emissions in the combustion chamber is recovered by a heat exchange system to heat the incoming odorous air (Mills, 1995). These systems operate at temperatures ranging from 800–1000°C and energy recoveries of 90–95% have been reported. However, despite this optimization effort, the energy consumption in regenerative incinerators has been reported to be more than 70 times higher than that of a chemical scrubber, more than 75 times higher than that of a biofilter and more than 160 times higher than that of a biotrickling filter (the most efficient technology in terms of energy consumption) (Estrada *et al.*, 2011). In addition, the combustion of the natural gas required causes a severe atmospheric impact in terms of global warming, generating more than 10 000 times more CO_2 equivalents than any other odour abatement technique according to a sustainability analysis using the IChemE sustainability metrics (Estrada *et al.*, 2011; IChemE, The Sustainability Metrics, 2002).

The other alternative available to reduce the energy input is catalytic oxidation, where the malodorous compounds are completely oxidized at moderate temperatures (which depend on the nature of the compound) in the presence of a solid catalyst. When comparing this technique to standard or regenerative incineration, the use of lower temperatures enables a reduction in the amount of fuel and the cost of the construction materials. The main drawback of this technology is again related to the nature of the odorous emissions: due to its complexity, a complete odorant oxidation requires an excess of oxygen and a combination of different catalysts in order to achieve the desired abatement efficiencies. In addition, some compounds (such as reduced sulfur odorants) can poison the catalysts, reducing their activity and lifespan (Lens, 2001).

Not much economic data are available for full scale incineration facilities treating odours. The European Commission Reference Document on Best Available Techniques in Common Waste Water and Gas Treatment/Management Systems in the Chemical Industry reported investment costs ranging from €8–52 $(m^3/h)^{-1}_{air\ treated}$ (European Comission, 2003). These figures support the widely accepted idea that incineration nowadays presents the highest investment costs among odour abatement technologies, and does not constitute a cost-effective technology to deal with most common malodours. Likewise, operating costs higher than €8 $(m^3/h)^{-1}$ are common in this type of technology.

When evaluating the social benefits of incineration in terms of welfare, health and safety, this technology provides reasonable results. However, the effect of the greenhouse gases derived from the combustion of the support fuel and the hazard derived from potential oxidation byproducts such as nitrous and sulfur oxides must be taken into account (Schlegelmilch *et al.*, 2005). In addition, the operational risks associated to these technologies are those inherent to a combustion process (high temperature, hot surfaces, high pressure, explosion) together with those derived from the handling and storage of the support fuel (Estrada *et al.*, 2011). Overall, incineration is not perceived as a sustainable technology for typical odour abatement, although it is still used in landfill and industrial applications and in highly sensitive scenarios (Burgess *et al.*, 2001; Delhoménie and Heitz, 2005).

4.4.2   Biological Technologies

Over the last decades, physical/chemical technologies for odour treatment have been gradually replaced by biological techniques such as biofiltration, biotrickling filtration, bioscrubbers or activated sludge diffusion. In the last few years, the use of biological methods has gradually increased based on their proven high efficiency and robustness, and on their lower operating costs when compared to their physical/chemical counterparts (lower energy requirements and no need of chemical reagents). In addition, biotechniques show lower environmental impacts, converting pollutants into substances with a lower odour potential (e.g. CO_2, H_2O, SO_4^{2-}, NO_3^- and biomass). For instance, in 2005, van Groenestijn and Kraakman estimated that there were more than 7500 biological waste gas treatment systems and related systems in Europe (half of them in sewage treatment and composting plants) (van Groenestijn and Kraakman, 2005). Although not all of them are probably related to odour

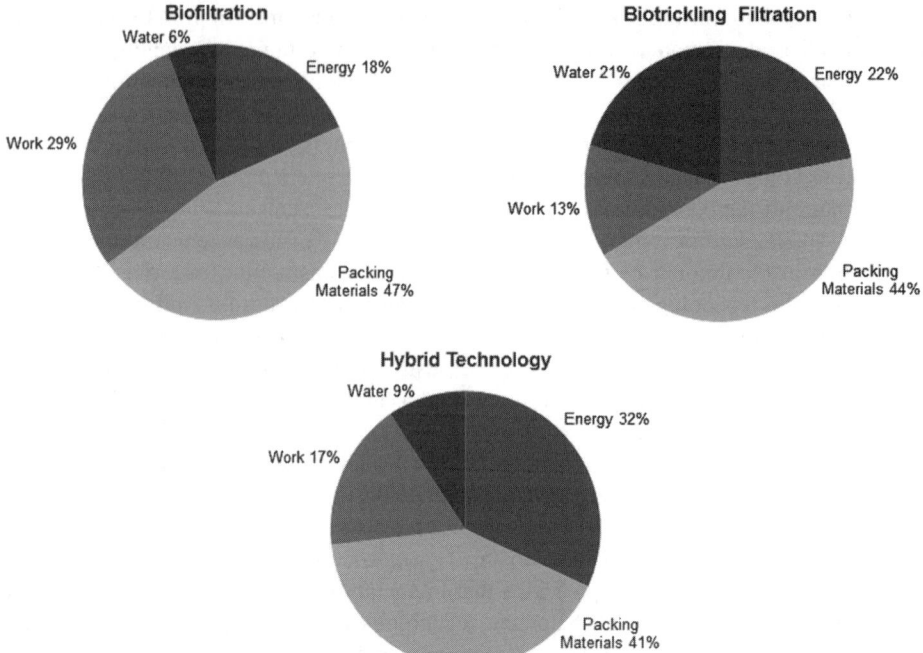

Figure 4.9 *Contribution of different design and operational parameters to the operating cost of biological technologies according to Estrada* et al. *(2012). Reprinted with permission from (Estrada et al., 2012) Copyright (2012).*

abatement, this figure highlights the increasing relevance of biological waste gas treatment techniques. Recent studies have also shown their relatively high robustness towards temperature and inlet load fluctuations for both industrial gaseous effluents and odorants (Kraakman, 2003; Lebrero *et al.*, 2010). However, despite their increasing use, these technologies are still subjected to an active research in order to improve their technological features (see Figure 4.9).

Biological techniques for odour abatement are based on the enzymatic oxidation of odorants once they have been transferred from the gaseous emission to an aqueous phase, where the microorganisms responsible for this enzymatic oxidation are present. The main reason for their lower operating costs is that these oxidation mechanisms take place at ambient temperature and pressure, and in the absence of an external supply of chemicals (only water and nutrients). The absence of extreme operating conditions and hazardous chemicals constitutes an additional advantage from a safety viewpoint for on-site staff and operators. On the other hand, their main disadvantage is the high land requirements caused by the high EBRT needed to achieve acceptable odour removal efficiencies (2–120 s for biological technologies versus 1–5 s for physical/chemical technologies) (Stuetz *et al.*, 2001).

4.4.2.1   Fundamentals of Biological Oxidation

Biological oxidation of odorants is based on the use of those compounds as a source of carbon and/or energy by microorganisms in a bioreactor (Revah and Morgan-Sagastume,

2005). Thus, VOCs such as terpenes, aldehydes, ketones or hydrocarbons are used as a carbon source for the build-up of new cellular material (maintenance and cellular replication) and as an energy source by means of their oxidation to H_2O and CO_2 using the oxygen diffusing from the emission as the electron acceptor (Kennes and Thalasso, 1998). Microorganisms can also use odorants such as H_2S or NH_3 as an energy source (by oxidation to sulfate and nitrate, respectively) to sustain both maintenance and/or cellular growth based on the assimilation of inorganic carbon from the environment. The energy produced in those oxidation reactions is either immediately used or stored under the form of high-energy phosphate bonds (Nielsen *et al.*, 2003). Thus, it is the ability/need of microorganisms to use odorants as a raw material for their growth and maintenance which makes possible the biological abatement of odours. There are two main requirements for these processes to occur: the presence of an aqueous medium to support all the metabolic reactions, and the availability of macronutrients (such as phosphorus, nitrogen, sulfur, potassium . . .) and micronutrients (generally trace amounts of metals needed for enzyme synthesis) (Deshusses, 1997).

Bacteria and fungi are the microorganisms responsible for odour destruction in most biological deodorization systems. Bacteria often exhibit high growth and biodegradation rates, a high resistance to toxicity and are able to degrade a wide range of odorants (Devinny *et al.*, 1999). However, their optimal pH is usually around 7 and a high water activity (moisture content) is necessary for their correct cellular functioning. On the other hand, fungi show a narrower range of degradable pollutants, but they tolerate low pH values (2–5), low humidity and nutrient limitation, which are common conditions in biological systems (van Groenestijn *et al.*, 2001). In addition, some fungi possess structural proteins in their cell wall known as hydrophobins, which facilitate the direct transport of hydrophobic compounds from the gas phase to the fungal hyphae, thus avoiding the mass transfer resistance in the liquid phase (Kennes and Veiga, 2004; Kraakman *et al.*, 2011).

Once the odorants have been transferred from the odorous emission to the aqueous phase containing the microorganisms (either in suspension or as a biofilm), they diffuse through the cellular walls/membranes (see Figure 4.10). It is inside the cells where pollutant destruction takes place. This destruction occurs via sequential catabolic reactions, which are globally known as metabolic routes. The result is the decomposition of odorants into metabolites that enter the central routes of the microbial metabolism. These metabolic routes are pollutant and microorganism specific.

Odorant biodegradation can take place via complete mineralization, partial oxidation or cometabolic degradation. Mineralization (complete oxidation) is the most favourable situation from an energetic viewpoint. On the other hand, partial oxidation (e.g. benzene to cathecol) generates secondary metabolites to be degraded and usually occurs at high pollutant concentration or in microbial populations with a limited spectrum of carbon sources. Cometabolism usually takes place for highly recalcitrant compounds (highly ramified or very stable ring structures) which are unable to induce the synthesis of the enzymes needed for their biodegradation. Halogenated hydrocarbons and polycyclic aromatics are considered extremely recalcitrant compounds due to their high resistance to microbial attack.

Biotechnologies are usually efficient for the treatment of biodegradable compounds at low to moderate concentrations although the development of two-phase partitioning systems has increased this maximum concentration threshold up to 20 000 ppm_v (Yeom and Daugulis, 2001; Muñoz *et al.*, 2007). Activated sludge is frequently used as inoculum

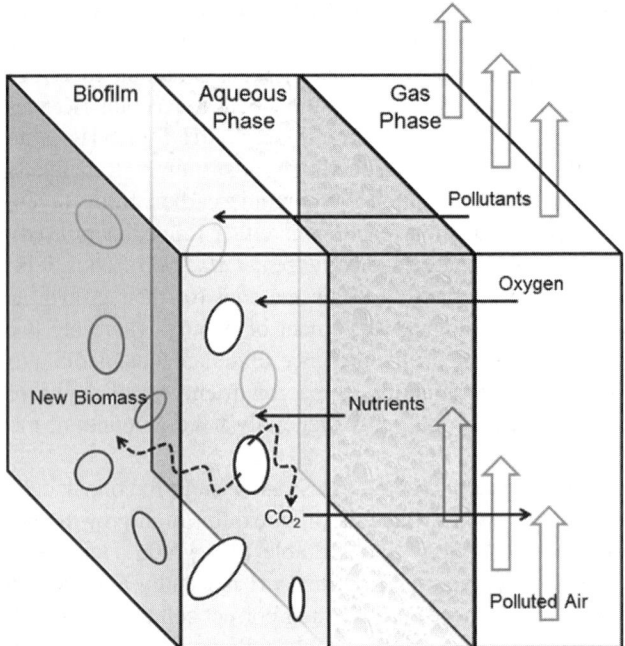

Figure 4.10   *Odorant biodegradation mechanisms in bioreactors.*

in biological odour abatement systems where no indigenous microflora is present (e.g. soil or compost biofilters) due to its high microbial diversity (Estrada *et al.*, 2012). Most of the microorganisms found in activated sludge are able to use alcohols, ketones, aldehydes, and so on. as their only carbon and energy source. On the other hand, recalcitrant odorants may require the presence of specialized microorganisms that can be externally added or used as customized inocula (Delhoménie and Heitz, 2005).

Prior to biodegradation, odorants must be transferred from the air emission to the aqueous phase surrounding the biofilm or where microorganisms are suspended. Overall, the volumetric mass transfer flowrate ($N_{g/l}$/kg m^{-3} h^{-1}) of an odorant from the gas to the aqueous phase is described in Equation (4.4):

$$N_{g/l} = K_L^{g/l} a \left(\frac{C_g}{K_{g/l}} - C_l \right) \tag{4.4}$$

Where: $\mathbf{K_L^{g/l} a}$(h^{-1}) is the volumetric transfer coefficient gas/water, $\mathbf{C_g}$ (g m^{-3}) is the odorant concentration in the gas phase, $\mathbf{C_l}$ (g m^{-3}) is the odorant concentration in the liquid phase and $\mathbf{K_{g/l}}$ [dimensionless] is the partition coefficient between the gas and the aqueous phase (Stuetz *et al.*, 2001).

Volumetric transfer coefficients are a function of the diffusivity of the odorant in water, the gas turbulence in the bioreactor and the area available for mass transfer in the system (size and type of packing in packed systems or bubble size in aerated systems). Gas/water partition coefficients can be estimated from the Henry's law constants since odorants are always present at very low concentrations, where the Henry's law is valid. It is important

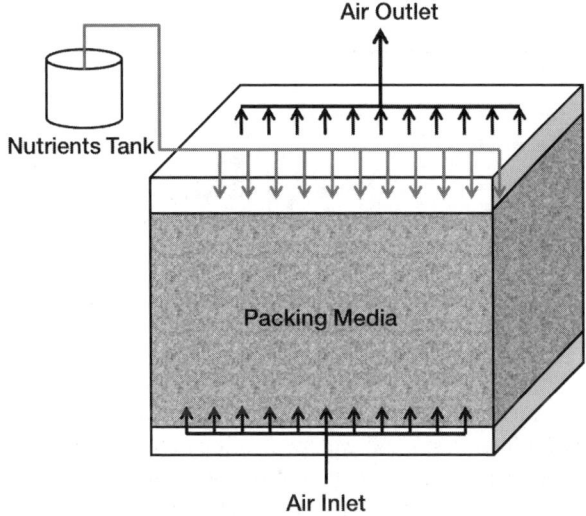

Figure 4.11   *Schematic design of a biofilter.*

to stress that the transport (and hence biodegradation) of hydrophobic compounds such as alkanes, terpenes or certain sulfur compounds is often limited by their low concentration gradients available for transport (low driving forces) due to their high partition coefficients (Muñoz *et al.*, 2007).

4.4.2.2   Biofiltration

Biofiltration is a biological odour abatement technique based on the biodegradation of odorants by a microbial community attached onto a fixed packing material in the presence of a stagnant water layer surrounding the biofilm. The odorous emissions are forced through this packed bed that hosts the microorganisms as a biofilm (see Figure 4.11) (Lebrero *et al.*, 2011). As described earlier, odorants must diffuse from the gas to the biofilm where biodegradation takes place, facing a mass transfer resistance first in the aqueous phase surrounding the biofilm and then in the biofilm itself. Common removal efficiencies reported in biofilters range from 75% for highly hydrophobic odorants (Henry's Law constant higher than 20 M atm^{-1}) to 99% for water soluble odorants (Henry's Law constant lower than 0.07 M atm^{-1}), these removal efficiencies being similar to those achieved by chemical scrubbers (Estrada *et al.*, 2011).

The selection of biofilter packing media is a key decision determining both the odorant removal efficiencies and the biofilter lifespan. Both inorganic and organic supports are used in biofiltration, the latter providing an extra carbon and nutrient source capable to maintain microbial activity when treating emissions with low carbon concentrations (as those found in odour abatement). Among inorganic materials, ceramics, plastic, lava rock or activated carbon are the most common packing materials. They provide an extra structural stability that increases the biofilter lifespan when used as the sole packing material or when mixed with organic materials such as soil, compost, and so on. However, both types of media must fulfil the same criteria in order to support an efficient and sustained biodegradation:

Table 4.5   *Recommended values for the design and operation of biofilters (Iranpour et al., 2005).*

Operating parameter	Range value	Units
Temperature	25–35	°C
pH	6–8	–
Moisture content	40–60	%
Porosity	40–80	%
Organic matter content	35–55	w/w%
Pressure drop	<10	cm water
Bed particles size	>4 mm (at least 60% of them)	mm

high specific surface areas and porosities, high buffer capacities (to maintain pH values near 7 for bacterial biofilters), a good water retention capacity and structural stability to prevent the formation of preferential pathways (Woertz *et al.*, 2002) (see Table 4.5). A recent compilation of data from full scale odour abatement facilities carried out by the authors, revealed that 87% of the biofilters used organic packing materials (either alone or mixed with inorganic media). Among the organic materials used wood chips and compost are the most popular (37 and 33%, respectively).

The high land requirements for its implementation, together with the frequent replacement of packing media, constitute the main drawback of biofiltration. The high EBRT needed (from 20 s to 2 min) and the low packing heights (1–2 m) to keep low pressure drops across the media result in high design areas. Another limitation often stressed by biofiltration detractors is the low control capacity over operational parameters such as pH, temperature, humidity, nutrients or degradation intermediates accumulated. Therefore, biofilters typically use a humidifier upfront to improve the moisture control in the biofilter media. At this point, it must be highlighted that fungal biofilters have some advantages such as the ability to grow at low pH and moisture contents or the mycelial growth, which increases the gas-biofilm surface area available for odorant mass transport (Vergara-Fernández *et al.*, 2011). Two different configurations can be found for biofiltration:

- Open biofilters, which require high areas for their installation as a result of the high EBRTs needed.
- Closed biofilters, with lower land needs and a better control over the operational parameters, which support a more flexible and efficient operation overcoming some of the drawbacks previously mentioned.

Biofilters are the most common biological technology for odour abatement, being intensively used in applications such as WWTPs or compost facilities. For instance more than 300 biofilters were operating in WWTPs in the USA in 2005 (Iranpour *et al.*, 2005). The accumulated experience over the last 60 years provides an extensive knowledge background of both design and operation parameters, making biofiltration a reliable technology for odour abatement when applied and operated properly provided its inherent limitations (Jorio and Heitz, 1999).

From an environmental viewpoint, biofiltration presents moderate to high environmental impacts. In terms of material consumption, biofiltration has been reported as the highest packing consuming odour abatement technology, with an annual consumption of around $4 \, kg \, (m^3/h)^{-1}_{air \, treated}$. This is mainly caused by the short lifespan of the packing material and the high EBRT of this technology. However, this consumption can be drastically reduced when using more durable media (e.g. mixtures of organic and inorganic materials) (Jorio and Heitz, 1999). In this regard, the variability of the commercial packing material lifespan is a key parameter: lifespans ranging from 1–15 years have been reported for nutrient-enriched inorganic media, and from 1–3 years for conventional organic media (Prado *et al.*, 2009). A recent sustainability study reported land requirements higher than 1.5×10^{-2} $m^2 \, (m^3/h)^{-1}_{air \, treated}$, while the rest of the odour abatement technologies exhibited land requirements at least three times lower (Estrada *et al.*, 2011). On the other hand, biofiltration presents a low energy consumption, mainly due to the low pressure drops maintained across the packing media and the lack of additional energy consuming processes such as water recycle, temperature control, process automation, etc. The annual water requirements of biofiltration are the lowest among the biotechniques $(220 \, l \, (m^3/h)^{-1}_{air \, treated})$ and its low water quality requirements make it possible to use recycled or partially treated water for periodical irrigation, this being a common advantage of all biological treatments.

When analysing the investment costs of biofiltration (Figure 4.4), it stands as the most economical biological alternative, although more expensive than physical/chemical technologies. This can be attributed to the accumulated design and construction experience, and its relative simplicity, which does not require high automation or specific construction material requirements. A compilation of recent full-scale biofiltration facilities investment costs reported costs ranging from €5 to 28 $(m^3/h)^{-1}_{air \, treated}$ depending on the design flowrate. The following equation, obtained from the above mentioned compilation, can be used to estimate the investment costs of biofilters (Estrada *et al.*, 2011):

$$Ic = 3168 \times F^{-0.58} \tag{4.5}$$

Where **Ic** is the investment cost $(€ \, (m^3/h)^{-1})$ and **F** the gas flowrate $(m^3 \, h^{-1})$.

The high land requirements for the installation of biofilters can however modify these costs in areas with low land availability, where the land pricing will entail an additional cost in the biofilter implementation.

In terms of operating costs, biofiltration presents moderate annual costs of €2.0 $(m^3/h)^{-1}$, ranking as the second most economical alternative for odour treatment, after biotrickling filtration. At this point, it is also worth highlighting the high variability of the packing material price, which is the main contributor to the total operating costs with a 47%, (Figure 4.5) ranging from €25 m^{-3} for compost to more than €500/m^3 for advanced media like Biosorbens^{TM} (Prado *et al.*, 2009). As mentioned earlier, the lifespan of the packing materials also exhibits a large variability, ranging from 1–15 years. Both parameters can drastically modify the final operating cost of biofiltration.

Figure 4.12 shows the relative impact of the most relevant parameters in terms of design, operation and commodity prices on the net present value at 20 years of lifespan. The three main parameters determining the cost of biofiltration are related to the packing material: EBRT, packing material lifespan and work cost (which itself is related to the media substitution and maintenance needed). The sensitivity found for all parameters evaluated is low when compared to that of physical/chemical technologies. Hence, increases of 25% in the parameters result in NPV20 responses ranging from 8–10%, while chemical scrubbers showed responses higher than 15% for H_2S concentration and chemicals cost, and

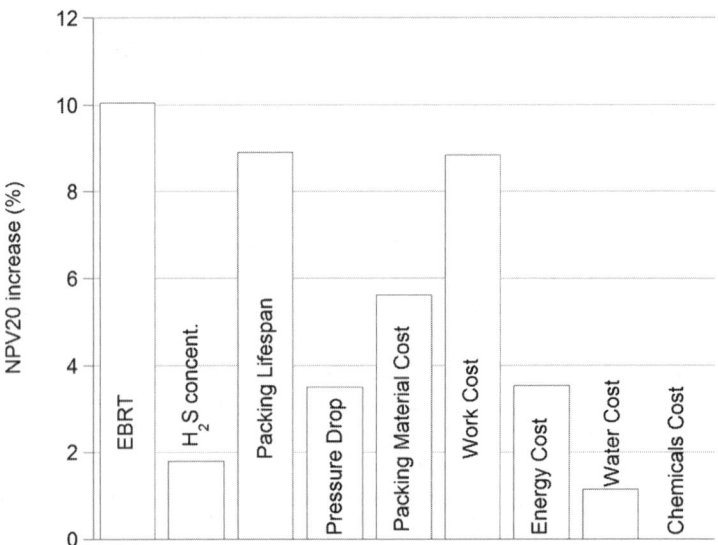

Figure 4.12 *Influence of an increase of 25 % in the selected parameters on the NPV20 of a biofiltration facility at a WWTP (gas flowrate 50 000 m³ h⁻¹, H₂S concentration 21 mg m⁻³, bed height 1 m, pressure drop 1000 Pa, packing material cost €72 m⁻³, EBRT 60 s).*

activated carbon adsorption exhibited NPV20 responses higher than 20% for a 25% H$_2$S concentration increase. In brief, the main advantages of biofiltration are the low dependence of process economic on H$_2$S (odorants) concentration and the low energy costs (Estrada *et al.*, 2012).

Despite not being one of the key parameters influencing the NPV20 (causing variations of 1% in the NPV20 when increased by 25%), the use of recycled water or secondary effluents can significantly reduce the operating costs of biofiltration systems in specific scenarios. For instance, in a WWTP under high tap water fares scenarios (such as Denmark), the use of the secondary effluent of the plant (available at zero cost) can reduce the annual operating costs by 29%. Overall, the possibility of using low-quality water allows for important savings in all biotechnologies (Estrada *et al.*, 2012).

The robustness of biofiltration is usually pointed out as one of its main limitations, although recent studies have shown that the performance of biofilters after common disturbances does not deteriorate as much as expected (Lebrero *et al.*, 2010). Empirical studies have shown that a decrease in the water content of the media (as a result of a failure in the water supply) and fluctuations of the inlet odorant concentrations cause the largest deterioration in biofilter performance. Good operational practice is often required to cope with these sometimes unavoidable issues, although a correct design can also help to mitigate the negative impact of these disorders by installing backup pumps or by combining foul air from different sources to buffer odour concentration fluctuations (Estrada *et al.*, 2012).

The lower abatement efficiencies achieved in most biofilters, compared to activated carbon filtration or incineration, significantly impact on the social benefits of biofiltration. However, the odour abatement efficiencies provided by biofilters are often sufficient for the most common applications or scenarios. In addition, biofiltration is perceived by society as a green and environmentally friendly technology and its implementation usually has a favourable acceptance. Biofiltration (like the rest of the biotechnologies) is a safe technology with very low risks associated to its operation due to the absence of high temperatures, fuels or hazardous chemicals.

In brief, biofiltration provides a reasonable odour abatement performance at moderate investment costs and low operating costs, which renders it as a good choice in non-sensitive scenarios with high space availability or in scenarios with limited economic and technological resources. State of the art biofilters are constructed enclosed with inorganic customized media and fully automatized in order to continuously monitor the operating variables to optimize odour removal, which increases both process robustness and the long-term operational stability. However, the implementation of these advanced systems entails a substantial increase in both investment and operating costs (Chitwood and Devinny, 2005).

4.4.2.3   Biotrickling Filtration

A biotrickling filter consists of a column packed with an inert packing material (usually plastic rings, resins, ceramic material, rocks...) where microorganisms grow attached. In addition to the nature of the packing support, the main difference from biofilters is the presence of a nutrient solution continuously recycling at rates ranging from $10\text{--}30\ \text{min}^{-1}\ (\text{m}^3{}_{\text{lecho}})^{-1}$ (Figure 4.13). These systems present high specific surface areas (between

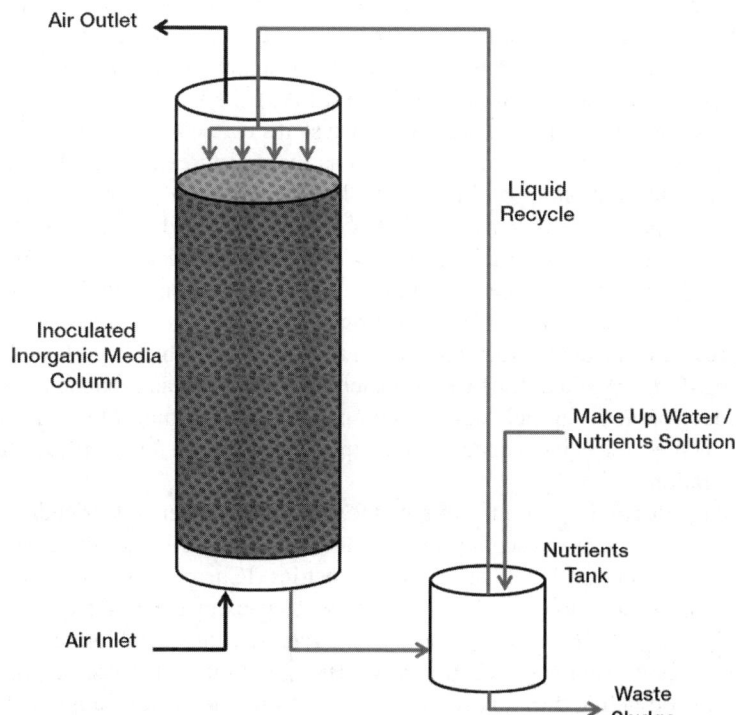

Figure 4.13   *Schematic representation of a biotrickling filter.*

$100–400 \ m^2 \ m^{-3}$), which allow for higher mass transfer in lower bed heights and at lower pressure drops (often ranging from $100–400 \ Pa \ m_{bed}^{-1}$) (Revah and Morgan-Sagastume, 2005). Odorants are initially absorbed in the aqueous film trickling over the biofilm and degraded afterwards by the microorganisms present in the biofilm. The high gas-liquid contact areas provided by the packed bed allow for the operation at gas EBRTs ranging from 2–15 s. Nevertheless, the mass transfer to the aqueous phase constitutes the main limitation of this technology for the removal of highly hydrophobic compounds, which accounts for 50–80% (Lebrero *et al.*, 2012). For water soluble compounds, such as H_2S, the removal efficiencies reported increase up to 99% (Delhoménie and Heitz, 2005). Indeed, H_2S removal efficiencies of 100% have been reported in systems operating at EBRTs similar to those applied in physical/chemical technologies (1.6–2.2 s) (Easter *et al.*, 2008; Gabriel and Deshusses, 2003). These experimental observations render biotrickling filtration as the best-performing biological alternative in scenarios with space limitations.

The use of inorganic packing materials requires the addition of nutrients to the system. Those nutrients are required for the complete mineralization of odorants (Revah and Morgan-Sagastume, 2005). The composition of this specific nutrient solution must be carefully designed since the low odorant concentrations commonly found in malodorous emissions, together with the poor mass transfer of highly hydrophobic compounds, might result in a poor microbial activity mediated by a combined detrimental effect of low nutrient and low odorant concentrations (Guieysse *et al.*, 2008). One or two stage designs are available, the latter being the only configuration providing odour abatement efficiencies of 95%. One stage biotrickling filters provide a good performance only when H_2S abatement is required and odour removal efficiencies lower than 90% are acceptable. Co- or counter-current configurations for the gas and aqueous phases do not present important differences in terms of odour abatement (Cox and Deshusses, 1999).

The main advantage of biotrickling filtration systems over conventional biofilters is the control over several operational variables such as the moisture content, pH, temperature and nutrients concentration. In addition, biotrickling filters allow for the continuous wash of bioconversion products out of the system, thus avoiding the media acidification problems that usually reduce the lifespan of biofilters' media and biological activity. As a rule of thumb, 2.5 l of water are supplied per g H_2S removed, which maintains a pH of 2, but is still optimum for acidophilic H_2S oxidizing communities. The use of inert packing materials provides a media lifespan up to more than 10 years, which significantly reduces the operating costs associated to packing material replacement and maintenance. Due to the recent developments in their design and operation, biotrickling filters are nowadays more implemented, especially treating high strength odours involving moderate to high H_2S concentrations.

Biotrickling filtration constitutes one of the most environmentally friendly technologies for odour abatement, presenting low to moderate impacts for all the environmental impacts analysed by the IChemE sustainability metrics (IChemE, The Sustainability Metrics, 2002). This technology presents the lowest yearly packing material requirements 0.04 $kg \ (m^3/h)^{-1}_{air \ treated}$ ($\approx$ 100 times lower than the requirements of a biofilter) mainly due to the lower packing volumes needed (lower EBRT) and the higher lifespan of the inorganic media. Biotrickling filtration is the technology with the lowest energy requirements due to the usually low pressure drops maintained across the bed and the limited liquid recycle rates (when compared to chemical scrubbers). In terms of land requirements, this technology also benefits from the lowest EBRTs among biotechnologies. Some biotrickling filters can provide an efficient treatment with land requirements comparable to those of chemical scrubbers (taking into account the extra areas for chemical storage) or activated carbon adsorption systems. The high water requirements are, a priori, the main drawback of biotrickling filtration. Hence, it presents the highest water needs among all odour abatement technologies as a result of the need for a continuous supply of water and nutrients as well as the continuous wash out of bioconversion products such as H_2SO_4. However, biotrickling filtration does not require high quality water, hence, the use of partially treated wastewater or secondary plant effluents (when available) offers an opportunity to simultaneously reduce both the impact by water consumption and the operating costs. Moreover, partially treated effluents can also supply the basic nutrients required for microbial growth (nitrogen, sulfur, phosphorus, etc.), which constitutes a further reduction in the operating costs (Estrada *et al.*, 2011).

Biotrickling filtration turns out to be the second most expensive odour abatement technology in terms of investments, after the hybrid technologies. This can be attributed to the relatively large gas contact time to the low experience gained over the years on this kind of systems and the higher mechanical complexity compared to other standard systems like biofilters or activated carbon filters. A compilation of recent full-scale biotrickling filter investment costs reported costs in the range of €10–41 $(m^3/h)^{-1}{}_{air\ treated}$ for two-stage designs. Equation (4.6) was obtained by fitting these data and can be used to estimate the investment costs of biotrickling filters as a function of the design gas flowrate (Estrada *et al.*, 2011):

$$Ic = 1794 \times F^{-0.47} \qquad (4.6)$$

Where **Ic** is the investment cost (€ $(m^3/h)^{-1}$) and **F** the gas flowrate ($m^3\ h^{-1}$)

Despite the relatively high investment costs of this technology, one of the main advantages of biotrickling filters is their low operating costs, which have been recently reported to be as low as €1.2 $(m^3/h)^{-1}$ on a yearly basis. 44% of these costs originate from the packing material, while water and energy contribute with 21 and 22%, respectively. The remaining 13% are derived from work costs according to Estrada *et al.* (Estrada *et al.*, 2012) (Figure 4.9).

When the total costs (NPV20) sensitivity is evaluated, the analysis reveals that biotrickling filtration is one of the most economical and less sensitive technologies towards changes on design/operational parameters and commodity prices. This technology presents a low sensitivity towards the EBRT, H_2S concentration, pressure drop or packing lifespan, which is due to the high lifespan of its packing material when compared to activated carbon adsorption or biofiltration. Although the analysis shows that the EBRT, the packing material lifespan and the packing material costs are the most important variables affecting the economics of biotrickling filtration, the variation of the NPV20 is always lower than 7% when these parameters are increased by 25%. An additional advantage in odour abatement scenarios with high fluctuations in the odorant concentrations is its low sensitivity towards variations in H_2S concentration, with only a 3% increase in the NPV20 when the H_2S loading increases by 25%. This low sensitivity entails stable operating costs for this technology and therefore accuracy in the yearly costs predictions (see Figure 4.14).

Water costs account for only 21% of the total operating costs (Figure 4.9) and are not a key economic parameter in terms of sensitivity, increasing the NPV20 by 3% when the

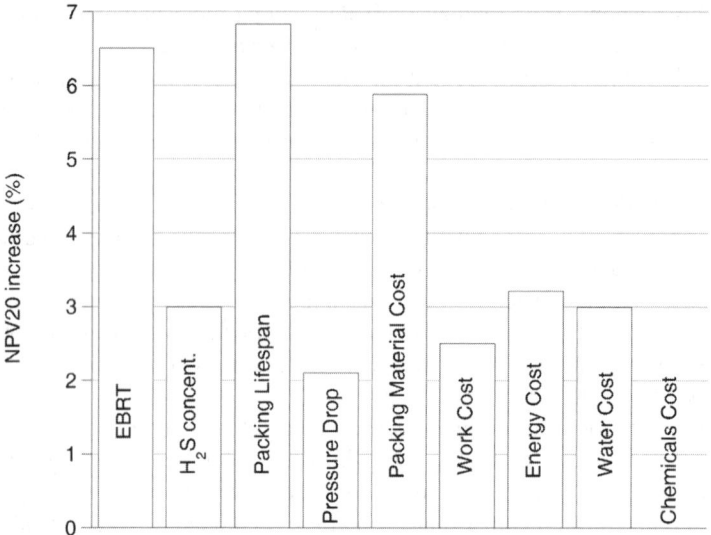

Figure 4.14 *Influence of an increase of 25% in the selected parameters on the NPV20 of a two-stages biotrickling filter at a WWTP (gas flowrate 50 000 m³ h⁻¹, H₂S concentration 21 mg m⁻³, bed height 1 m per stage, total pressure drop 500 Pa, packing material cost €1200 m⁻³, EBRT 15 s, 7.5 s per stage).*

water price is increased by 25%. However, the use of recycled water or secondary effluents (when available) can drastically reduce the operating costs of biotrickling filters under some specific scenarios. For instance, in a WWTP with high tap water fares, the use of secondary effluent (available at zero cost) can reduce the operating costs by more than 70% compared to those obtained with full fare water. The possibility of using low-quality water allows for savings even higher than those found in the case of biofiltration. This makes biotrickling filtration an even more attractive technology for odour abatement in wastewater treatment (Estrada *et al.*, 2012).

In terms of robustness, biotrickling filters perform better than conventional biofilters, ranking at the same level as chemical scrubbers. However their robustness is lower than that of activated carbon adsorption (Estrada *et al.*, 2012). The critical parameter affecting the removal efficiency in biotrickling filters is water supply. In this case, the negative effect of a failure in the water supply or water recycling system is higher than in biofilters due to both biofilm drying and the accumulation of the acidic degradation product at high H_2S loadings. This deterioration is also often faster than in biofilters due to their low EBRT of operation and generally lower water holding capacities. Thus, the installation of duty and standby pumps, backup water supply systems, flow and level transmitters and alarms is highly recommended for stable odour abatement, although this will directly impact on the investment costs. Biotrickling filters also present a moderate sensitivity towards electricity supply interruptions, although lab scale and full-scale studies have observed a good robustness when biotrickling filters are correctly designed and operated (Kraakman, 2003; Lebrero *et al.*, 2011).

The social benefits associated to biotrickling filters are limited by the moderate abatement efficiencies when dealing with highly hydrophobic odorants. Biotrickling filtration is still a relatively new applied technology, but will have a favourable acceptance due to its low environmental impacts. Biotrickling filtration is also a safe technology since it does not operate with hazardous chemicals, which is also advantageous from a social viewpoint (Estrada *et al.*, 2011).

Overall, biotrickling filtration provides outstanding economic benefits (lowest NPV20), even larger than biofiltration, and even in scenarios where packing material costs are high. This, together with the low environmental impact and efficient abatement performance (when two-stage systems are implemented) makes biotrickling filtration one of the most promising technologies in a near future. Studies have shown a good abatement performance for even hydrophobic VOCs and in addition, innovative bioreactor configurations, new packing materials and the addition of a non-aqueous phase have been successfully tested to cope with potential mass transfer limitations (Kraakman *et al.*, 2011; Lebrero *et al.*, 2012).

4.4.2.4   Hybrid Technology

This hybrid technology consists of a two-stage system combining a biotrickling filter and an activated carbon filter installed in series. In the first stage, the biotrickling filter removes most of the odorous compounds, especially water soluble odorants from the malodorous stream. Following this biological stage, an activated carbon adsorption system acts as a polishing step, removing the residual odours consisting mostly of the highly hydrophobic odorant fraction that escapes from the biotrickling filter. This kind of systems provides three major benefits:

1. The presence of this additional adsorption system increases the odorants removal efficiency to high levels (over 99.7% for all odorous compounds, regardless of their hydrophobicity), thus overcoming the main limitation of conventional biotrickling filters.
2. The combination of both systems provides an extra robustness for odour abatement, as one technology can back up the other one and minimize the effect of any potential malfunctioning in one of the systems.
3. The location of the biotrickling filter in the first stage results in the removal of most of the H_2S and other sulfur compounds, allowing for an increased lifespan of the activated carbon and therefore reducing the most important cost associated to adsorption processes.

The operating and design parameters for both technologies will be similar to those described for their standalone applications in terms of residence time, pressure drops, packing materials and irrigation or liquid recycling rates (the latter in the case of the biotrickling filter). For instance, a hybrid system designed to remove 99% of H_2S and 95% of odour will consist of a biotrickling filter operated at about 10 s of EBRT while the activated carbon filter will be designed at about 2 s. The main difference compared to standalone systems applies to the adsorption step, where the bed will face a much lower odorant load (as most odorants will be removed in the biotrickling filter). The lifespan of the activated carbon bed is much longer, and it's substitution is often performed on a time basis due to its polishing nature instead of on a load basis. When used as a backup polishing technology, the bed lifespan increases and its performance is not so critical. Thus bed replacement on a time basis is usually adopted and typically every two years when the biotrickling filter is properly designed and operated.

The hybrid technology presents a low annual material consumption (0.48 kg $(m^3/h)^{-1}$ air treated). This consumption is higher than that of biotrickling filtration in standalone applications due to the presence of the activated carbon filter. However, this configuration presents 40% less material consumption than a conventional activated adsorption filter, which constitutes a significant environmental advantage. The presence of two packed beds in series increases the pressure drop through the system, which implies an increase in the energy consumption. A hybrid system consumes almost four times the energy required for a two stage biotrickling filter and about 40% more than a standard activated carbon filter. This energy consumption is the highest among the most commonly used odour abatement technologies (excluding incineration or activated sludge diffusion processes) (Estrada *et al.*, 2011). The land requirements of a hybrid system are obviously a combination of those of both systems since at least two beds are required and depend on the operating parameters. Water requirements are only associated to the biotrickling filtration stage and can be estimated according to the empirical design parameter of 2.5 l_{H_2O} g^{-1} $_{H_2S\ removed}$. Thus, for the same emission and assuming that all H_2S is completely removed in the biological step due to its high water solubility, the annual water consumption will be the same of a standalone biotrickling filter ($5.2 \cdot 10^{-2}$ l $(m^3/h)^{-1}$ air treated). In this case, the substitution of high quality water by partially treated water or secondary effluent is also possible, with the associated savings in economic terms and the decrease in the environmental impact.

No investment costs of full scale hybrid systems have been so far reported in literature. However, the investment costs for this technology will presumably be slightly higher than for standalone biotrickling filters due to the extra cost of the carbon filter beds, which would make the hybrid technology the most expensive technology in terms of investment costs. On the other hand, the operating costs of this technology can be estimated from the regular operation of their components separately. A recent study reported annual operating costs of €2.7 $(m^3/h)^{-1}$ for hybrid systems, which ranks the costs of this technology right over those of biotrickling filtration, but still far below the high operating costs reported for chemical scrubbers or activated carbon adsorption systems. These costs are mainly due to the packing material (41%) and energy (32%) (Figure 4.9) (Estrada *et al.*, 2012). Both costs are related to the presence of two beds, which entail a higher cost for packing media replacement and an extra pressure drop that increases the energy requirements.

The NPV20 of the hybrid technology showed the highest sensitivity towards variations in the EBRT, packing lifespan and packing material price. On the other hand, the NPV20 presented the lowest sensitivity towards variations in H_2S concentration, with only a 1.75% increase when this parameter was increased by 25% (Figure 4.15). This supports the superior performance of this technology in applications with high odorant concentration fluctuations and is due to the fact that H_2S is often completely eliminated in the biological stage and do not influence the lifespan of the adsorption stage. When all parameters are increased by 25%, it is clearly shown that the energy cost and the pressure drop significantly influences the NPV20 (5–6%). This is caused by the extra pressure drop, and thus energy demand, that the two beds in series induce. However, none of the effects is higher than 8%, which renders the hybrid technology similar to biotrickling filtration in terms of economic stability while supporting a usually more robust odour abatement performance than standard biotechnologies and chemical scrubbers (Figure 4.11).

Like the other biotechnologies, the biotrickling filtration step benefits from the reduced environmental impact and operating costs mediated from the use of recycled or partially

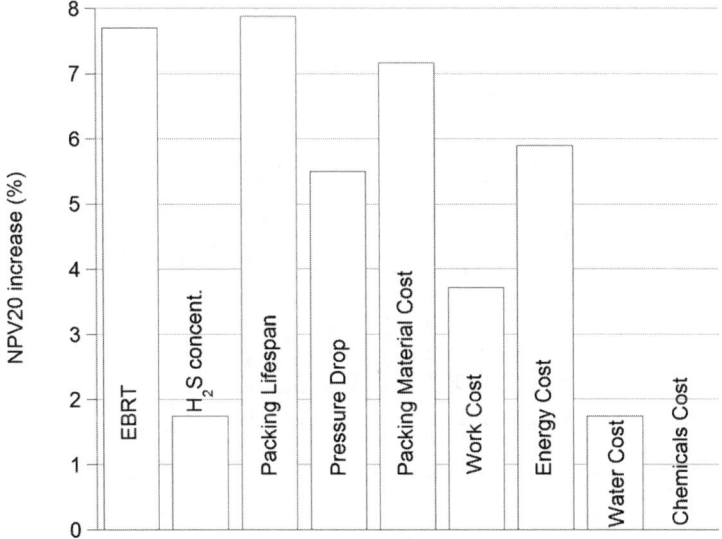

Figure 4.15 *Influence of an increase of 25% in the selected parameters on the NPV20 of a hybrid system at a WWTP (gas flowrate 50 000 m³ h⁻¹, H₂S concentration 21 mg m⁻³, biotrickling filter: EBRT = 9 s, packing height = 2 m, inert plastic packing cost of €1200 m⁻³, total ΔP = 500 Pa. Impregnated activated carbon bed: EBRT = 2.5 s, packing height = 0.5 m, ΔP = 1750 Pa, cost = €7.2 kg⁻¹).*

treated water. The use of secondary effluent (when available at nearly zero cost) can reduce the operating costs of this technology by more than 40%, and offers 32% savings in the NPV20 in expensive water scenarios such as in northern European countries. This means more than €1.5 million of savings in 20 years for a hybrid system treating 50 000 m³ h⁻¹.

The presence of two beds combining the benefits of biological oxidation and physical adsorption determines the robustness of this technology. In fact, the robustness of the hybrid technology is comparable to that of activated carbon adsorption in standalone applications. The presence of the activated carbon filter minimizes the effect of any deterioration in the biological activity, at the expenses of a reduced lifespan of the activated carbon. For instance, a water supply disorder is usually considered probable and with critical effects on the performance of biofilters and biotrickling filters. However, in a hybrid technology, the effect would be marginal due to the odour abatement of the adsorption stage while the problem in the biological step is being solved.

The high odour abatement efficiencies provided by the hybrid technologies rank them among those supporting the highest benefits from a social point of view. The high abatement efficiency will benefit both plant operators and the nearby population, this technology being highly recommended for odour control in sensitive areas. The use of a biological stage drastically reduces the material needs of the carbon filter while supporting removal efficiencies as high as standalone adsorption filter but at a lower environmental impact and economic costs.

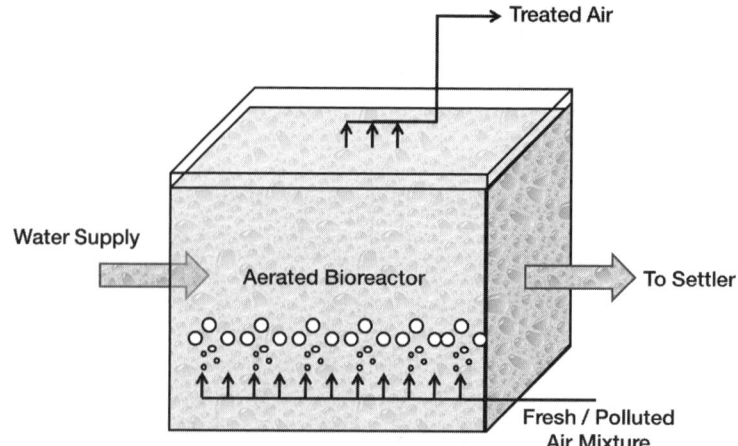

Figure 4.16   *Schematic representation of an activated sludge diffusion system.*

In summary, hybrid systems are probably one of the most expensive alternatives in terms of investment costs. However, despite their high capital costs, hybrid systems exhibit lower operating costs and environmental impacts than those of standalone activated carbon filters. In addition, they exhibit extremely high removal efficiencies and robustness, similar to those of adsorption filters. This technology can meet the strictest standards of odour removal in scenarios where the performance of biofilters or biotrickling filters is not sufficient, and represents the most cost-effective and sustainable alternative for highly sensitive scenarios. Further developments (mainly in the field of biotrickling filtration) will likely reduce the contribution of the adsorption stage to the overall deodorization and allow hybrid systems to be replaced by standalone biological technologies.

4.4.2.5   Activated Sludge Diffusion

At WWTPs, the malodorous emission can be directly sparged into the aeration tank of a wastewater treatment plant together with (or in substitution of) the air to satisfy the oxygen needs of the process (Hardy *et al.*, 2001) (Figure 4.16). Obviously, this activated sludge diffusion technology is only cost-effective in plants where aeration is provided via air diffusion. Malodorous compounds diffuse from the gas phase to the aqueous phase together with the oxygen needed for the oxidation of the organic matter and NH_4^+ of the wastewater and are then converted by the microbial community present in the activated sludge (Burgess *et al.*, 2001; Barbosa *et al.*, 2002). There are not many data available about this innovative technique for odour abatement, however large scale studies have reported odour removal efficiencies higher than 99% in WWTPs treating odours from compost plants (Ostojic *et al.*, 1992). Likewise, Bowker (2000) found deodorization efficiencies higher than 99.9% at H_2S concentrations up to 120 ppm, which confirms the potential of activated sludge diffusion system as an alternative for odour abatement.

In appropriately designed plants, the malodorous emission can cover 100% of the aeration needs of the activated sludge process. However, the malodorous emissions are usually higher than the aeration needs and additional odour treatment systems would be required (Bowker, 2000).

A recent sustainability analysis of the most commonly applied odour abatement technologies conducted by the authors confirmed that activated sludge diffusion systems are one of the most environmentally friendly techniques if the plant is provided with aeration via air diffusion. Hence, activated sludge diffusion presents no material needs, no water requirements and no additional land needs. The only relevant impact related to the process is the yearly energy requirements and the associated environmental impacts, which account for 556 MJ $(m^3/h)^{-1}_{air\ treated}$. However, it must be considered that this consumption does not represent an additional impact as this energy would be supplied for the aeration of the biological reactor and it is already included in the energy consumption of a WWTP.

Despite there not being much information about activated sludge diffusion systems for odour abatement so far, it is reasonable to assume that the investment costs of this technology are low because all the equipment required is already present in the wastewater treatment line (even the air blowers might be used for the dispersion of the malodorous emission). Additional costs can derive from the installation of moisture traps to avoid corrosion problems in the pipeline caused by H_2S. The installation should include also dust and grease aerosols filters and corrosion resistant materials in blowers and air piping. Corrosion is a major issue for this kind of systems, nevertheless, a survey collecting data from 30 WWTPs located in the USA where activated sludge diffusion was implemented showed that these concerns might be not well founded. The selection of adequate materials and protection equipment as described earlier can easily overcome these problems (Bowker and Burgess, 2001).

4.4.3   Technology Comparison Based on Case Studies

Despite being well established technologies with a wide experience gained in their design and operation, the supremacy of physical/chemical techniques in the field of odour treatment is being challenged by the increasing public concern on sustainability (including social, economic and environmental issues) (Lebrero *et al.*, 2011) (see Table 4.6). Activated carbon adsorption systems exhibit a high robustness and abatement efficiencies, which make this technology still competitive in highly sensitive scenarios. On the other hand, chemical scrubbers support removal levels comparable to those of conventional biotechniques but at a much higher environmental impact. In terms of robustness, chemical scrubbers are severely impacted by operational upsets such as liquid recycle failures and they are more sensitive in economic terms to variations in design parameters or commodity prices. Gabriel and Deshusses (Gabriel and Deshusses, 2004) demonstrated that chemical scrubbers are more expensive than biological methods in the long term. They estimated that the conversion of a full scale chemical scrubber into a biotrickling filter would save more than US$43 000 year^{-1} with a payback time of 1.5 years. In addition, this study confirmed that the biotrickling filter outperformed the chemical scrubber, providing increased odour removals at a lower operating cost (Gabriel and Deshusses, 2004). This confirms that this type of conversion is not only cost-effective but also mitigates the environmental impact of odour removal at increased odour removal efficiencies. Another full-scale example of the superior performance of biotechnologies was described for a WWTP located in Jacksonville (Florida,

Table 4.6 Odour abatement technologies comparison chart.

	Investment Cost	Operating Costs	Total Costs (NPV20)	Cost Sensitivity	Environmental Performance	Social Benefits	Abatement Robustness
Chemical Scrubber	+ +	−	−	−	−	−	+
Adsorption Systems	+ +	−	−	−	−	+ +	+ + +
Regenerative Incineration	−	−	−	N.D.	−	+	+ + +
Biofiltration	+	+ +	+	+	+ +	+	−
Biotrickling Filtration	−	+ + +	+ +	+	+ +	+ +	+
Hybrid Technology	−	+	−	+ +	+	+ + +	+ + +
Activated Sludge Diffusion	+ +	+ + +	+ + +	+ +	+ + +	+ +	+ +

+/positive aspect, benefit; −/negative performance, drawback; N.D. not determined.

US). In this case, a chemical scrubbing system and an advanced bioreactor system were compared for the treatment of 47 600 m^3 h^{-1}. The results showed that the operating costs of the biofilter were 10 times lower than those of the chemical scrubber at high removal efficiencies (99.6% for H$_2$S) and an acceptable robustness to common operating upsets (for instance, the H$_2$S removal efficiency never decreased below 75% and it recovered steady state performance two days after any upset) (Kraakman, 2006). In this context, the savings derived from the installation of bioreactors instead of chemical scrubbers was estimated as \$AUS660 000 year^{-1} at the Woodman Point WWTP (Perth, Australia) for the treatment of a malodorous emission of 74 250 m^{-3} h^{-1}. These savings could increase up to nearly \$AUS2 million year^{-1} at increased airflows in the plant (van Durme *et al.*, 2009). The use of standalone biotechniques for odour treatment is very common in WWTP, where most of the information on full scale deodorization systems is available.

The odorous emission is often separated and treated in standard size units operated in parallel. For instance, 10 biotrickling filters were installed at the Melbourne Western Treatment Plant (Australia) providing an overall treatment capacity of 126 000 m^3 h^{-1} in a parallel configuration with average odour removals of 98.5% (Cesca *et al.*, 2010) and 10 biotrickling filters were installed at Columbus (Ohio, US) treating approximately 170 000 m^3 h^{-1} with an average odour removal efficiency of 96% (Morton *et al.*, 2010). When further odour removal efficiencies are required, as in the Gippsland Water Factory (Victoria, Australia), a combination of technologies can be implemented. In this particular case, the malodorous stream was treated in an activated sludge diffusion system followed by an advanced biotrickling filtration system used as a polishing step (Trainor *et al.*, 2008).

References

Barbosa, V.L., Burgess, J.E., Darke, K. and Stuetz, R.M. (2002) Activated sludge biotreatment of sulfurous waste emissions, *Reviews in Environmental Science and Biotechnology*, **1**, 345–362.

Bowker, R.P.G. (2000) Biological odour control by diffusion into activated sludge basins, *Water Sci. Technol.*, **41** (6), 127–132.

Bowker, R.P.G. and Burgess, J.E. (2001) Activated sludge diffusion as an odour control technique, in *Odours in Wastewater Treatment: Measurement, Modelling and Control* (R. Stuetz and F.B. Frenchen, eds). IWA Publishing.

Bilsen, I. and De Fré, R. (2008) Evaluation of a neutralizing agent applied on a cooling tower at a citric acid production plant, in *3rd IWA International Conference on Odour and VOCs*, Barcelona, Spain.

Bruchet, A., Decottignies, V. and Filippi, G. (2008) Efficiency of masking agents: outcome of a 3 year study at pilot and full scales, in *3rd IWA International Conference on Odours and VOCs*, Barcelona, Spain.

Burgess, J.E., Parsons, S.A. and Stuetz, R.M. (2001) Developments in odour control and waste gas treatment biotechnology: a review, *Biotechnology Advances*, **19** (1), 35–63.

Card, T. (2001) Chemical odour scrubbing systems, in *Odours in Wastewater Treatment: Measurement, Modelling and Control* (R. Stuetz and F.B. Frenchen, eds). IWA Publishing, p. 308–344.

Cesca, J., McDonald, A., Rahardjo, A., *et al.* (2010) Case Study: Succesful Operation of Biotechnology for Odour Control at Western Treatment Plant, in *WEF/AWMA Odor and Air Pollutants conference*, Washington.

Chemel, C., Riesenmey, C., Batton-Hubert, M. and Vaillant, H. (2012) Odour-impact assessment around a landfill site from weather-type classification, complaint inventory and numerical simulation, *Journal of Environmental Management*, **93** (1), 85–94.

Chitwood, D.E. and Devinny, J.S. (2005) Commercial Applications of Biological Waste Gas Purification, in *Environmental Biotechnology*. Wiley-VCH Verlag GmbH & Co. KGaA, p. 427–438.

Cleugh, H. (1998) Effects of windbreaks on airflow, microclimates and crop yields, *Agroforestry Systems*, **41** (1), 55–84.

Cox, H.H.J. and Deshusses, M.A. (1999) Chemical removal of biomass from waste air biotrickling filters: screening of chemicals of potential interest, *Water Research*, **33** (10), 2383–2391.

Decottignies, V., Filippi, G. and Bruchet, A. (2007) Characterization of odour masking agents often used in the solid waste industry for odour abatement, *Water Sci. Technol.*, **55** (5), 359–364.

Delhoménie, M.-C. and Heitz, M. (2005) Biofiltration of Air: A Review, *Critical Reviews in Biotechnology*, **25** (1), 53–72.

Deshusses, M.A. (1997) Biological waste air treatment in biofilters, *Current Opinion in Biotechnology*, **8** (3), 335–339.

Devinny, J., Deshusses, M.A. and Webster, T.S. (1999) *Biofiltration for Air Pollution Control*. CRC Press, Boca Raton.

Easter, C., Witherspoon, J., Voig, R. and Cesca, J. (2008) An Odor Control Master Planning Approach to Public Outreach Programs. in *Proceedings of the 3rd IWA International Conference on Odour and VOCs*, Barcelona, Spain.

Edwards, F.G. and Nirmalakhandan, N. (1996) Biological treatment of airstreams contaminated with VOCs: an overview, *Water Science & Technology*, **34** (3–4), 565–571.

Estrada, J.M., Kraakman, N.J.R.B., Muñoz, R.L. and Lebrero, R. (2011) A Comparative Analysis of Odour Treatment Technologies in Wastewater Treatment Plants, *Environmental Science & Technology*, **45** (3), 1100–1106.

Estrada, J.M., Rodríguez, E., Quijano, G. and Muñoz, R. (2012) Influence of VOC concentration on the diversity and biodegradation performance of microbial communities, *Bioprocess and Biosystems Engineering* (In Press).

Estrada, J.M., Kraakman, N.J.R., Lebrero, R. and Muñoz, R. (2012) Torre, A Sensitivity Analysis of Process Design Parameters, Commodity Prices and Robustness on the Economics of Odour Abatement Technologies, *Accepted for Publication in Biotechnology Advances*.

European Comission (ed.) (2003) EC, *Integrated Pollution Prevention and Control*. Reference Document on Best Available Techniques in Common Waste Water and Gas Treatment/Management Systems in the Chemical Sector.

Frenchen, F.B. (2008) Emission of Odours from Sewer Systems–Countermeasures and Quantification of their Efficiency, in *NOSE 2008 – International Conference on Environmental Odour Monitoring and Control* Rome, Italy.

Gabriel, D. and Deshusses, M.A. (2003) Performance of a full-scale biotrickling filter treating H_2S at a gas contact time of 1.6 to 2.2 seconds, *Environmental Progress*, **22** (2), 111–118.

Gabriel, D. and Deshusses, M.A. (2004) Technical and economical analysis of the conversion of a full-scale scrubber to a biotrickling filter for odor control, *Water Science & Technology*, **50** (4), 309–318.

Guieysse, B., Hort, C., Platel, V., *et al.* (2008) Biological treatment of indoor air for VOC removal: Potential and challenges, *Biotechnology Advances*, **26** (5), 398–410.

Hainong, S., Xi Sheng, C., Hong Xiang, Z. *et al.* (2006) Study on Odor Formation Control during Kraft Pulping, in *Pan Pacific Conference Proceedings*.

Hardy, P., Burgess, J.E., Morton, S. and Stuetz, R.M. (2001) Simultaneous activated sludge wastewater treatment and odour control, *Water Sci. Technol.*, **44** (9), 189–96.

Hobbs, P.J., Pain, B.F., Kay, R.M. and Lee, P.A. (1996) Reduction of Odorous Compounds in Fresh Pig Slurry by Dietary Control of Crude Protein, *Journal of the Science of Food and Agriculture*, **71** (4), 508–514.

IChemE, The Sustainability Metrics (2002) The Institution of Chemical Engineers, Rugby.

INE (2009) *Spanish National Statistics Institute*, www.ine.es (accessed January, 2012).

Iranpour, R., Cox, H.H.J., Deshusses, M.A. and Schroeder, E.D. (2005) Literature review of air pollution control biofilters and biotrickling filters for odor and volatile organic compound removal, *Environmental Progress*, **24** (3), 254–267.

Jehlickova, B., Longhurst, P.J. and Drew, G.H. (2008) Assessing Effects of Odour: A critical review of assessing annoyance and impact on amenity, in *3rd. IWA International Conference on Odours and VOCs*, Barcelona, Spain.

Jorio, H. and Heitz, M. (1999) Traitement de l'air par biofiltration, *Revue Canadienne de Génie Civil*, **26** (4), 402–424.

Kaye, R. and Jiang, K. (2000) Development of odour impact criteria for sewage treatment plants using odour complaint history, *Water Science & Technology*, **41** (6), 57–74.

Kennes, C. and Thalasso, F. (1998) Review: Waste gas biotreatment technology, *Journal of Chemical Technology & Biotechnology*, **72** (4), 303–319.

Kennes, C. and Veiga, M.C. (2004) Fungal biocatalysts in the biofiltration of VOC-polluted air, *Journal of Biotechnology*, **113** (1–3), 305–319.

Koe, L. (2001) Process Covers for Odour Containment, in *Odours in Wastewater Treatment. Measuring, Modelling and Control* (R.M. Stuetz and F.B. Frenchen, eds). IWA Publishing, Cornwall.

Kraakman, N.J.R. (2003) Robustness of a full-scale biological system treating industrial CS2 emissions, *Environmental Progress*, **22** (2), 79–85.

Kraakman, N.J. (2006) Odour Emission Control with Advanced Biotechnology to Eliminate Disadvantages Associated with Chemical Scrubbers. (Odortool.com, Bioway). Available from: http://www.odortool.com/project_images/papers/12.%20Case%20study%20Jacksonville.pdf (accessed 22 December, 2011).

Kraakman, N., Rocha-Rios, J. and van Loosdrecht, M. (2011) Review of mass transfer aspects for biological gas treatment, *Applied Microbiology and Biotechnology*, 1–14.

Lebrero, R., Rodríguez, E., Martin, M. *et al.* (2010) H2S and VOCs abatement robustness in biofilters and air diffusion bioreactors: A comparative study, *Water Research*, **44** (13), 3905–3914.

Lebrero, R., Estrada, J.M. and Muñoz Torre, R. (2011) A comparative study of one and two-liquid phase biotrickling filters for odour removal in WWTP, in *Biotechniques for Air pollution control IV*, A Coruña, Spain.

Lebrero, R., Bouchy, L., Stuetz, R. and Muñoz, R. (2011) Odor Assessment and Management in Wastewater Treatment Plants: A Review, *Critical Reviews in Environmental Science and Technology*, **41** (10), 915–950.

Lebrero, R., Rodríguez, E., Estrada, J.M. *et al.* (2012) Effect of the EBRT on Methyl Mercaptan and Hydrophobic VOCs Removal, *Bioresource Technology*, **109**, 39–45.

Lens, P.B. (2001) Catalytic oxidation of odorous compounds from waste treatment processes, in *Odours in Wastewater Treatment: Measurement, Modelling and Control* (R.M. Stuetz and F.B. Frenchen, eds). IWA Publishing.

Lin, X.J., Barrington, S., Nicell, J. *et al.* (2006) Influence of windbreaks on livestock odour dispersion plume in the field, *Agriculture, Ecosystems & Environment*, **116** (3–4), 263–272.

Mills, B. (1995) Review of methods of odour control, *Filtration & Separation*, **32** (2), 146–152.

Morton, C.M., Pope, R.J., Reinhold, R.A., *et al.* (2010) Columbus Discovers a Greener Odor Control Approach in *WEF/AWMA Odor and Air Pollutants Conference*, Washington.

Mukhtar, S., Ullman, J.L., Carey, J.B. and Lacey, R.E. (2004) A Review of Literature Concerning Odors, Ammonia, and Dust from Broiler Production Facilities: 3. Land Application, Processing, and Storage of Broiler Litter, *The Journal of Applied Poultry Research*, **13** (3), 514–520.

Muñoz, R., Villaverde, S., Guieysse, B. and Revah, S. (2007) Two-phase partitioning bioreactors for treatment of volatile organic compounds, *Biotechnology Advances*, **25** (4), 410–422.

Muñoz, R., Sivret, E.C., Parcsi, G. *et al.* (2010) Monitoring techniques for odour abatement assessment, *Water Research*, **44** (18), 5129–5149.

Nielsen, J.H., Villadsen, J. and Lidén, G. (2003) *Bioreaction Engineering Principles.* Kluwer Academic/Plenum Publishers, New York.

Ostojic, N., Les, A.P. and Forbes, R. (1992) Activated sludge treatment for odor control, *Biocycle*, **33** (4), 74–78.

Prado, Ó.J., Gabriel, D. and Lafuente, J. (2009) Economical assessment of the design, construction and operation of open-bed biofilters for waste gas treatment, *Journal of Environmental Management*, **90** (8), 2515–2523.

Prado, Ó.J., Redondo, R.M., Lafuente, J. and Gabriel, D. (2009) Retrofitting of an Industrial Chemical Scrubber into a Biotrickling Filter: Performance at a Gas Contact Time below 1 s, *Journal of Environmental Engineering*, **135** (5), 359–366.

Revah, S. and Morgan-Sagastume, J. (2005) Methods for odor and VOC control, in *Biotechnology for Odour and Air Pollution* (Z. Shareefdeen and A. Singh, eds). Springer-Verlag, Heidelberg, Germany, p. 29–64.

Schlegelmilch, M., Streese, J. and Stegmann, R. (2005) Odour management and treatment technologies: An overview, *Waste Management*, **25** (9), 928–939.

Stuetz, R.M. and Frenchen, F.B. (eds) (2001) *Odours in Wastewater Treatments: Measurement, Modelling and Control.* IWA Publishing, Cornwall.

Stuetz, R.M., Gostelow, P. and Burgess, J.E. (2001) Odour Perception, in *Odours in Wastewater Treatment: Measurement, Modelling and Control* (R. Stuetz and F.B. Frenchen, eds). IWA Publishing.

Sucker, K., Both, R. and Winneke, G. (2008) Review of adverse health effects of odours in field studies, in *3rd IWA International Conference on Odour and VOCs*, Barcelona, Spain.

Sutton, A.L., Kephart, K.B., Verstegen, M.W. *et al.* (1999) Potential for reduction of odorous compounds in swine manure through diet modification, *J. Anim. Sci*, **77** (2), 430–439.

Tchobanoglous, G., Burton, F.L. and Stensel, H.D. (2003) *Wastewater Engineering: Treatment and Reuse*. McGraw Hill, New York.

Trainor, S., Cesca, J., Witherspoon, J. *et al.* (2008) An Innovative and Sustainable Odour Management Solution at the Gippsland Water Factory. (cited CH2M HILL Australia, Parson's Brinckerhoff, Gippsland Water). Available from: http://www.odortool.com/project_images/papers/602/Paper_Case%20study%20Gippsland%20Water%20Factory.pdf (accessed 22 December, 2011).

Turk, A. and Bandosz, T.J. (2001) Adsorption systems for odour treatment, in *Odours in Wastewater Treatment: Measurement, Modelling and Control* (R. Stuetz and F.B. Frenchen, eds). IWA Publishing.

Tyndall, J. and Colletti, J. (2007) Mitigating swine odor with strategically designed shelterbelt systems: a review, *Agroforestry Systems*, **69** (1), 45–65.

Vane, L.M. and Giroux, E.L. (2000) Henry's Law Constants and Micellar Partitioning of Volatile Organic Compounds in Surfactant Solutions, *Journal of Chemical & Engineering Data*, **45** (1), 38–47.

Van Broeck, G., Bogaert, S. and De Meyer, L. (2008) Monetary valuation of odour nuisance as a tool to evaluate cost effectiveness of possible odour reduction techniques, in *3rd. IWA International Conference on Odour and VOCs*, Barcelona, Spain.

van Durme, G.P., Nichols, C. and Cadee, K. (2009) Biotrickling filters reduce odour control costs Woodman Point WWTP, in *OZ-Water 2009*, Melbourne.

van Groenestijn, J.W., van Heininge, W.N. and Kraakm, N.J. (2001) Biofilters based on the action of fungi, *Water Sci. Technol.*, **44** (9), 227–232.

van Groenestijn, J.W. and Kraakman, N.J.R. (2005) Recent developments in biological waste gas purification in Europe, *Chemical Engineering Journal*, **113** (2–3), 85–91.

Varel, V.H. and Yen, J.T. (1997) Microbial perspective on fiber utilization by swine, *J. Anim. Sci*, **75** (10), 2715–2722.

Vergara-Fernández, A., Hernández, S. and Revah, S. (2011) Elimination of hydrophobic volatile organic compounds in fungal biofilters: Reducing start-up time using different carbon sources, *Biotechnology and Bioengineering*, **108** (4), 758–765.

Vincent, A.J. (2001) Sources of Odours in Wastewater Treatment, in *Odours in Wastewater Treatment: Measurement, Modelling and Control* (R.M. Stuetz and F.B. Frenchen, eds). IWA Publishing.

Water Environmental Federation (ed.) (2004) WEF, Control of Odour and Emissions from Wastewater Treatment Plants, in *Manual of Practice 25*.

Woertz, J.R., van Heiningen, W.N., van Eekert, M.H. *et al.* (2002) Dynamic bioreactor operation: Effects of packing material and mite predation on toluene removal from off-gas, *Appl. Microbiol. Biotechnol.*, **58** (5), 690–694.

Wright, D.W., Eaton, D.K., Nielsen, L.T., *et al.* (2005) Multidimensional Gas Chromatography—Olfactometry for the Identification and Prioritization of Malodors from Confined Animal Feeding Operations, *Journal of Agricultural and Food Chemistry*, **53** (22), 8663–8672.

Yeom, S.-H. and Daugulis, A.J. (2001) Development of a novel bioreactor system for treatment of gaseous benzene, *Biotechnology and Bioengineering*, **72** (2), 156–165.

Zarra, T., Naddeo, V., Belgiorno, V. *et al.* (2008) Odour monitoring of small wastewater treatment plant located in sensitive environment., *Water Science & Technology*, **58** (1), 89–94.

Zhu, J.Y., Chai, X.S., Pan, X.J. *et al.* (2002) Quantification and Reduction of Organic Sulfur Compound Formation in a Commercial Wood Pulping Process, *Environmental Science & Technology*, **36** (10), 2269–2272.

Part 5

Dispersion Modelling for Odour Exposure Assessment

M. Piringer[1] and G. Schauberger[2]
[1]*Section Environmental Meteorology, Central Institute for Meteorology and Geodynamics, Vienna, Austria*
[2]*WG Environmental Health, Department for Biomedical Sciences, University of Veterinary Medicine Vienna, Vienna, Austria*

5.1   Introduction

Part 5 of this handbook contains an overview on the state-of-the-art of odour dispersion modelling for environmental impact assessment. It starts with describing odour perception as a biological reaction of humans, whereby odour intensity, frequency and duration are the dominant parameters. This explains why dispersion models (see Section 5.3 for the types of models usually used to calculate odour dispersion), which usually calculate time averages of concentrations, have to be adapted somehow to parameterize short-term odour concentrations relevant in the time scale of a single human breath (Section 5.4). The short Section 5.5 explains why the same level of odour concentrations can cause different grades of annoyance whether it is felt pleasant or unpleasant. This is followed by a section on odour impact criteria defined as a composite of odour concentration threshold and exceedance probability; criteria usually differ according to the protection level. Section 5.7 deals with the meteorological input data necessary to run odour dispersion models; besides wind data, information on atmospheric stability and the mixing height is important, usually in the form of time series over at least one year. Finally in Section 5.8, the important issue of model evaluation is demonstrated for the case of an odour experiment. This whole

Odour Impact Assessment Handbook, First Edition. Edited by Vincenzo Belgiorno, Vincenzo Naddeo and Tiziano Zarra.
© 2013 John Wiley & Sons, Ltd. Published 2013 by John Wiley & Sons, Ltd.

Part V is intended to provide up-to-date advice for the proper treatment of questions of odour dispersion in practical applications.

5.2   Odour Perception

Odour perception is a biological reaction which can be expected if the concentration lies above the odour concentration threshold. Therefore odour sensation cannot be expressed by the concentration itself; instead, a biologically-effective parameter, the odour intensity, has to be used. This approach is similar as for many other biological reactions (e.g. brightness, loudness.) In many cases the odour sensation is expressed in terms of odour intensity which is described by the relationship between the magnitude of a physical stimulus (odour concentration) and its perceived odour intensity.

The sensation of odour is an instantaneous reaction to a certain odour concentration. With the known odour emission flow rate and the necessary meteorological information, the ambient odour concentration can be calculated by dispersion models (Section 5.3). Annoyance caused by environmental odour can only be assessed by including the temporal dimension. From this it follows that the ambient odour concentration has to be known at least on a 1-h basis over a one year period (Yu *et al.*, 2009). Dispersion models provide the ambient odour concentration for each time step (mainly half-hour or 1-h mean values of the ambient concentration). Evaluating time series of ambient odour concentrations, the probability of annoyance can be assessed.

The quantification of annoyance depends on various predictors which can be summarized by the FIDO factors (frequency, intensity, duration and offensiveness) which were suggested by (Watts and Sweeten, 1995). In New Zealand (Lohr *et al.*, 1996) and several countries in Europe, a fifth factor, the location, is additionally in use. This last factor describes the nuisance with regard to the sensitivity of the receiving environment. The location factor can directly be compared to the factor reasonableness. Reasonableness of odour sensation is denided as odour causing fewer objections within a community where odour is traditionally part of the environment, for example rural smells as part of the rural environment and for industrial smells in industrial areas. Problems also often arise when incompatible activities are located near each other. For example, complaints about existing intensive farming operations often occur when land use in the vicinity is changing. Personal knowledge of the operator of the livestock unit, long term residency, economic dependence on farming, familiarity with livestock farming and awareness of the agricultural-residential context are related to a reduced incidence of formal complaints (Ministry of the Environment, 2003). An assessment of this factor is often done by the land use (zoning) category of the neighbours, for example, a pure residential area has a higher protection level as a rural site.

5.2.1   Odour Intensity

The intensity of odour perception is one of the FIDO factors, which can be calculated by non-linear functions describing the relationship between the odour concentration and an intensity scale (e.g. 0 to 6 point scale) (Figure 5.1). Several functions are in use to determine this relationship (e.g. Weber Fechner Law (exponential function), Stevens Law (power function) (Sarkar and Hobbs, 2002; Sarkar *et al.*, 2003; Guo *et al.*, 2001).

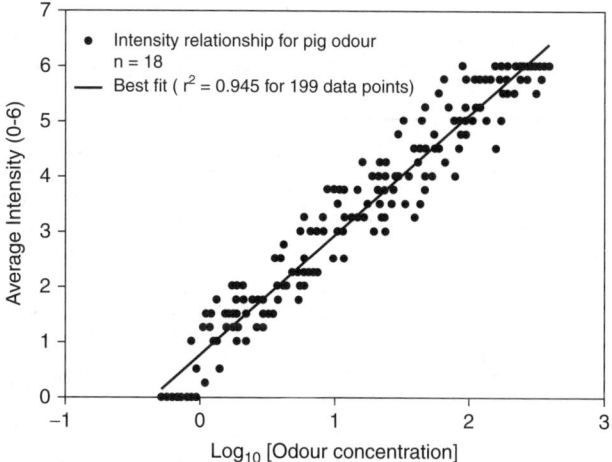

Figure 5.1 *Relation between measured odour concentration C and perceived odour intensity I for a 0 to 6 point scale for pig odour given by I = 2.19 log C + 0.736. (Reprinted with permission from Sheridan et al., 2004, Copyright (2004) Elsevier Ltd).*

In Table 5.1, the coefficients of the Weber-Fechner Law are summarized for various odour sources. If necessary, the coefficients were re-calculated for the selected 0 to 6 point scale. For this scale the intensity can also be expressed verbally: (0) <very weak odour (1) < weak odour (2) <distinct odour (3) < strong odour (4) < very strong odour (5) < extremely strong odour (6). Other intensity scales (0–5, 0–7, or 0–8) can be found in (Guo et al., 2001).

In jurisdiction, the following categories are in use to verbalize odour intensity for ambient odour concentrations C (Wallis and Oma, 2009):

- *Non-detectable* ($C < 2$ ou_E/m^3): Odours are non-detectable by most, if not all, individuals and acceptance of ambient air quality by the community is very high.
- *Acceptable* (2 ou_E/m^3 < C < 5 ou_E/m^3): Odours, if detected at all, are so faint and infrequent that they give rise at most to a few spasmodic complaints. The level of acceptance of the ambient air quality by the community is high.
- *Annoyance* (5 ou_E/m^3 < C < 15 ou_E/m^3): Moderate to strong odours are detected sufficiently often by a proportion of the community to give rise to regular odour complaints. Many members of the community will consider the ambient air quality as unacceptable.
- *Severe annoyance* (C > 15 ou_E/m^3): Strong odours are regularly detected by a large proportion of the community and give rise to many frequent odour complaints. Most members of the community will be annoyed and agitated about the ambient air quality.

As an orientation based upon laboratory experiments on perceived intensity, the following suggestions can be found (Environment Agency, 2009):

- 1–5 ou_E/m^3: the odour is recognizable;
- 5 ou_E/m^3 is a faint odour;
- 10 ou_E/m^3 is a distinct odour.

Table 5.1 *Coefficients of the Weber-Fechner Law describing the relationship between odour intensity I for a (0 to 6 point scale) and the odour concentration C by $I = k_1 \log C + k_2$ for various odour sources. For other than a 6 point intensity scale, the coefficients were re-calculated.*

Odour source	k_1	k_2	Source
Pigs	2.19	0.736	(Sheridan et al., 2004)
Pigs[a]	1.884	−0.509	(Nicolai et al., 2002)
Pig manure storage	1.61	0.45	(Misselbrook et al., 1993)
Pig manure storage[a]	1.932	−0.623	(Nicolai et al., w2002)
Pigs	0.615	0.270	(Zhang et al., 2002)
Pigs (including manure storage)[b]	1.073	0.585	(Feddes, 2006)
Pigs (including manure storage)[b]	0.934	−0.003	(Zhang et al., 2005)
Pigs (including manure storage)[a]	1.114	−2.364	(Guo et al., 2001)
Broiler	2.35	0.30	(Misselbrook et al., 1993)
Poultry	2.21	0.82	(Hayes et al., 2006)
Cattle[a]	1.106	−2.506	(Guo et al., 2001)
Municipal solid waste[c]	1.365	0.216	(Sarkar et al., 2003)

[a]originally calculated for a 5-point scale.
[b]originally calculated for an 8-point scale.
[c]originally calculated for a 7-point scale.

The values for normal background odours such as from traffic, grass cutting, plants, and so on, amount to anything from 5–40 ou_E/m^3 (Environment Agency, 2009).

In general, odour concentrations are measured in the laboratory by an olfactometer according to (EN 13725, 2003). Using this methodology the air samples are diluted by odour-free air. This means that the detection threshold of 1 ou_E/m^3 can only be perceived in an odour-free environment (laboratory). Therefore the perceived odour concentration in the field must be higher than 1 ou_E/m^3 to distinguish it from the background concentration. In field experiments, the distinguishable-ness of an odour source against the background odour is demanded (e.g. (VDI 3940 Part 1, 2006; VDI 3940 Part 2, 2006). The discrepancy between the definition of 1 ou_E/m^3 in the laboratory by using odour-free air and the situation in the field is solved by introducing the sniffing unit (Defoer and van Langenhove, 2003; van Langenhove and De Bruyn, 2001; van Langenhove and van Broeck, 2001). Sniffing team methods have some advantages over instrumental and olfactometric measurements. The main advantage is that they involve field measurements, by which the global impact of the source is evaluated, allowing to consider diffuse, surface and less clear sources, such as waste handling or transportation. Furthermore, these methods reflect the actual perceptibility of the odour in the environment (Nicolas *et al.*, 2006). Nevertheless the relationship between odour unit from the laboratory and the sniffing unit in field measurements is not yet known.

5.2.2   Temporal Dimension

Besides odour intensity, also time aspects have an important influence on the annoyance potential of environmental odours. In the FIDO factors the frequency as well as the duration of odour sensation is taken into account. A second point of view is the time pattern of the behaviour of the neighbours. This influence is not included in the odour impact criteria which are applied to calculate the separation distances. It is obviously not the same with respect to odour annoyance if odour episodes occur e.g. around sunset in summertime or during night-time in winter. It has to be discussed if the odour impact criteria, defined solely by a probability of exceedance of the selected odour threshold, are sufficient to guarantee protection with respect to the time of the day or the season of the year (Schauberger *et al.*, 2009).

The two parameters *frequency and duration* cover the temporal dimension of the FIDO factors. The frequency defines the fraction of time, when odour can be perceived. To archive odour sensation, the odour concentration has to be at least as high as the odour threshold (1 ou_E/m^3).

An important parameter describing the annoyance potential of odour is the duration of consecutive odour episodes (Yu *et al.*, 2009; Schauberger *et al.*, 2006). For a well-ventilated site in the North of Austria (Wels), the duration of consecutive odour episodes can be seen exemplarily (Figure 5.2). The duration of odour episodes is investigated in relation to the hour of the day and the day of the year. It is expressed by the size of the circles in the graph. The lines, marking the time of sunset and sunrise, separate daytime from night-time, which changes the character of the dispersion process in the atmosphere, expressed by the stability class. For all four wind directions, a distinct pattern can be seen. The two wind directions influenced by the valley wind system (N and S) show a distinct diurnal pattern: For North wind, episodes occur predominantly during daytime (Figure 5.2a), for South wind, during night-time (Figure 5.2c). The occurrence of long lasting odour episodes is much smaller than for the prevailing wind directions West and East. For these directions, the influence of solar radiation on the occurrence of odour episodes is less pronounced but still present (Figure 5.2b and d) and shows a minimum at midday in summer.

To find a better fit between odour complaint statistics and model calculations of the odour sensation by the use of dispersion models, the weighting of the odour sensation by the time of the day and the season of the year is proposed by (Schauberger *et al.*, 2009). The necessity of different protection levels, depending on the sensitivity of the neighbours, can be shown for environmental noise. During daytime the protection level is lower with a limit value of 55 dB, during night time, the limit value is 45 dB. This difference of 10 dB means that the sound level during daytime is allowed to be 10 times higher compared to night time.

By a weighting procedure each half hour value is multiplied by a weight depending on the time of the day and time of the year. For the time of the day the solar height is used according to a sigmoid weighting function w_t given by

$$w_t = w_0 + \tfrac{1}{2}(1 - w_0)[\tanh(2s_t(h - h_0)) + 1] \qquad (5.1)$$

with the off-set of the weighting function w_0 selected as 0.5, the shape factor $s_t = 0.25$, and the mid-point of the function $h_0 = -6°$. The slope can be calculated at the mid-point h_0 by $Sh(1 - w_0)$. The midpoint is taken as the solar height of the civil twilight, which is defined when the centre of the sun has a depression angle of 6° below an ideal horizon. At this solar height $h = -6°$, the weighting function becomes $w_t = \tfrac{1}{2}(w_0 + 1) = 0.75$.

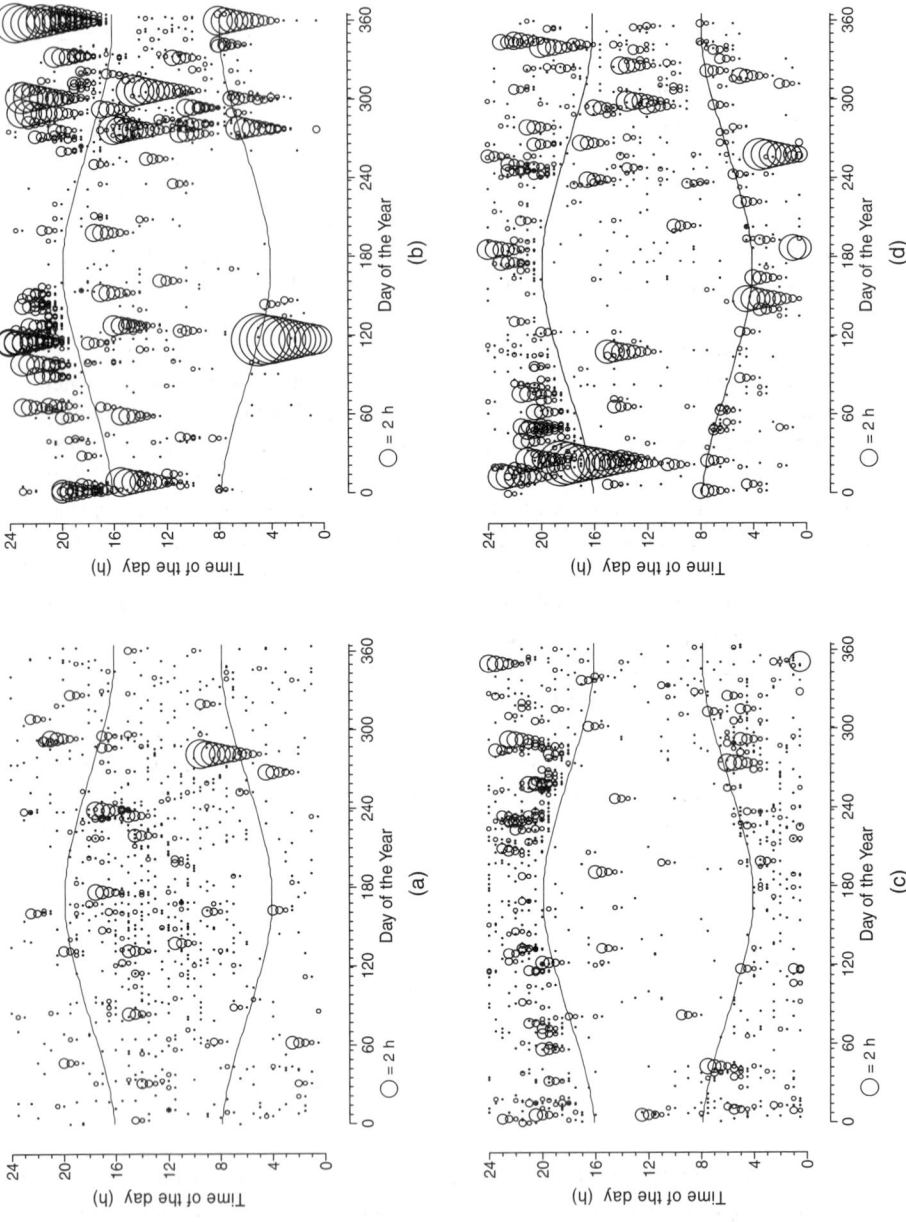

Figure 5.2 Duration of odour episodes as a function of the day of the year and the time of the day. The diameter of the circles is proportional to the duration of odour episodes. (a) North wind, (b) East wind, (c) South wind, and (d) West wind (all 45° sectors). The lines mark sunrise and sunset for Wels (geographic latitude: 47.17°N). (Reprinted with permission from Schauberger et al., 2006, Copyright (2006) Elsevier Ltd).

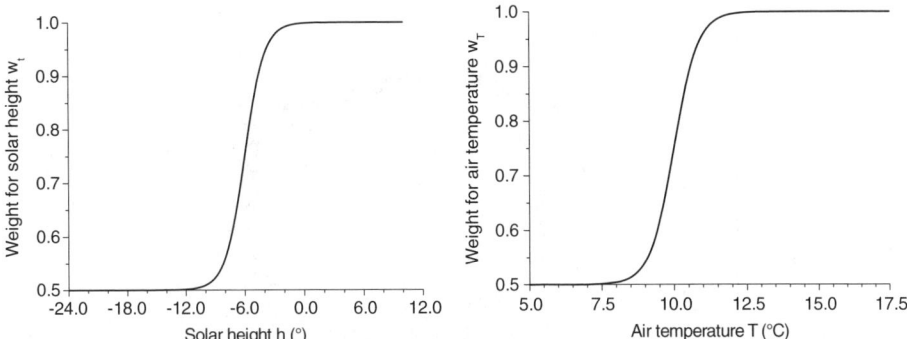

Figure 5.3 *Weighting functions for the time of the day w_t by using the solar height h (left) and for the time of the year w_T by using the ambient air temperature T (right). (Reprinted with permission from Schauberger et al., 2009, Copyright (2009) Kassel University Press).*

The weighting function for the time of the year is using the air temperature T as a predictor. The function is defined by

$$w_T = w_0 + \tfrac{1}{2}(1 - w_0)\left[\tanh\left(2\,s_T\,(T - T_0)\right) + 1\right] \tag{5.2}$$

with the shape factor $s_T = 0.6$ and the mid-point $T_0 = 10°C$.

The two weighting functions for the time of the day w_t and the time of the year w_T are depicted in Figure 5.3. w_t, as an alias for the time of the day, is reaching 0.50 for the nautical twilight ($h = -12°$), 0.75 at the civil twilight ($h = -6°$), and 1.00 for sunrise and sunset ($h = 0°$). w_T gives 0.50 for $T = 7.5°C$, 0.75 for the midpoint $T = 10.0°C$ (which equals approximately the annual mean temperature in the Austrian flatlands), and 1.00 for $T = 12.5°C$.

The total weight w is calculated by the product of the two weighting functions w_t and w_T. The maximum of the weighting function $w = w_t\, w_T$ gives 1.0, the minimum 0.25, which results in a range of 4.

The total weighting function w, defined by the product of the two individual weights, is calculated for a two year period of meteorological data (ambient air temperature) and presented in Figure 5.4. The isopleth $w = 0.95$ circumscribes the area with a higher impact of odour sensation (summer and daytime).

This procedure, weighting the sensation distance with w leads to the desired modification of the ambient odour concentration. For lower distances than the sensation distance, the ambient odour concentration is modified to be higher than 1 ou$_E$/m^3. The relationship between the ambient concentration C and the distance d can be described by a power function according to $d \propto C^a$. The exponent a lies in the range of 0.3–0.6, depending on the exceedance probability. For an exceedance probability of 3% the exponent can be assumed with $a = 0.6$ (Schauberger et al., 2012). During day time and in summer, the threshold value is in the range of 1 ou$_E$/m^3 meaning that the sensation distance is not reduced. For night time and low air temperatures, when the weighting function is close to the minimum of 0.25, the threshold will be in the range of 10 ou$_E$/m^3, reducing the sensation distance in the range of 4. Therefore the possible modification of the ambient concentration due to the weighting function lies in the range of about 10 which is the same range as for environmental noise.

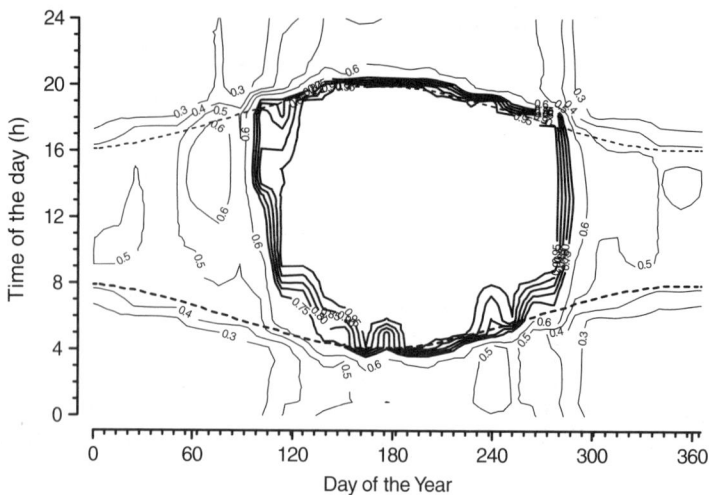

Figure 5.4   *Weight w of the sensation distances as a function of the day of the year and the time of the day. The dashed lines mark sunrise and sunset. Intervals are 0.05 in the range between 0.75–1 (thick lines) and 0.10 in the range between 0.3–0.6 (thin lines). (Reprinted with permission from Schauberger et al., 2009, Copyright (2009) Kassel University Press).*

5.3   Overview on Types of Odour Dispersion Model[1]

Two classes of dispersion models are currently used for topics of (regulatory) odour dispersion, namely Gauss and Lagrange models. Both model classes belong to the so-called non-CFD (computational fluid dynamics) models. Non-CFD models do not explicitly resolve fluid-dynamics equations but physical processes are parameterized. Generally, different grades of approximations and simplifications to the primitive equations are used when calculating concentrations. For example, these models do not calculate the flow around a single building or obstacle when applied to an urban-like geometry, but the effect of a group of obstacles is taken into account through an increased surface roughness value or by a coarse resolution of the buildings (Schatzmann *et al.*, 2010). Non-CFD dispersion models are in general less complex and easier to run than typical CFD models and require much shorter calculation time. One main advantage is that non-CFD models may be run over a long series of input data to represent different meteorological conditions.

For short-range local problems (up to about 5 km) at stake here, simple Gaussian type models have been and are still being used. These models are applicable for pollutant emissions into stationary and uniform atmospheric flows (for example, tall stack releases in flat, unobstructed terrain, Figure 5.5).

Inherent to all Gaussian type models (even if a boundary-layer parameterization is included such as within the ADMS model family) is their inability to allow for meteorological changes (direction, speed, atmospheric stability) within space and the time interval in which the concentration field is calculated (usually half an hour or one hour) so that a

---

[1]   This Section is partly taken from Schatzmann M., Olesen H.R. and Franke J. COST 732 model evaluation case studies:approach and results (2010). COST Office, Brussels.

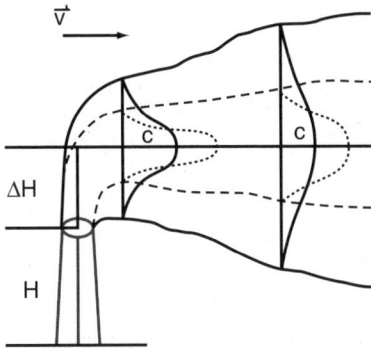

Figure 5.5 *Schematic view of the Gaussian plume (H: stack height; ΔH: effective stack height; c: plume concentration). Source: ZAMG.*

steady state for this period is assumed. Gaussian type models might incorporate chemical models of various complexities and may allow for a large number of sources making them suitable for large air quality studies.

It is accepted that over rougher surfaces these models can still be applied when clouds of pollutants disperse above the buildings or when the dimensions of the cloud are much bigger than the dimensions of the obstacles. Within the urban canopy and in the near field of the source where the flow conditions are usually far from uniform or motionless their usefulness is restricted to screening purposes, that is, to decide (e.g. by analysing the concentration levels involved) whether the application of a more complex model to the study case is appropriate or necessary.

Examples of Gauss models applicable to odour pollution are the ADMS model family and the AODM, the Austrian Odour Dispersion Model based on the Austrian regulatory Gauss model.

ADMS-Urban ((CERC, 2007); http://www.cerc.co.uk) is an advanced atmospheric pollution dispersion model for calculating concentrations of atmospheric pollutants emitted continuously from point, line, volume and area sources or intermittently from point sources. It is able to describe in details what happens in a range of scales, from the street scale to the city-wide scale. It incorporates the latest understanding of the boundary layer structure. The model uses advanced algorithms for the height-dependence of wind speed, turbulence and stability to produce improved predictions. The model also includes algorithms which take into account: downwash effects of nearby buildings within the path of the dispersing pollution plume; effects of complex terrain; wet deposition, gravitational settling and dry deposition; short term fluctuations in pollutant concentration; chemical reactions; pollution plume rise as a function of distance; averaging time ranging from very short to annual. The system also includes a meteorological data input pre-processor. Included in the model is also a linearized perturbation model which allows for the calculation of flow fields over changing terrain or roughness. The treatment of odour in ADMS is described briefly in Section 5.4.

The Austrian odour dispersion model (AODM, (Piringer *et al.*, 2007; Schauberger *et al.*, 2000, 2002)) estimates mean ambient concentrations by the Austrian regulatory dispersion model (Kolb, 1981; ÖNorm, 1992) and transforms these to instantaneous values depending on the stability of the atmosphere (Section 5.4). The model has been validated internationally

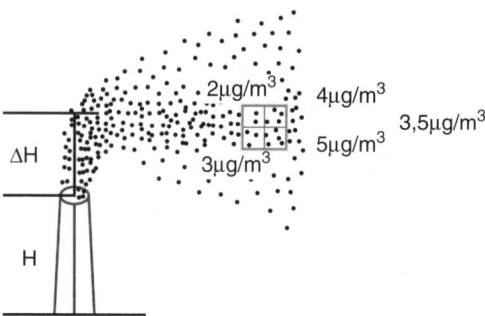

Figure 5.6   *Schematic view of a cloud of particles (Lagrange simulation) including the grid to determine ambient concentrations (H: stack height; ΔH: effective stack height).*
Source: ZAMG.

with generally good results (Pechinger and Petz, 1997, 1995; Piringer and Baumann-Stanzer, 2009, 2011). The regulatory model is a Gaussian plume model applied for single stack emissions and distances from 100 m up to 15 km. Plume rise formulae used in the model are a combination of formulae suggested by (Carson and Moses, 1969) and (Briggs, 1975). The model uses a traditional discrete stability classification scheme with dispersion parameters developed by (Reuter, 1970).

In Lagrangian models, the plume consists of individual plume parcels (Figure 5.6), and their paths are modelled on the basis of a random walk process. They need a complete mean and turbulent flow field as model input, which is usually delivered in form of 3D-gridded fields by either a diagnostic or prognostic model. The important influence of pressure gradients and forces on the flow development can only indirectly (= empirically) be taken into account. These models start with a first guess of the three-dimensional flow field that is subsequently modified until the divergence of the flow falls below a chosen limit. For a given obstacle array many different mass conserving flow fields can be found, depending on the particular choice of the initial flow field and the 'tuning of knobs' inside the model which, for example, determines whether the fluid at a specific position moves over or goes around an obstacle. Thus meandering and other shear effects as well as topographical effects can in principle be simulated. Diagnostic models may be helpful in analysing known cases, in particular when a good set of observed data is available for the area of interest.

Examples of Lagrange models available in Central Europe which are able to calculate odour concentrations are LASAT and GRAL.

The dispersion model LASAT ((Janicke Consulting, 2011) http://www.janicke.de) simulates the dispersion and the transport of a representative sample of tracer particles utilizing a random walk process (Lagrangian simulation). It computes the transport of passive trace substances in the lower atmosphere (up to heights of about 2000 m) on a local and regional scale (up to distances of about 150 km). A number of physical processes, including time dependencies, are simulated, such as transport by the mean wind field, dispersion in the atmosphere, sedimentation of heavy aerosols, deposition on the ground (dry deposition), washout of trace substances by rain and wet deposition, first order chemical reactions. The quality of the results achievable by Lagrangian models mainly depends on the wind field they are based on. A simplified version of LASAT is offered free of charge (AUSTAL2000, http://www.austal2000.de) which is favoured by German guide lines (GOAA, 2008; TA Luft, 2002).

GRAL (http://www.umwelt.steiermark.at/cms/beitrag/11023486/19222537/) is a Lagrange particle model especially developed for dispersion in complex terrain and in low wind speed conditions (Oettl and Uhrner, 2010). It is specially apt to take into account large horizontal wind direction fluctuations ('meandering') in low wind speeds (Oettl *et al.*, 2005).

5.4 Algorithms to Estimate Short-Term Odour Concentrations[2]

Contrary to most air borne pollutants odour is not a feature of a certain chemical species but a physiological reaction of humans. The sensation and perception of odorants depends on sniffing as an active stage of stimulus transport. The sniff volume is inversely proportional to the concentration of an odorant. The mean breathing rate during odour perception is in the range of 4 s (Kleemann *et al.*, 2009).

For the assessment of peak values, describing the biologically relevant exposure, often the so called peak-to-mean concept is used. This is a way to adopt dispersion models to short-term odour concentrations. The goal of the use of peak-to-mean factors is to mimic the human nose in a better way as it can be achieved by long term mean values.

The step from the one-hour mean value (as output of the dispersion model) to an instantaneous odour concentration is shown in Figure 5.7. For the one-hour mean value, the threshold for odour perception (here taken as 1 ou_E/m^3) is not exceeded. Taking mean values over 10 minutes, one concentration value exceeds the threshold. For the short term mean values of 12 s, concentrations in the range of 5–6 ou_E/m^3 can be expected, which means a distinct odour perception over several breaths. Figure 5.7 shows that the shorter the selected time interval, the higher the maximum concentration. For the shortest period of 12 s, a new feature of the time series can be seen. Besides 12 s intervals with odour concentrations above zero, a certain percentage of zero observations can be expected. The frequency of non-zero intervals is called intermittency γ defined by the conditional probability $\gamma = prob\{C|C > C_D\}$ with the concentration of the detection limit C_D (Chatwin and Sullivan, 1989).

Given a mean concentration over one hour, the mean value of a shorter period can be calculated using the well-known relationship (cited by (Smith, 1973)

$$\frac{C_p}{C_m} = \left(\frac{t_m}{t_p}\right)^u \tag{5.3}$$

with the mean concentration, C_m, calculated for an integration time of t_m and the peak concentration C_p, for an integration time of t_p. The evidence of this relationship by measurements was shown by (Hinds, 1969).

---

[2] This section is partly published in: Schauberger, G., Piringer, M., Schmitzer, R., *et al.*: Concept to assess the human perception of odour by estimating short-time peak concentrations from one-hour mean values. Reply to a comment by Müller *et al. Atmospheric Environment* 2012; **54**: 624–628.

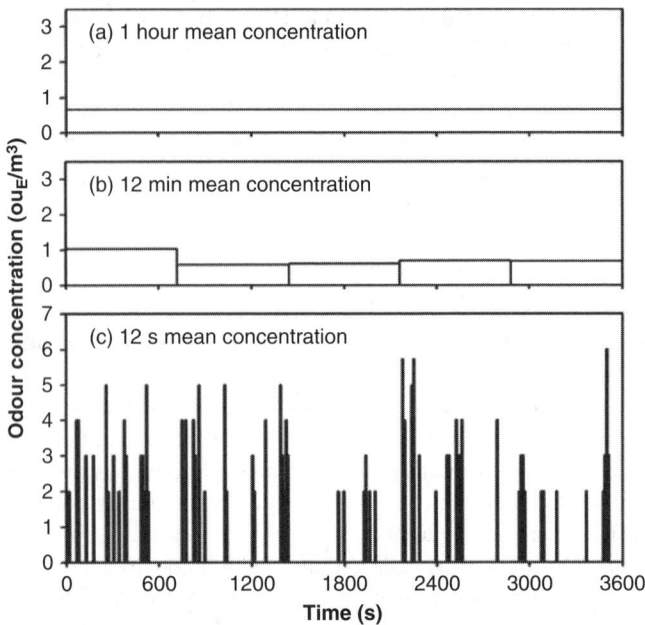

Figure 5.7 *Time course of the odour concentration (ou_E/m^3) for three time intervals. (a) one-hour mean value (e.g. output of a dispersion model), (b) 12-min and (c) 12-s mean odour concentrations observed at a single receptor point during a field study. The 12-s mean values were recorded and subsequently used to calculate 12-min and one-hour mean concentrations. (Reprinted with permission from Schauberger et al., 2012, Copyright (2012) Elsevier Ltd, modified from (Nicell, 2009).*

According to the relationship above, the peak-to-mean factor is defined by $F = C_p/C_m$. The open question is the definition of the peak value C_p. It can be defined manifold (Gross, 2001; Klein and Young, 2010). The following definitions are used frequently:

1. $C_p = C_m + \sqrt{\sigma}$, that is, the peak value is defined by the mean value and the standard deviation $\sqrt{\sigma}$. The quotient between $\sqrt{\sigma}$ and the mean value C_m is called fluctuation intensity $i = \sqrt{\sigma}/C_m$, therefore the peak-to-mean factor on the basis of the fluctuation intensity is $F_i = i + 1$.
2. The peak value is defined by the 90-percentile C_{p90}, so $F_{90} = C_{p90}/C_m$.
3. The peak value is defined by the 98-percentile C_{p98} or the 99-percentile C_{p99} (Klein and Young, 2010).
4. The peak value is defined by the maximum C_{max}, then $F_{max} = C_{max}/C_m$ (Klein and Young, 2010).

Especially for Germany, the peak value C_p is well-defined by the comparison between empirical field measurements (VDI 3940 Part 1, 2006) and dispersion model calculations. If we assume that the assessor sniffs every 10 s to decide if the sample smells, then 360 breaths (sample size) during one hour are obtained. In the German jurisdiction an hour is counted as a so called odour hour if at least 10% of the 360 breaths can be evaluated as odorous. For practical reasons, only a period of 10 min (60 breaths) is used as a sample to judge a certain hour. If 6 out of 60 periods (10 min) are assessed as odorous by a panellist,

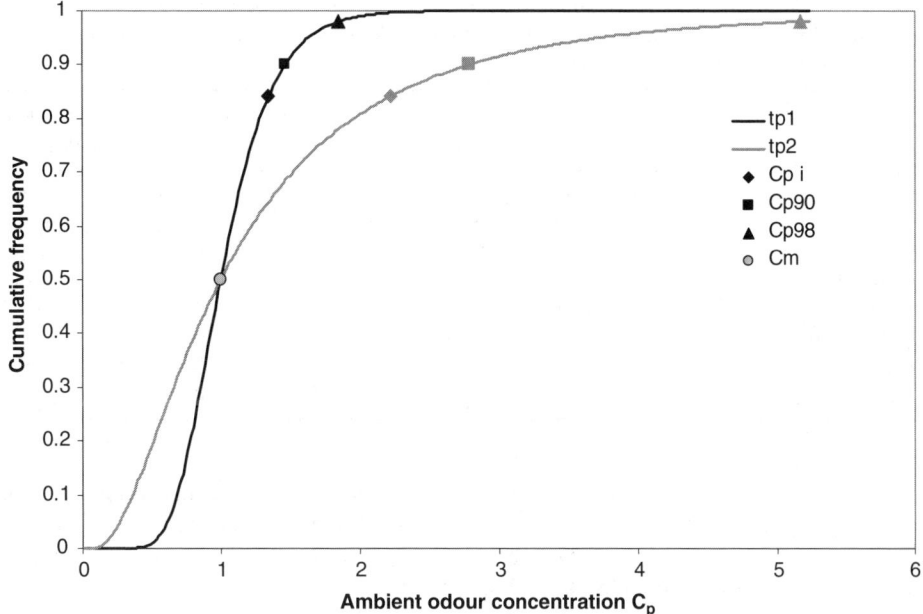

Figure 5.8 *Schematic diagram showing the cumulative distribution function (log-normal) for two different integration intervals t_p with $t_{p1} > t_{p2}$ of the peak concentration C_p. The mean value (one hour mean) $C_m = 1$ ou_E/m^3 is the same for both CDFs. The peak values are defined by the fluctuation intensity i (C_{pi}), the 90-percentile (C_{p90}) and the 98th-percentile (C_{p98}), showing the influence of the integration interval t_p and the determination of the peak value C_p due to a certain percentile. (Reprinted with permission from Schauberger et al., 2012, Copyright (2012) Elsevier Ltd.).*

this defines an odour hour. Therefore the 90th-percentile is used to define the peak value C_p with $t_p = 1$ s to assess the incidence of an odour hour.

The influence of the integration interval t_p on the peak concentration C_p is shown schematically in Figure 5.8. The shorter the integration interval, the higher the variance of the ambient concentration C. On the other hand, the graph shows the sensitivity of the definition of the peak value, which is defined by a certain percentile of the cumulative distribution function CDF. In this example the peak-to mean factor $F = C_p/C_m$ varies for the two integration intervals t_{p1} and t_{p2} between $F_i = 1.35$, $F_{90} = 1.47$ and $F_{98} = 1.85$ and $F_i = 2.23$, $F_{90} = 2.79$ and $F_{98} = 5.17$, respectively. This example shows the importance of a proper definition of the determination of the peak value C_p.

The assessment of maximum values for shorter periods than one hour is not only relevant for environmental odour but also for toxic and inflammable pollutants (Hanna, 1984; Hilderman *et al.*, 1999; Mylne, 1988). This estimation by one-hour mean values can lead to an underestimation of the impact of the ambient concentration. This error depends on the observed impact of the ambient concentration. In many cases the health impact H is described by a non-linear dose response function (Hilderman and Wilson, 1999). Especially health related phenomena show such a relationship with the ambient concentration C which can be described by a power function $H \propto C^\alpha$ with an exponent α (Miller *et al.*, 2000) in the range between 1.0 and 3.5. Some chemicals show an exponent between 2.0 and 3.0

Table 5.2   *Maximum peak-to-mean factors $F = C_p/C_m$ calculated by Equation (5.3) for integration times $t_p = 5$ s and $t_m = 3600$ s. (Reprinted with permission from Schauberger et al., 2012. Copyright (2012) Elsevier Ltd).*

Stability class	Smith (1973)		Trinity Consultants (1976) (Beychock, 1994)	
	Exponent u	F	Exponent u	F
unstable	0.64	67.4	0.68	87.7
slightly unstable	0.51	28.7	0.55	37.3
neutral	0.38	12.2	0.43	16.9
slightly stable	0.25	5.2	0.30	7.2
stable	0	1.0	0.18	3.3
very stable	0	1.0	0.18	3.3

for the toxicity and fatalities. Only if the exponent $\alpha = 1$ then the concentration can be determined by a mean value. However, if $\alpha > 1$, then the use of the mean concentration will underestimate the impact of the substances. The health effects of toxic gases in this context are described in detail by (Hilderman, 1997).

The assessment of a concentration value for a shorter integration time on the basis of a 1-h mean can be calculated by a peak-to-man factor. This conversion depends on the dilution process in the atmosphere which is predominantly influenced by turbulent mixing. The following predictors are discussed, which influence the concentration fluctuation (Hanna and Insley, 1989; Olesen *et al.*, 2005):

1. Stability of the atmosphere
2. Intermittency
3. Travel time or distance from the source
4. Lateral distance from the axis of the wake
5. Geometry of the source (emission height and source configuration)

Turbulent mixing in the atmosphere depends strongly on the stability of the atmosphere. The stability can be determined for example by discrete stability classes or by the Monin–Obukhov length (see Part 5.7.2 for details). The influence of the stability of the atmosphere on the peak-to-mean value is calculated for example by Equation (5.3), using $t_m = 3600$ s (calculated one-hour mean) and $t_p = 5$ s (duration of a single breath). The peak-to-mean factors F, depending on atmospheric stability, are then derived either by taking values for the exponent u proposed by (Smith, 1973) or by Trinity Consultants (1976) (cited in (Beychock, 1994), Table 5.2). From measurements ($t_p = 1$ s, C_p defined to be the maximum) in the near field of the source, a peak-to-mean factor in the range of $4 < F_{max} < 99$ was found (Lung *et al.*, 2002). The influence of the stability of the atmosphere on the exponent u of Equation (5.3) was shown by (Santos *et al.*, 2009).

As it was shown before, the fluctuation intensity $i = \sqrt{\sigma}/C_m$ can be used to assess the variance and therefore the peak-to-mean factor F. The relationship between the fluctuation intensity i and the peak-to-mean factor F can be described as (Lung *et al.*, 2002)

$$F = 1 + \alpha\, i^2 \tag{5.4}$$

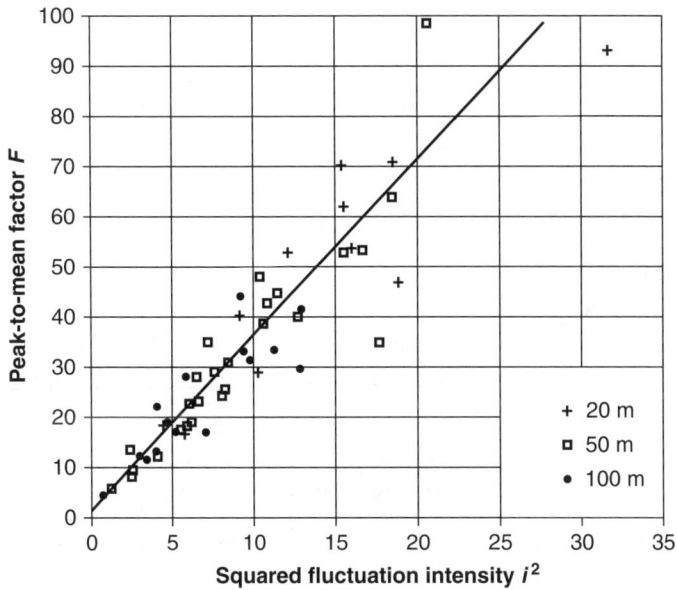

Figure 5.9 *Relationship between the squared fluctuation intensity i² and the peak-to-mean factor F measured in the vicinity of a point source (distance between 20–100 m). (Reprinted with permission from Lung et al., 2002, Copyright (2002) Liebniz-Institut fur Agrartechnik Potsdam-Bornim).*

with $\alpha = 3.6$ and the fluctuation intensity $i = \sqrt{\sigma}/C_m$ (Figure 5.9). Here, the peak-to-mean factor is related to a peak concentration measured for an integration time of 1 s and the 1-h mean. Various peak-to mean factors F (according to the definition of the peak concentration by the standard deviation, the 98th-percentile or the maximum) were measured by (Klein and Young, 2010), showing the increase of the factor F with the selected percentile (see also Figure 5.8).

The fluctuation intensity i is also related to the intermittency γ by

$$i^2 = \frac{\beta}{\gamma} - 1 \tag{5.5}$$

The parameter β is determined to lie between $\beta = 2$ (Hanna, 1967), $\beta = 3$ (Best *et al.*, 2001), and $\beta = 3.6$ (Lung *et al.*, 2002). Other functions describing this relationship can be found in (Klein and Young, 2010).

The lateral distance y, normalized by the lateral dispersion parameter σ_y from the plume centre line shows a strong influence on the fluctuation intensity (Hinds, 1969; Best *et al.*, 2001; Katestone Scientific, 1998; Løfstrøm *et al.*, 1996). The following function describes this relationship:

$$i(x, y) = i(x) \, \exp\left(\frac{y^2}{a \, \sigma_y^2}\right) \tag{5.6}$$

with $a = 2$ suggested by (Best *et al.*, 2001) and $a = 4$ by (Katestone Scientific, 1998).

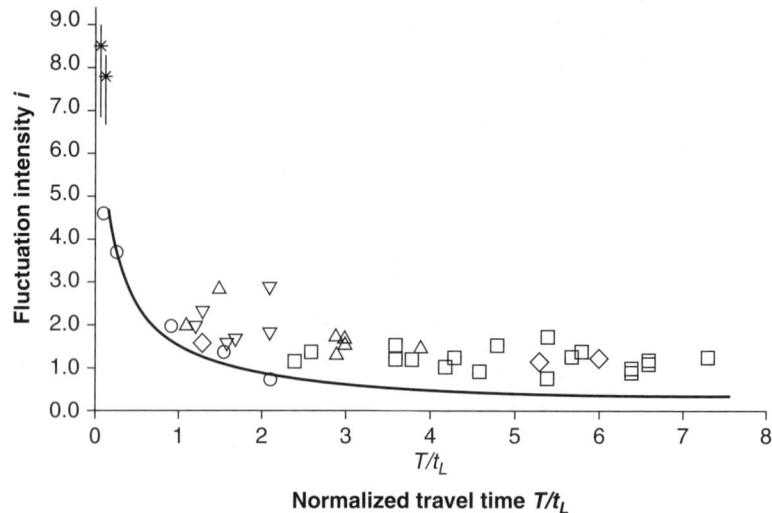

Figure 5.10   *Relationship between the fluctuation intensity i and the normalized travel time T/t_L. (Reprinted with permission from Mylne, 1990, Copyright (1990) Crown Copyright, The Met Office).*

The reduction of the fluctuation intensity with the distance is shown exemplarily by empirical data provided by (Mylne, 1990) (Figure 5.10). The fluctuation intensity seems to approach a constant value of 1.0 ± 0.3.

The influence of the geometry of the source (elevated point source versus area source) is discussed in detail by (Katestone Scientific, 1998), showing lower peak-to-mean ratios for area sources compared to elevated point sources. Exponents $u = 6/17$ and $u = 3/14$ for point and line sources, respectively, have been suggested by (Best *et al.*, 2001). Higher fluctuations in elevated sources are found by (Fackrell and Robins, 1982) and (Mylne, 1993). For area sources the reduction of the fluctuation intensity with distance is lower compared to elevated sources.

The distance dependent reduction of the fluctuation intensity can be calculated by a model published by (Best *et al.*, 2001) and (Katestone Scientific, 1998) as a function of atmospheric stability and the geometry of the emitting source. References (Hanna, 1984) and (Løfstrøm *et al.*, 1996) suggested in each case a model for the fluctuation intensity i as a function of the distance x and the lateral distance y. These models are based on the dispersion parameters σ_y and σ_z, which are known from the Gaussian dispersion model. The previous two models and the following model (Schauberger *et al.*, 2000) for the decrease of the peak-to-mean factor with distance from the source can be used as a post-processing tool for dispersion calculations.

$$F = 1 + (F_0 - 1) \, \exp\left(-0.7317 \frac{T}{t_L}\right) \tag{5.7}$$

where the peak-to-mean factors of Table 5.2 are used as F_0. Further downwind they are modified by an exponential attenuation function of T/t_L, where $T = x/v$ is the time of travel with the distance x and the mean wind velocity v, and t_L is a measure of the Lagrangian time scale (Mylne, 1992), which involves knowledge of the standard deviations of the three wind components. The calculation of the Lagrangian time scale in this context was improved by (Piringer *et al.*, 2007). This model can be used as a post-processing tool for dispersion calculations.

An example of the decrease of the peak-to-mean factor with distance resulting from the application of equation (5.7), which is in use with the AODM model in Austria, is shown in Fig. 5.11. In addition, the constant factor 4 applied by AUSTAL2000 is shown by a thin straight line. When applied to the Gauss model, F values for distances larger than 100 m are relevant. At shorter distances the implicit assumption in Gaussian plume models that the longitudinal diffusion is negligible compared to the lateral and vertical diffusion is no longer valid (Piringer *et al.*, 2007). F depends strongly on the stability class. For the very unstable class 2, F, starting at rather high values near the source, approaches 1 within about 50 meters from the source. For the unstable class 3, this is the case within about 100 m. This is in agreement with ideas that vertical turbulent mixing in weak winds can lead to short periods of high ground-level concentrations, whereas the ambient mean concentrations are low. For classes 4 and 5, the decrease of the peak-to-mean factor is more gradual with increasing distance, because vertical mixing is reduced and horizontal diffusion is dominating the dispersion process. For classes 6 and 7, F is always 1 in the Smith scheme (Fig. 5.11a) and below 2 according to Trinity Consultants (5.11b).

It is obvious that a constant factor $F = 4$ as with AUSTAL2000 will lead to an underestimation in the near field compared to the far field and vice versa.

As indicated in Figure 5.11, also constant peak-to-mean factors are in use. Some examples: $F = 10$ for the pervious regulatory Gaussian dispersion model in Germany (Rühling and Lohmeyer, 1998), the Danish model with $F = 7.8$ (Olesen *et al.*, 2005), and $F = 4$ for AUSTAL2000. A peak-to-mean factor $F = 1$ means that the long term mean value (e.g. 1-h mean value) is selected to evaluate the sensation of environmental odour.

Also with the ADMS model (CERC, 2007), the number of exceedences of particular concentrations in a year can be calculated. For up to 10 (odour) concentrations and pre-selected exceeding probabilities, the model calculates the frequency as well as the number of periods of exceeding the concentrations. This procedure is similar to that of AUSTAL2000 when an odour concentration of 0.25 ou_E m^{-3} is taken. A meteorological dependence on this factor is not taken into account.

In an extensive study of odour perception in agricultural areas (Both, 2006), the observed exceedance probability from field measurements (VDI 3940 Part 1, 2006) was compared to calculations with the German regulatory dispersion model AUSTAL2000. Using the data for three agriculturally dominated villages (Hartmann, 2006), the dispersion calculations show a trend for overestimations for small exceedance probabilities (which can be expected at larger distances) (Figure 5.12). For exceedance probabilities of about 2%, the regression model shows an overestimation up to a factor of 3. For higher exceedance probabilities (which means in general lower distances), the overestimation is reduced to quite an extent. The overestimation of separation distances by the dispersion model AUSTAL2000 for a low exceedance probability of 2% (irrelevance criterion) resulting from using a constant factor 4 is discussed by (Hartmann and Hölscher, 2007) and (Schauberger *et al.*, 2012) in detail. In general, a constant peak-to-mean factor (even if $F = 1$) will underestimate the

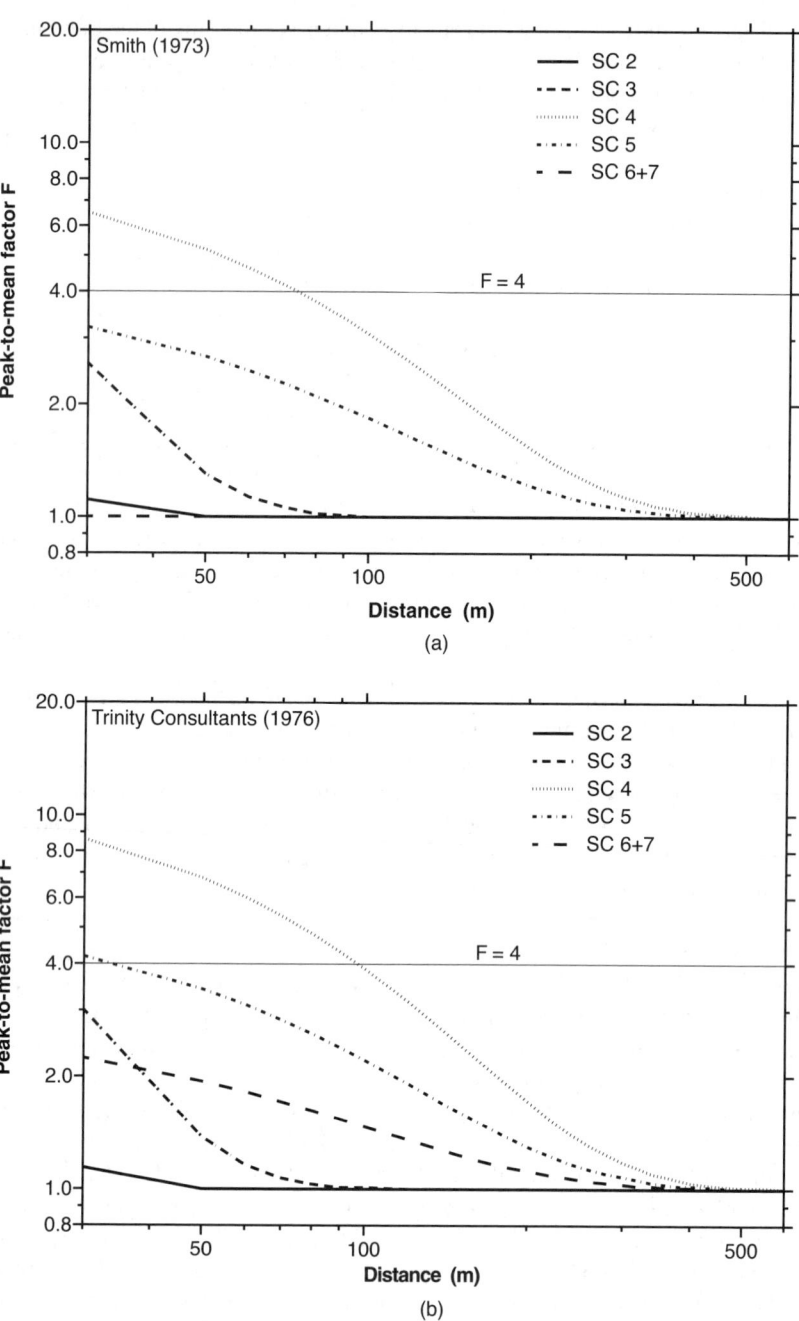

Figure 5.11 *Distance depending peak-to-mean factor F depending on stability classes SC. 2 (a): Maximum peak-to-mean factors according to Smith (1973) (b): Maximum peak-to-mean factors according to Trinity Consultants (1976; cit. in (Olesen et al., 2005)); see Table 5.2. The constant factor F = 4 applied by AUSTAL2000 is shown as a thin straight line.*

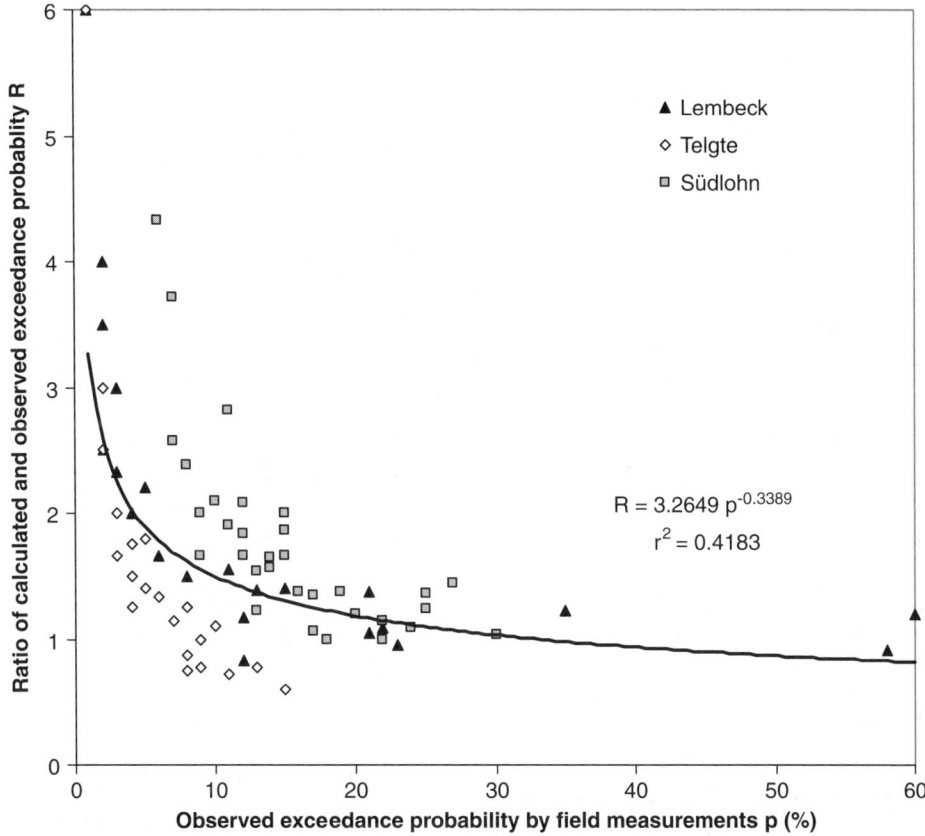

Figure 5.12   *Ratio of the calculated (AUSTAL2000) and the observed exceedance probability from field measurements (VDI 3940 Part 1, 2006) as a function of the observed exceedance probability for three agriculturally dominated villages (Lembeck, Telgte, and Südlohn) with several odour emitting livestock buildings (data taken from (Hartmann, 2006)).*

predicted odour sensation in the near field. In the far field the peak-to-mean factor reaches the value $F = 1$, shown in Figure 5.11.

5.5   Annoyance

For practical use separation distances are calculated to reduce or avoid odour annoyance depending on a certain protection level. At such a distance the frequency of odour sensation over a certain odour concentration threshold C_T does not exceed a pre-selected level, called the exceedance probability p_T. The exceedance probability can be defined as a conditional probability $p_T = prob\{C|C > C_T\}$ This concept is based on investigations of (Miedema and Ham, 1988) and (Miedema *et al.*, 2000) who found a strong relationship between the odour concentration threshold for an exceedance probability of $p_T = 2\%$ $C_{2\%}$ (respectively

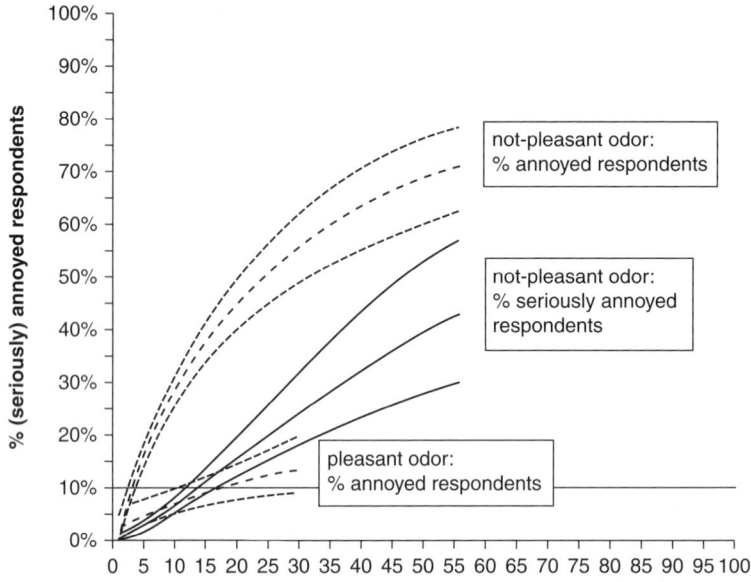

Figure 5.13 *Percentage of 'annoyed' (thermometer scale: 1–10) and 'seriously annoyed' (thermometer scale: 7–10) persons related to the frequency of both non-pleasant and pleasant odour exposure. (Reprinted with permission from Sucker et al., 2008, Copyright (2007) Springer Science + Business Media).*

the 98 percentile) and the percentage of the highly annoyed neighbours *HA*, using an integration time of 1 hour ($F = 1$)

$$HA = K \ \log \ C_{2\%} \qquad (5.8)$$

with a constant $K = 9.25$ (Miedema *et al.*, 2000) or $10 < K < 12$ for pigs (Nicolas *et al.*, 2008).

The dose–response relationship between the odour frequency and the percentage of 'annoyed' and 'seriously annoyed' people (annoyance thermometer scale from 7–10) can be graphically represented by an S-shaped curve as illustrated in Figure 5.13. It is obvious that a hedonic tone has an abundantly clear effect on the dose–response relationship between odour frequency and annoyance. Pleasant odours have a significantly lower annoyance potential than unpleasant/neutral odours (Sucker *et al.*, 2008). The exposure level described by the frequency of odour sensation is based on field measurements according to (VDI 3940 Part 2, 2006).

The reasonableness of odour sensation has a strong influence on the annoying potential. Taking the 10% value of annoyed people, the corresponding concentration of the 2% exceedance probability (1-h mean vale) shows $C_{2\%} = 1.3$ ou$_E$/m^3 for the general public. In areas dominated by agricultural land use the 'acceptable' concentration reaches $C_{2\%} = 3.2$ ou$_E$/m^3. If pig odour is a historical feature of the environment, then $C_{2\%} = 6.3$ ou$_E$/m^3. For those inhabitants which are directly involved in livestock husbandry, the concentration is determined to $C_{2\%} = 13$ ou$_E$/m^3 (Irish Environmental Protection Agency, 2001). These findings are good arguments that impact criteria can be adapted according to zoning and to the acceptance of a certain odour level by residents. This relates also to the location factor mentioned in Section 5.2.

5.6 Odour Impact Criteria for Use in Dispersion Modelling

The determination of the separation distance depends on the protection level, which is established in various jurisdictions in different ways. The odour impact criterion for a certain protection level depends on three parameters. The first parameter defines the integration interval of the ambient odour concentration. In many cases this is calculated by the peak-to-mean factor F which fixes the integration time of the ambient concentration in relation to the outcome of the dispersion model. The other two parameters are the exceedance probability p_T of a certain odour concentration threshold C_T which together form an odour impact criterion. In Figure 5.14 a schematic diagram shows the interaction between odour impact criteria and the selected integration time for the ambient concentration; the lower the integration time, the weaker the odour impact criteria.

The odour impact criteria can be adapted to the required protection level in two ways. (1) The exceedance probability varies, keeping the odour threshold constant. For example, in Germany the exceedance probability $p_T = 10\%$ for pure residential areas and $p_T = 15\%$ for rural sites, whereas the odour concentration threshold remains constant with $C_T = 1\ ou_E/m^3$. (2) The odour concentration threshold is variable, whereas the exceedance probability is kept constant (e.g. Australia).

For various national jurisdictions, these criteria are compared and summarized in Table 5.3 for European countries and in Table 5.4 for Australia/New Zealand and North America.

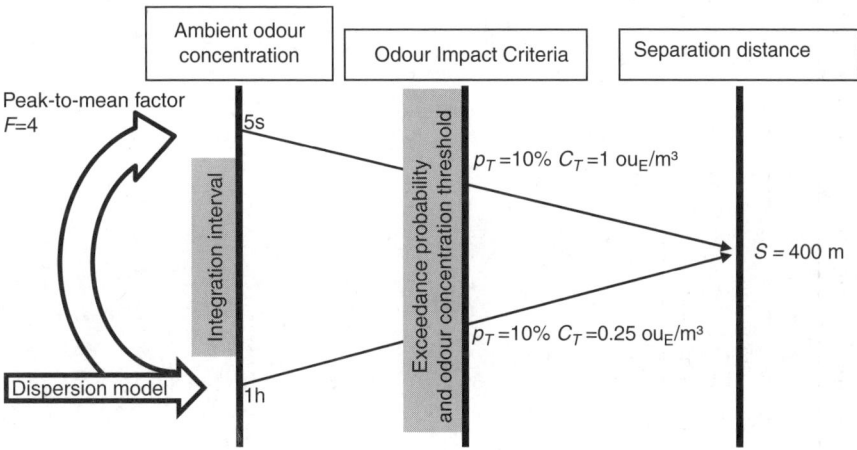

Figure 5.14 *Relationship between the integration interval of the ambient odour concentration (between a 1-h mean value and the duration of a single breath), the odour impact criteria (exceedance probability and odour concentration threshold) and the separation distance to avoid odour annoyance. The transition from a 1-h mean value to shorter integration intervals is done by peak-to-mean factors. The relationship is shown for a constant peak-to-mean factor $F = 4$ as it is done by the German odour impact criteria.*

Table 5.3 Odour impact criteria defined by the odour concentration threshold C_T (in ou_E/m^3) and the exceedance probability p_T (in %) for various European countries. The ambient odour concentration is determined either by the integration time or the peak-to-mean factor. The impact criteria are applied to this (modified) concentration.

Country	Ambient odour concentration		Odour impact criteria (C_T / p_T)	Protection level	Source
	Integration time of the ambient concentration	Peak-to-Mean factor			
Germany	1 s	4	1 ou_E/m^3/2%	irrelevance criterion	(GOAA, 2008)
				residential areas:	
			1 ou_E/m^3/6.7%	poultry	
			1 ou_E/m^3/13.3%	fattening pigs	
			1 ou_E/m^3/20%	milking cows	
			1 ou_E/m^3/10%	other animals	
				rural areas:	
			1 ou_E/m^3/10%	poultry	
			1 ou_E/m^3/20%	fattening pigs	
			1 ou_E/m^3/30%	milking cows	
			1 ou_E/m^3/15%	other animals	
				farm land, sparsely populated:	
			1 ou_E/m^3/16.7%	poultry	
			1 ou_E/m^3/33.3%	fattening pigs	
			1 ou_E/m^3/50%	milking cows	
			1 ou_E/m^3/25%	other animals	
Austria	5 s	variable[a]	1 ou_E/m^3/8% and 5 ou_E/m^3/3%	residential areas	(Austrian Academy of Sciences, 1994)
Ireland	1 h	1	4.3 ou_E/m^3/2%	residential areas, pig	(Sheridan et al., 2004)
			6 ou_E/m^3/2%	residential areas, pig, old farms	(Irish Environmental Protection Agency, 2001) (Hayes et al., 2006)

Table 5.3 (Continued)

Country	Ambient odour concentration		Odour impact criteria (C_T/p_T)	Protection level	Source
	Integration time of the ambient concentration	Peak-to-Mean factor			
			9.7 ou$_E$/m³/2%	residential areas, poultry	(Sheridan et al., 2004)
			6 ou$_E$/m³/2%	residential areas, poultry	(Nicolas et al., 2008)
Belgium	1 h	1	6 ou$_E$/m³ /2%	pigs	
			10 ou$_E$/m³/2%	poultry	
The Netherlands	1 h	1	10 ou$_E$/m³/2%	serious annoyance	(Yang and Hobson, 2000)
			1–5 ou$_E$/m³/2%	generally accepted	
			1 ou$_E$/m³/2%	no serious annoyance	
			1 ou$_E$/m³/0.5%	safe target for new sources	
			10 ou$_E$/m³/0.01%	highly intermittent sources	
	1 h	1	0.5 ou$_E$/m³/2%	new facilities, densely populated	(Wallis and Cadee, 2008)
			1.5 ou$_E$/m³/2%	existing facilities, densely populated	
			1.0 ou$_E$/m³/2%	new facilities, sparsely populated	
			3.5 ou$_E$/m³/2%	existing facilities, sparsely populated	
Denmark	1 min	7.8	5–10 ou$_E$/m³/1%	residential areas	(Olesen et al., 2005; Mahin, 2001)
			10–30 ou$_E$/m³/1%	industrial and rural areas	(British Columbia, Ministry of Water, Land and Air Protection, 2005)
Hungary	1 h	1	0.6–1.2 ou$_E$/m³/2%		(Cseh et al., 2010)

[a] peak-to-mean factor depends on the distance and the atmospheric stability (Piringer et al., 2007; Schauberger et al., 2000).

Table 5.4 Odour impact criteria defined by the odour concentration threshold C_T (in ou_E/m^3) and the exceedance probability p_T (in %) for Australia, New Zealand and North America. The ambient odour concentration is determined either by the integration time or the peak-to-mean factor. The impact criteria are applied to this (modified) concentration.

Country	Ambient odour concentration		Odour impact criteria (C_T / p_T)	Protection level	Source
	Integration time of the ambient concentration	Peak-to-Mean factor			
Australia Queensland		10 5	5 ou_E/m^3/0.5% 5 ou_E/m³/0.5%	stacks ground-level or down-washed plumes	(Environmental Protection Agency (EPA), 2004)
Australia New South Wales	3 s	a	$C_T = f(D)/1\%$	C_T (ou_E/m^3) depends on the population density D (1/km²); $C_T = -(\log D - 4.5)/0.6$	(Wallis and Cadee, 2008)
Australia West Australia	3 min	a	2 ou_E/m^3/0.5% 4 ou_E/m^3/0.1%		(Wallis and Cadee, 2008)
Australia Victoria	3 min	a	4 ou_E/m^3/0.1%		(Wallis and Cadee, 2008)
Australia Queensland		2	2.5 ou_E/m^3/ 0.5%	residential areas	(Wallis and Cadee, 2008)
Australia South Australia	3 min	a	$C_T = f(D)/0.1\%$	C_T (ou_E/m^3) depends on the population density D (1/km²); $C_T = -(\log D - 4.5)/0.6$	(Wallis and Cadee, 2008)

Table 5.4 (Continued)

Country	Ambient odour concentration		Odour impact criteria (C_T / p_T)	Protection level	Source
	Integration time of the ambient concentration	Peak-to-Mean factor			
New Zealand	1 h	1	1 $ou_E/m^3/0.5\%$	high sensitivity/unstable and semi unstable	(Ministry of the Environment, 2003)
			2 $ou_E/m^3/0.5\%$	high sensitivity/stable	(Ministry of the Environment, 2003)
			5 $ou_E/m^3/0.5\%$	moderate sensitivity	
			5 to 10 $ou_E/m^3/0.5\%$	low sensitivity	
USA Pennsylvania	2 min	2	4 $ou_E/m^3/0.57\%$	residential with highway	(Ministry of the Environment, 2003)
USA California	1 h	1	4 $ou_E/m^3/1.1\%$	industrial with some residential and highway	(Ministry of the Environment, 2003)
USA Pennsylvania	1 h	1	20 $ou_E/m^3/1.1\%$	residential	(Ministry of the Environment, 2003)
USA California	5 min	2.29	4 $ou_E/m^3/0.5\%$	plant fence-line	(Ministry of the Environment, 2003)
Canada City of Calgary	1 h	1	20 $ou_E/m^3/1.1\%$	rural with growing residential	(Ministry of the Environment, 2003)

[a] No guidelines are given to determine the peak-to-mean factor for an integration time which deviates from the 1 h mean.

In Europe (Table 5.3), in Germany, Denmark and Austria a factor $F > 1$ is used to assess the expected odour concentration in the range of several seconds. All other European countries use a factor $F = 1$. In Austria, a method is used to calculate the factor F (here a peak-to-mean factor dependent on meteorological conditions) as a function of the distance to avoid the disadvantages of a constant factor which was discussed above. Ireland, The Netherlands, Belgium and Hungary do not modify the odour concentration before applying their odour impact criteria ($F = 1$). The largest variation of odour impact criteria (dependent on the annoyance level and/or population density) is used in the Netherlands. The consideration of the hedonic tone (pleasant/unpleasant) for the odour impact criteria can be seen for Germany, Ireland, and Belgium with increasing protection needs for poultry, pigs, and milking cows.

In the Australian states indicated in Table 5.4, either an integration time is in the range of 3 min (with the exception of New South Wales where it is 3 s) or a constant factor is in use. For several states in Australia, no information on the factor F is available. New Zealand uses four classes of odour impact criteria, dependent on land use. Pennsylvania and California in the USA use two odour impact criteria each, depending on land use. The city of Calgary in Canada has also one odour impact criterion.

From Tables 5.3 and 5.4 it can be concluded that the protection level of the surrounding areas is often categorized by zoning these areas, mainly dependent on land use (e.g. for agricultural use). In New South Wales (Australia), for example, the protection level depends on the population density (Department of Environment and Conservation, 2006). The odour concentration threshold C_T is obtained from the population density D (inhabitants/km^2) by $C_T = -(\log D - 4.5)/0.6$ (equals $C_T = 2$ ou$_E$/m^3 for urban areas (about $D = 2000$/km^2) and $C_T = 4$ ou$_E$/m^3 for rural sites (about $D = 125$/km^2)).

Besides single values describing an odour impact criterion also functions are in use to describe the relationship between exceedance probability p_T and odour concentration threshold C_T for a certain protection level. The function $C_T = k_1/p_T$ for the protection level of rural ($k_1 = 8$) and pure residential (urban) areas ($k_1 = 4$) is suggested by (Watts and Sweeten, 1995). A function $C_T = k_2 \, p_T^{-0.4}$ to divide the odour impact into non-detectable, acceptable, annoyance, and severe annoyance using the coefficients $k_2 = 2/3$, $k_2 = 2$, and $k_2 = 6$ is suggested by (Wallis and Cadee, 2008). These functions are depicted in Figure 5.15. Additionally, empirical data for a pig farm with an odour emission rate of 20 000 ou$_E$/s are added (Nicolas *et al.*, 2008). For this farm, separation distances between 100–400 m with protection level increasing were calculated. The slope of these data are in good agreement with the functions proposed by (Watts and Sweeten, 1995). These functions as well as the empirical data can be interpreted as isopleths of constant protection levels. The national odour impact criteria depicted in Figure 5.15 are in the range of the isopleths of (Wallis and Cadee, 2008) whereas the empirical data of (Nicolas *et al.*, 2008) are more related to the slope of (Watts and Sweeten, 1995).

If we assume, that the functions of (Watts and Sweeten, 1995) as well as (Wallis and Cadee, 2008) in Figure 5.15 depict constant protection levels, then two odour impact criteria, which lie on the same isopleth should result in identical separation distances. An impact criterion with a high odour concentration threshold and a low exceedance probability (e.g. $C_T = 4$ ou$_E$/m^3 and $p_T = 1\%$ with $F = 1$) is predominantly influenced by very stable meteorological situations which cause a low dilution (Yu *et al.*, 2009; Schauberger *et al.*, 2006). For inhabitants this means that high odour intensity will

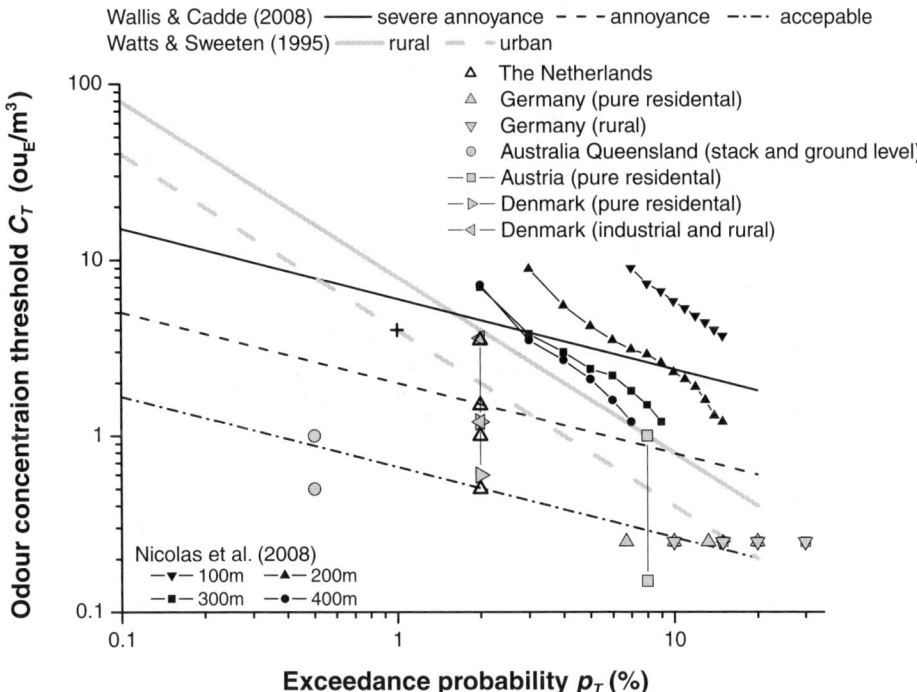

Figure 5.15 *Relationship between exceedance probability p_T and odour concentration threshold C_T. The functions of (Watts and Sweeten, 1995) and (Wallis and Cadee, 2008) describe isopleths of a certain annoyance potential. Additionally, empirical data are added for a pig farm with an odour emission rate of 20 000 ou_E/s (Nicolas et al., 2008). National odour impact criteria are shown for several countries (Tables 5.3 and 5.4). If the peak-to-mean factor $F > 1$ (filled symbols), then the odour concentration threshold C_T is converted to a 1-h mean value. The two crosses mark the impact criteria for Figure 5.16.*

be perceived in only few hours of the year (e.g. 87 h for an exceedance probability of $p_T = 1\%$).

If an impact criterion with a low odour concentration threshold and a high exceedance probability (e.g. Germany $C_T = 0.25\ ou_E/m^3$ and $p_T = 15\%$ with $F = 1$) is selected, weak odour intensities will be perceived over a higher fraction of time (e.g. 1324 hours for an exceedance probability of $p_T = 15\%$). The separation distances for these two impact criteria are depicted in Figure 5.16.

The two odour impact criteria selected for the comparison in Figure 5.16 lie on the same isopleth of (Watts and Sweeten, 1995) (two crosses in Figure 5.15) indicating a constant protection level for urban areas. The calculated separation distances, contrary to expectation, are very different and show a stronger dependence on wind direction for higher exceedance probabilities (grey line in Figure 5.16). From Figure 5.16 it must be concluded that the concept of odour impact criteria, used in various jurisdictions should be improved. It is obvious that separation distances, calculated for the identical protection level, should be similar.

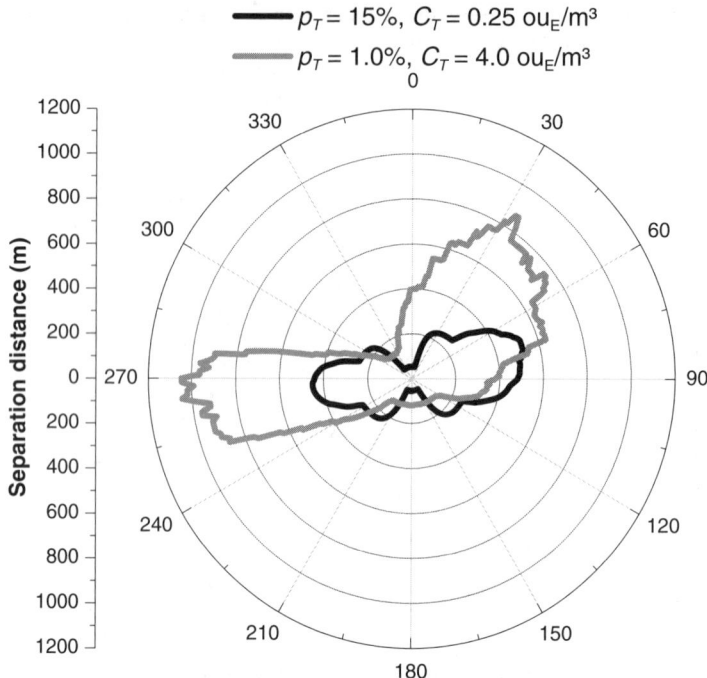

Figure 5.16 *Separation distances for two different odour impact criteria for the same protection level for urban areas according to the functions of (Watts and Sweeten, 1995) (Figure 5.15). Low odour intensity and higher exceedance probabilities ($p_T = 15\%$, $C_T = 0.25$ ou$_E$/m³) and high odour intensity combined with a low exceedance probability ($p_T = 1.0\%$, $C_T = 4$ ou$_E$/m³). The calculation is performed for a livestock building with 2000 fattening pigs (13 000 ou$_E$/m³) for Wels, Austria.*

5.7 Meteorological Input to Odour Dispersion Models

Once pollutants and thus also odorous substances are set free into the atmosphere, they are transported with the wind and diluted by turbulent diffusion. Meteorological input to dispersion models therefore consists of wind and stability information as well as a measure of vertical plume extent, the so-called mixing height. The availability and quality of these data are crucial for the representativeness of the model results.

5.7.1 Wind Information

Measurements of wind direction and wind speed in form of a continuous time series seems to be readily available and is done world-wide due to the WMO guide line which, for rural stations, declares wind measurements as representative if placed 10 m above ground without close obstacles (temperature and humidity measurements have to be conducted at 2 m). For urban areas, no guide lines for proper siting exist, although a report of COST-Action 715[3]

---

[3] COST (*CO*-operation in the fields of *S*cience and *T*echnique) is an intergovernmental European framework for international cooperation between nationally funded research activities.

Figure 5.17 *Change of plume direction with height (Image provided courtesy of August Kaiser, ZAMG).*

(Piringer and Joffre, 2005) gives some recommendations (representative measurements are achieved if sensors are placed 2–5 times above the average building height).

Odour pollution is often a local-scale problem, and in this scale small-scale wind fluctuations become important: the wind may change with height and time (Figure 5.17); with low stacks, downwash might occur, leading to high concentrations in the wake of a building (Figure 5.18).

In Gauss models, flow-disturbing features like building influence or topography can only be treated via simple empirical relations and assumptions (e.g. flow around or across an isolated hill via the dividing streamline concept). In Lagrange models including a diagnostic wind field model, a more realistic simulation of the flow field due to topography or buildings is possible. A simple assessment of the ambient odour concentration in the near field of buildings can be found in (Schauberger and Piringer, 2004). For all types of models a meteorological station representative for the area of interest has to be chosen or erected

Figure 5.18 *Example of building downwash (Image provided courtesy of August Kaiser, ZAMG).*

to deliver the desired time series of meteorological parameters, at least over one whole year, in the form of hourly or half-hourly values. In areas with a high frequency of weak winds, like valleys or basins, operation of a three-dimensional ultrasonic anemometer is to be preferred over a conventional wind vane.

5.7.2   Information on Atmospheric Stability

Besides wind data, dispersion models need parameters describing the vertical structure of the boundary layer (the atmospheric stability) as input. This information can be provided in its simplest form by discrete stability classes or, more advanced, by direct measures of atmospheric turbulence from 3D sonics.

For the purposes of dispersion modelling, discrete stability classes SC are still widely used. They are usually determined on the basis of routine meteorological observations representing rural conditions. As an example, the stability classification scheme used in Austria (with dispersion parameters developed by (Reuter, 1970)) is presented here. The details of the two schemes are given in Part 4.6 of (Piringer and Joffre, 2005).

Stability classes with the Reuter-Turner scheme (Reuter, 1970) are determined as a function of half-hourly mean wind speed and a combination of sun elevation angle, cloud base height and cloud cover; alternatively, the radiation balance (net radiation) or the vertical temperature gradient is used in combination with the mean wind speed. A three-dimensional statistics of stability classes contains the percentage frequency of each combination of wind direction (36 categories), wind speed (12 categories), and stability class (six categories), for example over the whole year. The calculation of stability classes is necessary to determine the vertical dispersion parameters σy and σz as discussed by (Hanna and Chang, 1992).

In practice, within the scheme explained in (Reuter, 1970), stability classes 2–7 can occur in Central Europe. Stability classes 2–3 occur during daytime in a well-mixed boundary layer, class 3 allowing also for cases of high wind velocity and moderate cloud cover. SC 4 is representative for cloudy and/or windy conditions including precipitation or fog and can occur day and night. Stability classes 5–7 occur at night, static stability increasing with class number. With the scheme based on cloudiness data, classes 2 and 3 can occur only during daytime, classes 5–7 only during night-time.

As an example, a comparison for two sites in Austria is presented. The area of the town of Wels is situated in the well-ventilated Austrian North-Alpine foreland, the village Frauental lies in a flat valley in Central Styria south of the main Alpine chain with generally lower wind speeds. Both at Frauental and Wels, discrete stability classes are determined as a function of half-hourly mean wind speed and a combination of sun elevation angle, cloud base height and cloud cover ('cloudiness method'); at Frauental, in addition, the radiation balance in combination with the mean wind speed has been used ('net radiation method').

The statistics of stability classes differs with methodology and site (Figure 5.19). At Frauental, a comparison between the net radiation -based and cloudiness-based statistics is possible. With in-situ net radiation, class 6 representative for stable situations domi-nates, comprising more than 35 % of all cases year-round. Compared to the statistics with cloudiness data where class 4 dominates, differences occur mainly for classes 3, 4, and 6, while classes 2 and 7 occur with about the same probability. Apparently the large abun-dance of class 6 when using net radiation data is mainly transformed to classes 3 and 4 in the cloudiness scheme, as is also revealed by a direct statistical comparison (not shown).

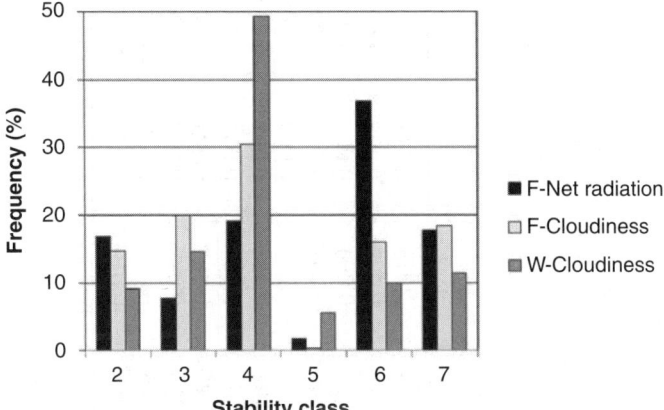

Figure 5.19 *Relative frequency (%) of stability classes 2–7 for the sites Frauental: (F) and Wels (W) using the net radiation and cloudiness methods (for details see text).*

This deviation between net radiation and cloudiness based statistics is systematic, that is, it occurs independent of site, and can be explained by the fact that in the cloudiness-based statistics, in contrast to net radiation, unstable situations occur during daytime, stable situations during night-time only. In areas with orographic modifications like Frauental, a negative net radiation and therefore stable situations can occur still after sunrise and already before sunset, while the cloudiness method can only calculate classes 2–4 by definition, which is the main cause for this difference.

The cloudiness-based statistics for Wels show an even more pronounced dominance of stability class 4 than for Frauental. Class 4 takes 50% of all cases year-round. Other classes occur at about 15% at most. The reason for the dominance of class 4 in the Austrian flatlands is the more frequent occurrence of higher wind speeds and especially wintertime cloudiness compared to Alpine sites. Both effects strengthen class 4 and weaken especially class 2 at daytime (which, in wintertime, is practically non-existent at Wels) and classes 6 and 7 at night with frequencies around 10% each only.

Alternatively, atmospheric stability can be deduced from three-dimensional (3D) ultrasonic anemometer measurements. Sonic anemometers measure the along-path velocity component from the travel time of acoustic waves between transducers separated about 10–20 cm. In addition to the three-dimensional wind vector, the sound velocity is derived, from which the so called 'sonic temperature' is calculated. The measurement of sonic temperature fluctuations is necessary to calculate the sensible heat flux. Other quantities which are derived from sonic measurements are the means, standard deviations, and co-variances of the wind components and the momentum flux, the Monin-Obukhov stability parameter, and the friction velocity. Sonic anemometers usually sample at 10 Hz, and the data are usually stored as averages over 10 min or half an hour.

As an example for the determination of atmospheric stability from 3D sonics, the scheme of the German guide line ('TA Luft', 2002) is shown in Table 5.5. The table shows the Monin-Obukhov stability parameter MOS (1/m) depending on the roughness length z_0 (m) and the stability class after Klug-Manier. In this way, stability classes are obtained without additional sensors or data.

Table 5.5 *Dependence of Monin-Obukhov stability parameter (m^{-1}) from the roughness length z_0 (m) and the stability class (scheme, Klug-Manier) according to the German guideline (TA Luft, 2002).*

Stability class		Roughness length z_0 in m								
	Klug-Manier	0.01	0.02	0.05	0.1	0.2	0.5	1	1.5	2
I	very stable	0.143	0.111	0.077	0.059	0.042	0.025	0.015	0.011	0.008
Class limit		**0.063**	**0.050**	**0.035**	**0.026**	**0.019**	**0.011**	**0.007**	**0.005**	**0.004**
II	stable	0.040	0.032	0.023	0.017	0.012	0.007	0.004	0.003	0.002
Class limit		**0.020**	**0.016**	**0.011**	**0.008**	**0.006**	**0.004**	**0.002**	**0.002**	**0.001**
III.1	neutral night	0.000	0.000	0.000	0.000	0.000	0.000	0.000	0.000	0.000
Class limit		**-0.020**	**-0.016**	**-0.011**	**-0.008**	**-0.006**	**-0.004**	**-0.003**	**-0.002**	**-0.002**
III.2	neutral day	-0.040	-0.031	-0.022	-0.017	-0.012	-0.008	-0.005	-0.004	-0.003
Class limit		**-0.057**	**-0.044**	**-0.031**	**-0.024**	**-0.017**	**-0.011**	**-0.007**	**-0.005**	**-0.004**
IV	unstable	-0.100	-0.077	-0.053	-0.040	-0.029	-0.018	-0.012	-0.009	-0.007
Class limit		**-0.143**	**-0.111**	**-0.077**	**-0.057**	**-0.042**	**-0.026**	**-0.017**	**-0.013**	**-0.010**
V	very unstable	-0.250	-0.200	-0.143	-0.100	-0.071	-0.045	-0.029	-0.022	-0.018

Figure 5.20 *Position of the ultrasonic anemometer at the confluence of the river Traun into the Danube east of the city of Linz (rural site).*

Klug-Manier classes are numbered from I to V and are classified according to atmospheric stability as follows:

- Stability classes V and IV comprise very unstable and unstable conditions, meaning good vertical mixing in the boundary layer. They do not occur during night-time. Class V occurs only between May and September.
- Stability classes III/2 and III/1 are classified as neutral. III/2 occurs predominantly at daytime, III/1 predominantly at night-time and during sunrise and sunset. These classes are typical for cloudy and/or windy conditions.
- Stability classes II and I occur with stable and very stable conditions, mostly, but not exclusively at night. They occur with reduced vertical mixing; horizontal transport over long distances is possible.

The method is very sensitive to the surroundings (surface conditions, roughness elements) around the measurement site. This enables to discern stability in different environments, for example, to distinct between rural and urban conditions, as shown in the following example from the Austrian industrial city of Linz. Three-dimensional sonics have been operated at two sites in autumn 2006, one situated at the confluence of the river Traun into the Danube

Figure 5.21   *Position of the ultrasonic anemometer on top of a school building in the downtown area of Linz (urban site) (Image provided courtesy of Franz Freytag, Voestalpine).*

east of the city (Figure 5.20), the other on top of a school building in the downtown area (Figure 5.21).

From the average daily course of the sensible heat flux over this three months period, the expected urban–rural differences are clearly visible (Figure 5.22). At the urban site, due to the built-up area with sealed surfaces and anthropogenic activity (traffic, industry, domestic heating), the sensible heat flux shows on average smaller negative values at night and larger positive values during daytime, as expected. The resulting differences in atmospheric stability, explained by the MOS parameter, are shown in Figure 5.23. During the night-time, the rural site shows a strong tendency towards stable conditions (positive MOS); however, also the urban site is partly strongly stable, especially during the first half of the night. Daytime unstable conditions at the urban location start earlier in the morning and last longer in the afternoon.

The resulting frequencies of Klug-Manier stability classes are displayed in Figure 5.24. Differences are observed for classes V, III/1 and I. Very unstable conditions represented by class V are more than twice as frequent at the urban than at the rural site; this result is easily explained by the aforementioned anthropogenic activities in urban areas. Night-time differences in atmospheric stability need more explanation in this specific case. The rural site experiences more neutral and less very stable cases than the urban site. This at first sight surprising result can be caused by higher wind speeds at the undisturbed rural site. However, the very reason for this somewhat unexpected result can be explained by looking

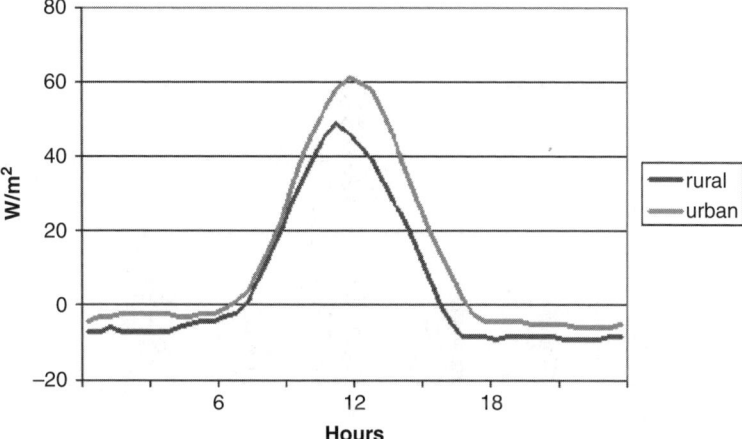

Figure 5.22   *Average daily course of the sensible heat flux (W m^{-2}) during autumn 2006 at two sites at Linz, Austria.*

at the stability-dependent wind roses at the urban location (Figure 5.25): The wind direction distribution for stable conditions show a marked occurrence of northerly winds. These are subject to night-time outflow from tributary valleys of the river Danube North of the city, coupled with a very stable boundary layer. These winds are lifted over the built-up area of Linz and registered well by the ultrasonic anemometer positioned above roof-top and thus situated in this undisturbed flow, which is probably not at all registered at street level. A similar phenomenon of up-lifting of a night-time valley wind could be observed and documented at Graz (Piringer and Baumann, 1999).

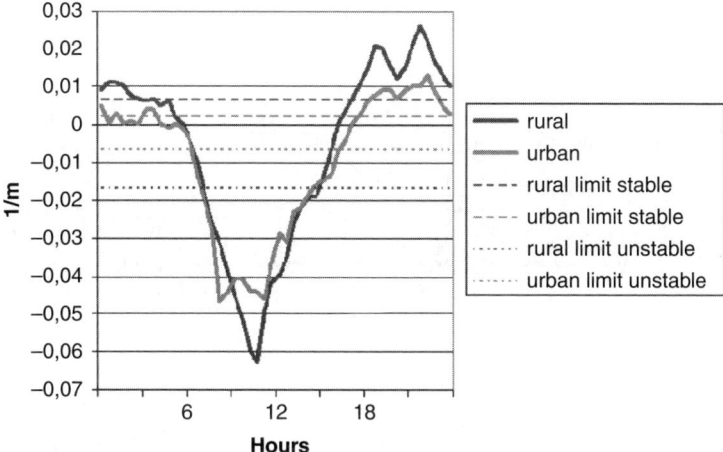

Figure 5.23   *Average daily course of the MOS parameter (m^{-1}) during autumn 2006 at two sites at Linz, Austria.*

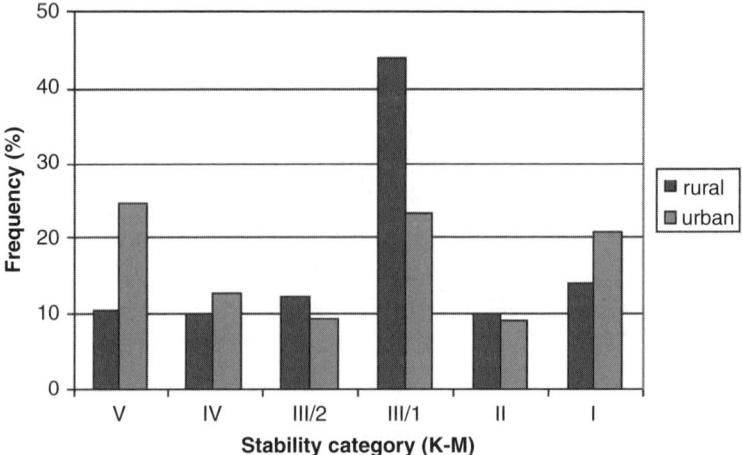

Figure 5.24   *Frequency distribution of stability classes (Klug-Manier) during autumn 2006 at two sites at Linz, Austria.*

5.7.3   Information on the Mixing Height

The mixing height (MH) is an important parameter for practically all air pollution applications except probably the street scale; it is still one major uncertainty for most air quality models. It is the height of the layer adjacent to the ground over which pollutants emitted within this layer or entrained into it become vertically dispersed by convection or mechanical turbulence within a time scale of about 1 h (Seibert *et al.*, 2000). An overview on methods how to derive MH from measurements as well as parameterizations in meteorological pre-processors is given in (COST-710, 1998); a review of methods for urban areas is given in (Piringer *et al.*, 2007). Here we focus on state-of-the-art methods to derive continuous time series of MH for the purpose of (odour) dispersion modelling.

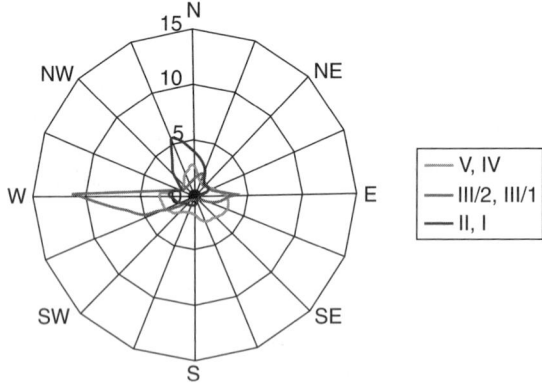

Figure 5.25   *Average wind roses depending on groups of Klug-Manier stability classes at the urban site during autumn 2006 at Linz, Austria.*

Table 5.6   *Examples of mixing heights (m) dependent on atmospheric stability for use in (regulatory) models in Austria, Germany and the UK.*

		Very unstable	Unstable	Neutral	Slightly stable	Stable	Very stable
Austria (Kolb, 1981)	Summer	2500	2000	1100			
	Winter	1100	1100	800			
ADMS (CERC, 2007)		1300	900	850	800	400	100
Germany (VDI 3782 Part 1, 2009)		1100	1100	800	800	250	250
LASAT (Janicke Consulting, 2011)		1100	1100	800		102	

The simplest way is to assign to each stability class a fixed value of MH which is often done for regulatory dispersion modelling (Table 5.6). The values are intended to be used for specific models as indicated or regulations like the Austrian or German ones. The Austrian regulation is the only one using a seasonal dependence and needs no specification of MH for stable conditions due to the assumed narrowness of the plume. LASAT uses fixed values only for negative Monin-Obukhov lengths. For positive values, the approach in (Seinfeld and Pandis, 1998) is used, MH depending on the Monin-Obukhov length, the friction velocity and the Coriolis parameter.

The seasonal dependence of MH is supported by an investigation undertaken in Austria during the COST-Action 715 (Piringer and Joffre, 2005). Figure 5.26 displays average MH and their standard deviations (both in m above ground) dependent on stability class for an urban and a rural site in the area of Linz, Austria (hourly values over a period of five

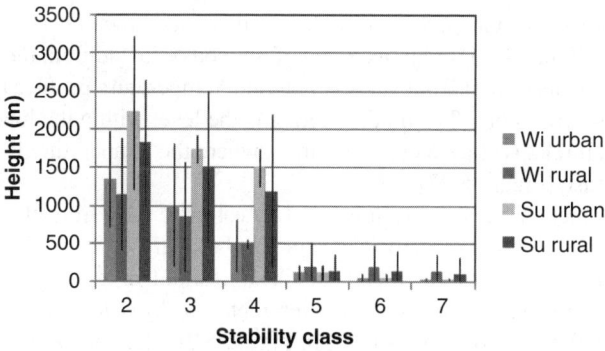

Figure 5.26   *Average mixing heights and standard deviations (thin lines, both in m above ground) dependent on stability classes 2–7 (Reuter-Turner, cloudiness scheme) for summer (Su) and Winter (Wi) for the area Linz, Austria.*

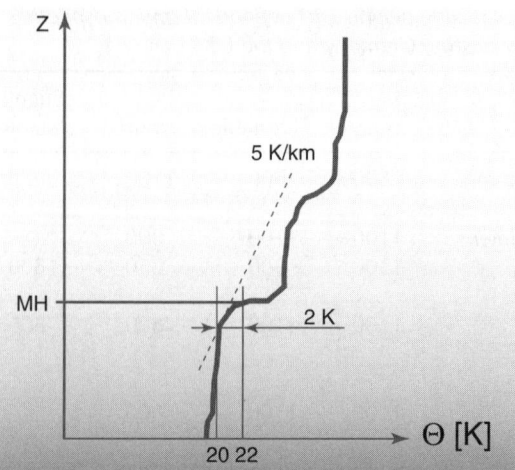

Figure 5.27   *Scheme to determine the mixing height (MH) with the Heffter method (Heffter, 1980) (Image provided courtesy of Christoph Lotteraner, ZAMG).*

years), calculated with the OML meteorological pre-processor (Piringer *et al.*, 1998). For unstable and neutral conditions (stability classes 2 to 4), on average lower mixing heights in winter as well as a decrease in average MH with increasing atmospheric stability is clearly visible. Urban-rural differences, with larger MH in the urban area, are most pronounced for unstable classes and neutral conditions in summer. Larger mixing heights in urban areas have to be expected due to enhanced mixing resulting from both the large surface roughness and increased surface heating. During night-time (stability classes 5 to 7), mixing heights are in the order of 100 m, and the calculation shows a tendency towards larger mixing heights in the rural area, attributable to the on-average larger wind speeds at the rural site in this specific case. From direct measurements, the opposite has to be expected, that is, larger night-time mixing heights at an urban site, due to the reasons given earlier.

In principle, all instruments providing a vertical temperature profile (e.g. radio- and tethersondes, RASS) can be used to deduce the mixing height (MH). It is retrieved from these profiles mostly by the parcel or Richardson number method (see (Seibert *et al.*, 2000) and (COST-710, 1998) for an overview of methods). In addition, the Heffter method (Heffter, 1980) diagnoses MH from vertical potential temperature profiles by searching for so-called 'critical' inversions. The mixing height is the level within the lowest layer with a potential temperature lapse rate $\Delta \Theta \geq 5 \, \text{K km}^{-1}$, where the temperature is 2 K higher than at the inversion base (Figure 5.27).

Recently, ceilometers are increasingly used to estimate MH from the analysis of the aerosol distribution. Heights of aerosol layers can be analysed from the optical vertical backscatter profiles using several methods, like threshold or gradient methods (Emeis *et al.*, 2004). As an example, the aerosol layers detected by a Jenoptik ceilometer at the Vienna airport on 3 June, 2010, are shown in Figure 5.28; the course of MH deduced via an algorithm developed at ZAMG is shown in Figure 5.29. The algorithm uses mainly the lowest aerosol layer as MH and includes smoothing and extrapolations in case of interruptions or large changes in aerosol layers.

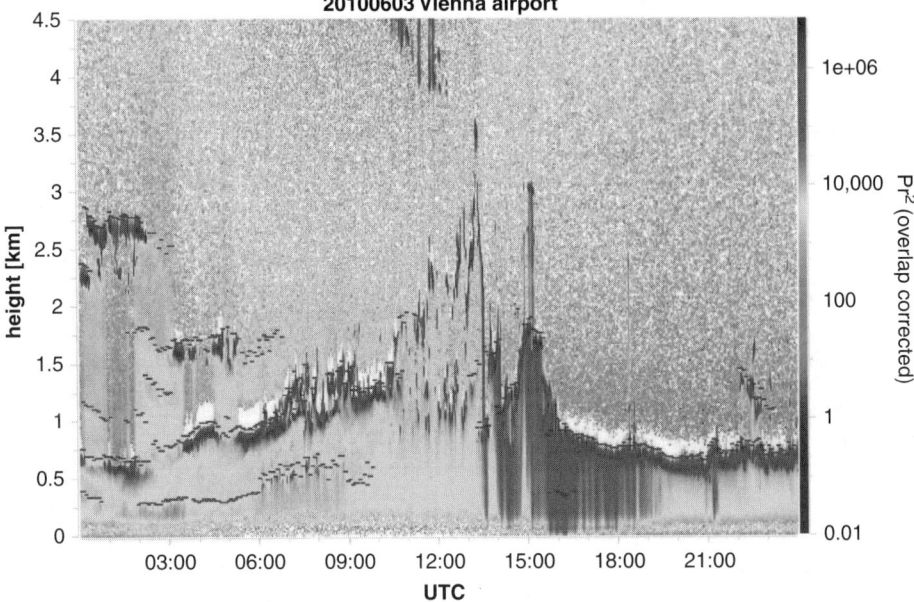

Figure 5.28   *Time series of aerosol layer height detected with a Jenoptik ceilometer at Vienna airport on 3 June, 2010 (Image provided courtesy of Jenoptik, Germany).*

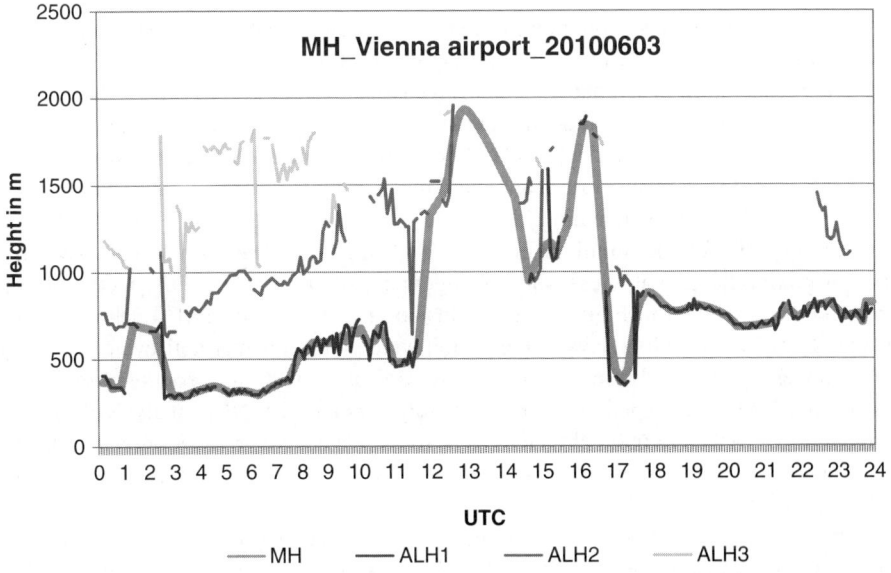

Figure 5.29   *Time series of the mixing height (MH, m) deduced from the aerosol layer heights (ALH) of the Jenoptik ceilometer at Vienna airport on 3 June, 2010 (Image provided courtesy of Christoph Lotteraner, ZAMG).*

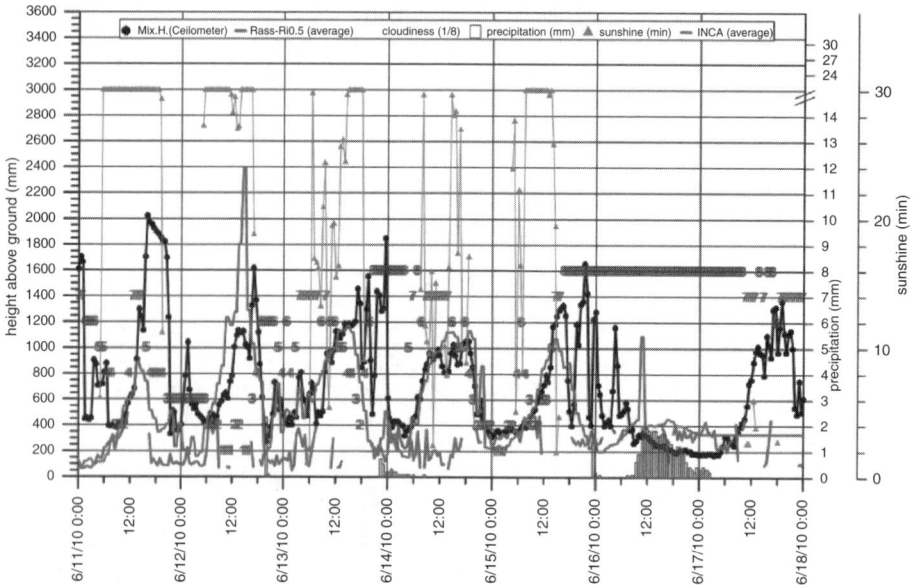

Figure 5.30   *Time series (June 11–17, 2010) of the mixing height (in m) from RASS (dotted line), ceilometer (black line) and INCA (black, thin) at Vienna airport. Precipitation (mm) as bars; sunshine duration (min) as thin line with the symbol of the sun; cloudiness (1/8) marked as numbers (Image provided courtesy of Gabriele Rau, ZAMG).*

MH can also be obtained from nowcasted meteorological fields. At ZAMG, INCA (Integrated Nowcasting through Comprehensive Analysis; (Haiden *et al.*, 2007) was developed. This is an observation-based analysis and forecasting system (+6 h). ALADIN forecasts are used as first guess fields and downscaled to 1 km resolution (operationally) taking into account high resolution terrain data and online available meteorological measurements. To derive MH from the 3D temperature and wind fields provided by INCA, the same methods are used as for the vertical sounding instruments.

As an example for the simultaneous use of all methods described, a one week time series of mixing heights at Vienna airport together with a few explaining meteorological parameters is displayed in Figure 5.30. From looking at the course of sunshine duration and cloudiness, the first five days of the period were predominantly anti-cyclonic, the last two days were cloudy and rainy. On the sunny days, a distinct daily course of the mixing height can be seen, most apparent from the ceilometers and the INCA analysis. The RASS is limited to a few 100 m in height and thus is able to detect MH mostly only at night-time and during the morning transition; on fair weather days, MH usually increases well above the RASS range.

From a close look at Figure 5.30, agreement and discrepancies in mixing heights from the different methods applied are clearly visible. There is often agreement in all three methods in the morning hours, when MH starts to increase; a good daytime coincidence between INCA and ceilometer is seen especially on 13, 14 and partly 15 June. Large discrepancies between these two methods occur predominantly in the late afternoon and evening hours on clear days (e.g. 11, 13 and 15 June), when the ceilometer tends to deliver large mixing heights; in these cases, it cannot detect the newly evolving stable boundary layer which INCA is frequently able to diagnose better. It is interesting that during the cloudy and rainy period on 16 June, all methods agree quite well in MH for most of the time; on 17 June, INCA apparently produced no data. Figure 5.30 indicates that an operational MH could be quite easily obtained by taking the lowest estimate of the three methods during night-time and of INCA and the ceilometer during daytime.

5.8   Evaluation of Odour Dispersion Models

Air pollution models especially in the micro- and meso-scale are nowadays commercially available and widely used in environmental impact studies. The systematic evaluation of these models is necessary to increase confidence in the modelled results. The quality assurance and improvement of micro-scale meteorological models was the focus of COST-Action 732 (Schatzmann *et al.*, 2010); http://www.mi.uni-hamburg.de/Home.484.0.html).

This COST-Action validated different CFD and non-CFD models in boundary layers with a number of obstacles. Basic test data are available from the internet platform CEDVAL (http://www.mi.uni-hamburg.de/CEDVAL-Validation_data.427.0.html). The Action set up a number of validation objectives. For non-CFD models that assume a straight centreline for any plume it is a priori known that they are unable to predict the position of a plume correctly if the geometry is complex. Therefore, non-CFD models may be used to predict the magnitude of concentrations. An evaluation objective for such models can therefore be based only on a comparison of data *unpaired* in space.

Evaluation tools for Gaussian models are therefore:

- Quantile-quantile and residual plots (no scatter plots)
- Plume width and maximum arcwise concentrations
- Metrics: bias and difference of standard deviations

In Lagrangian models the dispersion calculation is based on a diagnostic wind field model. For these models it makes sense to evaluate flow and concentration results in the same way (*paired* in space) as for the CFD flow simulations. Evaluation tools for Lagrangian models comprise all plots and metrics in the COST 732 Excel Workbook (http://www.dmu.dk/International/Air/Models/Background/MUST). The statistical measures for comparison are analogous to the BOOT software (Chang and Hanna, 2004); http://www.harmo.org/kit/BOOT_details.asp.

As an example, we report on an evaluation exercise conducted at ZAMG using the data set of the so-called OROD (Odour release and odour dispersion) project (Bächlin *et al.*, 2002) available from http://www.lohmeyer.de/eng/. The source is a pig fattening unit in fairly flat terrain. Concentration data sets from 14 SF6 tracer releases and measurements of 10-min duration, meteorological data, odour intensity data estimated by a panel (not used for evaluation) and the exact source values (source flow rate, volume flux and emission velocity at the source) are available. All tracer releases took place under neutral conditions and south-westerly wind directions, wind speeds ranging from 2.5 to 7.9 m s^{-1}. The experimental set-up is shown in Figure 5.31.

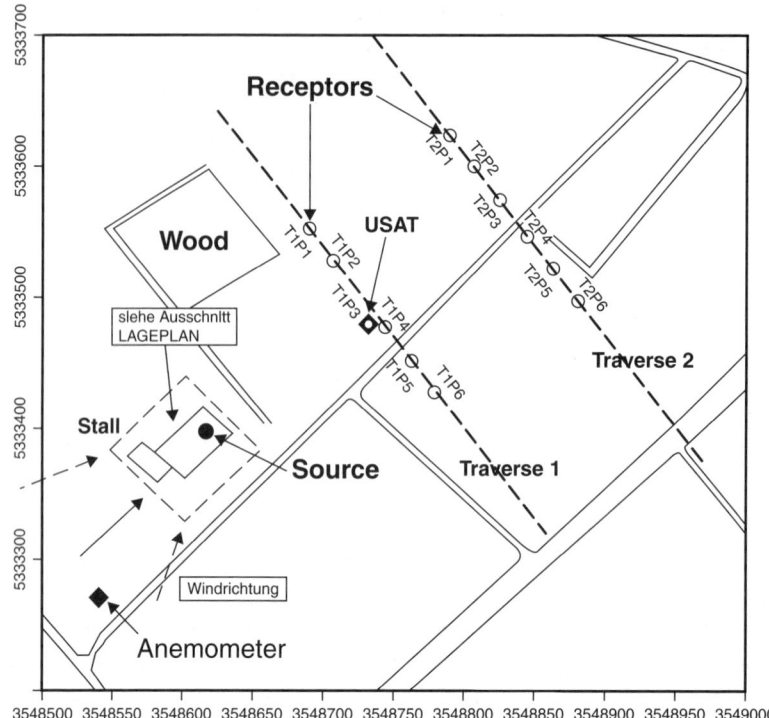

Figure 5.31   *Set-up of the OROD experiment indicating the source, the main wind direction and the two rows of receptors perpendicular to the main wind direction. (Reprinted with permission from Piringer and Baumann-Stanzer, 2009, Copyright (2009) Martin Piringer).*

The quantile-quantile plots for the models AODM (Gaussian, see Part 5.3) and LASAT (Lagrangian, see also Part 5.3) are shown in Figure 5.32. More results are presented in (Piringer and Baumann-Stanzer, 2009) and (Baumann-Stanzer and Piringer, 2011). The results of the Gauss model AODM and the more complex Lagrange model LASAT are comparable in this case. AODM is even slightly better in reproducing single very high concentrations, whereas LASAT seems to cut off already below 30 μg m^{-3}. Statistical performance measures for all models applied in this exercise, together with their range of acceptability according to (Chang and Hanna, 2004), are given in Table 5.7. The best values are highlighted in bold. For all runs except ADMS, MG, VG and NMSE are within the proposed range of acceptability. For ADMS, the statistical measures are somewhat higher than the proposed limits. FB is acceptable for AODM and the two LASAT runs only. LASAT performs best, on this statistical basis (three best values without, two best values with terrain). AODM and LASAT-Terrain perform best with respect to MG.

Numerous other validation exercises (e.g. (Pechinger and Petz, 1997, 1995; Hirtl and Baumann-Stanzer, 2007; Lohmeyer *et al.*, 2002; Oettl *et al.*, 2005; Oettl *et al.*, 2001), to name a few) have shown that there is no 'best' model, that is, no model alone performs best in all the statistical or graphical measures applied. This is a well-known outcome of such model evaluation exercises. They are however important as they reveal that it is of utmost

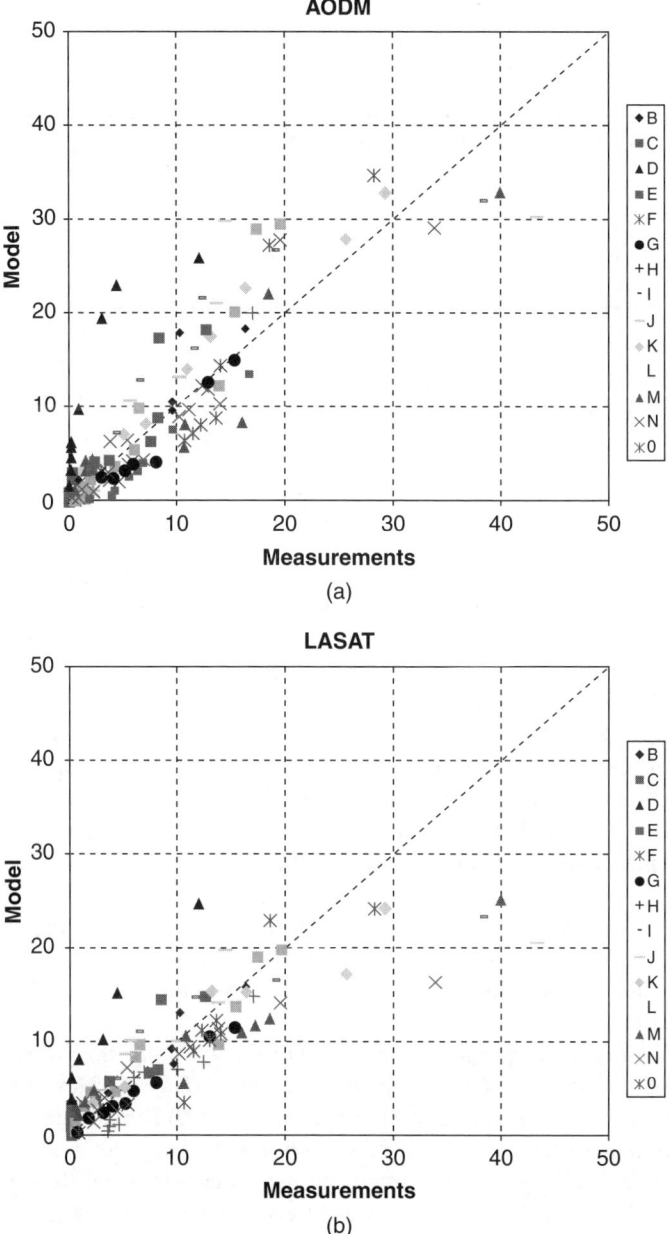

Figure 5.32 *Comparison of measured and modelled concentrations ($\mu g\ m^{-3}$) in the form of quantile-quantile plots for the OROD experiment. (a) AODM; (b) LASAT. (Reprinted with permission from Piringer and Baumann-Stanzer, 2009, Copyright (2009) Martin Piringer).*

Table 5.7   *Statistical performance measures and their range of acceptability for all models applied in the OROD experiment. MG: geometric mean; VG: geometric variance; FB: fractional bias; NMSE: normalized mean square error (Reprinted with permission from Piringer and Baumann-Stanzer, 2009, Copyright (2009) Martin Piringer).*

	ONGAUSSplus	AODM	ADMS	LASAT	LASAT Terrain	'acceptable'
MG	1.3	**1.0**	1.6	0.9	**10**	$0.7 < MG < 1.3$
VG	1.5	1.5	1.6	**1.3**	1.4	$VG < 1.6$
FB	0.4	−0.2	0.8	**0.01**	**0.01**	$-0.3 < FB < 0.3$
NMSE	1.7	1.2	5.8	**1.1**	1.3	$NMSE < 4$

importance to apply a model according to its physics and to take care for a representative meteorological input and appropriate boundary conditions.

References

Austrian Academy of Sciences. Umweltwissenschaftliche Grundlagen und Zielsetzungen im Rahmen des Nationalen Umweltplans für die Bereiche Klima, Luft, Geruch und Lärm (1994) Kommission für Reinhaltung der Luft der Österreichische Akademie der Wissenschaften, Schriftenreihe der Sektion I . Band 17. Bundesministeriums für Umwelt, Jugend und Familie, Wien.

Baumann-Stanzer K. and Piringer M. Validation of regulatory micro-scale air quality models: modelling odour disperison and built-up areas. *World Review of Science, Technology and Sust. Development.* 2011; **8**: 203–213.

Bächlin W., Rühling A. and Lohmeyer A. Bereitstellung von Validierungsdaten für Geruchsausbreitungsmodelle – Naturmessungen (2002) Forschungsbericht FZKA-BWPLUS, p. 183.

Best P., Lunney K. and Killip C. Statistical elements of predicting the impact of a variety of odour sources. *Water Science & Technology.* 2001; **44**: 157–164.

Beychock, M.R. Fundamentals of stack gas dispersion. (1994) M.R. Beychock, Newport Beach, CA.

Both R. *Geruchsbeurteilung in der Landwirtschaft – Vorstellung des Projekts und Ergebnisse der Immissionsmessungen. Emissionen der Tierhaltung. Messung, Beurteilung und Minderung von Gasen, Stäuben und Keimen.* (2006) Kuratorium für Technik und Bauwesen in der Llandwirtschaft. KTBL-Schrift 449 Darmstadt, pp. 150–158.

British Columbia, Ministry of Water, Land and Air Protection. Final Report Odour Management in British Columbia: Review and Recommendations (2005).

Briggs G. Plume rise predictions. Lectures on air pollution and environmental impact analysis (1975). American Meteorological Society.

Carson J.E. and Moses H. Validity of several plume rise formulas. *Air Pollution Control Assn-J.* 1969; **19**: 862–866.

Chatwin P.C. and Sullivan P.J. The intermittency factor of scalars in turbulence. *Physics of Fluids A.* 1989; **1**: 761–763.

CERC. *ADMS 4 Atmospheric Dispersion Modelling System User Guide*, Version 4.0 (2007) Cambridge Environmental Research Consultants.

Chang J.C. and Hanna S.R. Air quality model performance evaluation. *Meteorology and Atmospheric Physics*. 2004; **87**: 1–3.

COST-710. Harmonisation of pre-processing of meteorological data for atmospheric dispersion models. Final report (Fisher, B. *et al.*, eds). (1998) European Commission.

Cseh M, Nárai KF, Barcs E, Szepesi DB, Szepesi DJ, Dicke JL. Odor setback distance calculations around animal farms and solid waste landfills. *Idojaras*. 2010; **114**: 303–318.

Department of Environment and Conservation NSW. Technical Note: Assessment and management of odour from stationary sources in NSW. (2006) Department of Environment and Conservation, NSW, Australia.

Defoer N. and van Langenhove H. Variability and repeatability of olfactometric results of n-butanol, pig odour and a synthetic gas mixture. *Water Science and Technology*. 2003; **50** (4): 65–73.

Emeis S., Münkel C. and Vogt S. *et al.* Atmospheric boundary-layer structure from simultaneous SODAR, RASS and ceilometer measurements. *Atmospheric Environment* 2004; **38**: 273–286.

Environmental Protection Agency (EPA) Queensland. Guideline - Odour Impact Assessment from Developments. (2004) Queensland Government, Parks and Wildlife Service, Brisbane, Australia.

Environment Agency (UK). H4 – *Odour Management Environment Agency Technical Guidance Note, UK.* (2009).

EN 13725. *Air quality – Determination of odour concentration by dynamic olfactometry* (2003) Austrian Standards.

Fackrell J.E. and Robins A.G. Concentration fluctuations and fluxes in plumes from point sources in a turbulent boundary layer. *Journal of Fluid Mechanics*. 1982; **117**: 1–26.

Feddes J.J.R. Development of odour monitoring procedures for Alberta livestock operations: Measuring odour with confidence. (2006) Livestock Industry Development Fund and Alberta Agricultural Research Institute, Edmonton, Alberta, Canada.

GOAA. Guideline on Odour in Ambient Air (GOAA). Detection and Assessment of Odour in Ambient Air (2008).

Gross G. Estimation of annual odour load with a concentration fluctuation model. *Meteorologische Zeitschrift*. 2001; **10**: 419–10425.

Guo H., Jacobson L.D. and Schmidt D.R. Calibrating INPUFF-2 model by resident-panelists for long-distance odor dispersion from animal production sites. *Applied Engineering in Agriculture*. 2001; **17**: 859–868.

Guo H., Yu Z. and Lague C. Livestock Odour Dispersion Modeling: A Review. CSBE/SCGAB 2006 Annual Conference. 2006:170.

Hanna S.R. Concentration fluctuations in a smoke plume. *Atmospheric Environment*. 1967; **18**: 1091–1106.

Hartmann U. *Geruchsbeurteilung in der Landwirtschaft – Ermittlung der Belastungssituation in NRW durch Modellrechnung, Emissionen der Tierhaltung. Messung, Beurteilung und Minderung von Gasen, Stäuben und Keimen.* (2006) Kuratorium für Technik und Bauwesen in der Landwirtschaft, KTBL-Schrift 449. Darmstadt.

Hartmann U. and Hölscher M. *Ausbreitungsrechnungen für Geruchsimmissionen. Vergleich mit Messdaten in der Umgebung von Tierhaltungsanlagen* (2007) LANUV-Fachbericht 5. Landesamt für Natur, Umwelt und Verbraucherschutz Nordrhein-Westfalen (LANUV NRW), Recklinghausen.

Hanna S.R. and Chang J.C. Boundary-layer parameterizations for applied dispersion modelling over urban areas. *Boundary-Layer Meteorology*. 1992; **58**: 229–259.

Haiden T., Stadlbacher K., Steinheimer M. and Wittmann C. Integrated Nowcasting through Comprehensive Analysis (INCA) – System overview (2007) Central Institute for Meteorology and Geodynamics (ZAMG), Vienna, 49 pp.

Hayes E.T., Curran T.P. and Dodd V.A. A dispersion modelling approach to determine the odour impact of intensive poultry production units in Ireland. *Bioresource Technology*. 2006; **97**: 1773–1779.

Hanna S.R. The exponential probability density function and concentration fluctuations in smoke plumes. *Boundary-Layer Meteorology*. 1984; **29**: 361–375.

Hanna S.R. and Insley E.M. Time series analysis of concentration and wind fluctuations. *Boundary-Layer Meteorology*. 1989; **47**: 131–147.

Heffter J.L. Transport layer depth calculations. AMS, *Proc. 2nd Conf. on applications of Air Pollution Modelling*. 1980; 787–791.

Hinds W.T. Peak-to-mean concentration ratios from ground-level sources in building wakes. *Atmospheric Environment*. 1969; **3**: 145–156.

Hilderman T.L., Hrudey S.E. and Wilson D.J. A model for effective toxic load from fluctuating gas concentrations. *Journal of Hazardous Materials*. 1999; **64**: 115–134.

Hilderman T.L. and Wilson D.J. Simulating concentration fluctuation time series with intermittent zero periods and level dependent derivatives. *Boundary-Layer Meteorology*. 1999; **91**: 451–482.

Hilderman T.L. Stochastic simulation of concentration fluctuations for an effective toxic load model. *Department of Mechanical Engineering*. 1997; 146.

Hirtl M. and Baumann-Stanzer K. Evaluation of two dispersion models (ADMS-Roads and LASAT) applied to street canyons in Stockholm, London and Berlin. *Atmospheric Environment*. 2007; **41**: 5959–5971.

Irish Environmental Protection Agency. Odour impacts and odour emission control measures for intensive agriculture. (2001) Environmental Protection Agency, Ireland.

Janicke Consulting. Dispersion model LASAT. Reference/Working Book for Version 3.2 (2011).

Katestone Scientific. Peak-to-mean ratios for odour assessments (1998) A report from Katestone Scientific to Environment Protection Authority of New South Wales.

Kleemann A.M., Kopietz R., Albrecht J., *et al.* Investigation of breathing parameters during odor perception and olfactory imager. *Chemical Senses*. 2009; **34**: 1–9.

Klein P.M. and Young D.T. Concentration fluctuations in a downtown urban area. Part I: analysis of Joint Urban 2003 full-scale fast-response measurements. *Environmental Fluid Mechanics*. 2010; **11**: 23–42.

Løfstrøm P., Jørgensen H., Lyck E *et al. Test of a New Concentration Fluctuation Model for Decision Makers* (1996) 21st NATO/CCMS Int. Techn. Meeting, Plenum Publishing Corp, New York.

Lohmeyer A., Müller W.J. and Bächlin W. A comparison of street canyon concentration predictions by different modellers: final results now available from the Podbi-exercise. *Atmospheric Environment.* 2002; **36**: 157–158.

Lohr L. Perceptions of rural air quality: What will the neighbors think? *Journal of Agribusiness.* 1996; **14**: 109–128.

Kolb H. Ein normatives physikalisches Modell zur Simulierung der Ausbreitung von Schadstoffen in der Atmosphäre mit besonderer Berücksichtigung der Verhältnisse in Österreich (1981). University of Vienna, Austria.

Lung T., Müller H.J., Gläser M. and Möller B. Measurements and modelling of full-scale concentration fluctuations. *Agrartechnische Forschung.* 2002; **8**: E5–E15.

Mahin T.D. Comparison of different approaches used to regulate odours around the world. *Water Science and Technology.* 2001; **44**: 87–102.

Miedema H.M.E. and Ham J.M. Odour annoyance in residential areas. *Atmospheric Environment.* 1988; **22**: 2501–2507.

Miedema H.M.E., Walpot J.I., Vos H. and Steunenberg C.F. Exposure-annoyance relationships for odour from industrial sources. *Atmospheric Environment.* 2000; **34**: 2927–2936.

Misselbrook T.H., Clarkson C.R. and Pain B.F.U. Relationship Between Concentration and Intensity of Odours for Pig Slurry and Broiler Houses. *Journal of Agricultural Engineering Research.* 1993; **55**: 163–169.

Ministry of the Environment. Good practice guide for assessing and managing odour in New Zealand. (2003) Ministry of the Environment, Wellington, New Zealand.

Miller F.J., Schlosser P.M. and Janszen D.B. Haber's rule: A special case in a family of curves relating concentration and duration of exposure to a fixed level of response for a given endpoint. *Toxicology.* 2000; **149**: 21–34.

Mylne K.R.U. Concentration fluctuation measurements of a tracer plume at up to 1 km range in the atmosphere. Proceedings of the 9th Symposium on Turbulence and Diffusion. 1988; 168–171.

Mylne K.R.U. Concentration fluctuation measurements of a tracer plume at up to 1 km range in the atmosphere. *Ninth Symposium on Turbulence and Diffusion.* 1990: 168–171.

Mylne K.R. The vertical profile of concentration fluctuations in near-surface plumes. *Boundary-Layer Meteorology.* 1993; **65**: 111–136.

Mylne K.R. Concentration fluctuation measurements in a plume dispersing in a stable surface layer. *Boundary-Layer Meteorology.* 1992; **60**: 15–48.

Nicolai R., Clanton C. and Guo H. Modelling the relationship between detection threshold and intensity of swine odours. *Proceedings of the Second International Conference: Air Pollution from Agricultural Operations.* 2002: 296–304.

Nicolas J., Craffe F. and Romain A.C. Estimation of odor emission rate from landfill areas using the sniffing team method. *Waste Management.* 2006; **26**: 1259–1269.

Nicolas J., Delva J., Cobut P. and Romain A. Development and validating procedure of a formula to calculate a minimum separation distance from piggeries and poultry facilities to sensitive receptors. *Atmospheric Environment.* 2008; **42**: 7087–7095.

Nicell J.A. Assessment and regulation of odour impacts. *Atmospheric Environment.* 2009; **43**: 196–206.

Nicolas J., Romain A., Delva J. *et al.* Odour annoyance assessment around landfill sites: methods and results. (2008) *NOSE: International Conference on Environmental Odour Monitoring and Control Rome*, Italy, pp. 29–36.

Oettl D. and Uhrner U. Documentation of the Lagrangian Particle Model GRAL (Graz Lagrangian Model) Vs. 4.10 (2010).

Oettl D., Sturm P. and Almbauer R. Evaluation of GRAL for the pollutant dispersion from a city street tunnel portal at depressed level. *Environmental Modelling and Software.* 2005; **20**: 499–504.

Oettl D., Kukkonen J., Almbauer R.A. *et al.* Evaluation of a Gaussian and a Lagrangian model against a roadside data set, with emphasis on low wind speed conditions. *Atmospheric Environment.* 2001; **35**: 2123–2132.

Olesen H., Løfstrøm P., Berkowicz R. and Ketzel M. *Regulatory Odour Model Development: Survey of Modelling Tools and Datasets with Focus on Building Effects* (2005) NERI Technical Report No 541.

Olesen H.R., Løfstrøm P., Berkowicz R. and Ketzel M. *Regulatory Odour Model Development: Survey of Modelling Tools and Datasets with Focus on Building Effects* (2005) Ministry of the Environment, Denmark.

Oettl D., Goulart A., Degrazia G. and Anfossi D. A new hypothesis on meandering atmospheric flows in low wind speed conditions. *Atmospheric Environment.* 2005; **39**: 1739–1748.

ÖNorm M 9440. Ausbreitung von luftverunreinigenden Stoffen in der Atmosphäre; Berechnung von Immissionskonzentrationen und Ermittlung von Schornsteinhöhen (1992). Austrian Standards Institute.

Pechinger U. and Petz E. Model evaluation of the Austrian Gaussian plume model on M 9440: Comparison with the Copenhagen and the Liliestrøm datasets. *International Journal of Environment and Pollution.* 1997; **8**: 287–294.

Pechinger U. and Petz E. Model evaluation of the Austrian Gaussian plume model on M 9440: Comparison with the Kincaid dataset. *International Journal of Environment and Pollution.* 1995; **5**: 338–349.

Piringer M. and Joffre S. The urban surface energy budget and mixing height in European cities: data, models and challenges for urban meteorology and air quality–Final report of Working Group 2 of COST Action 715 (2005) ISBN 954-9526-29-1. Demetra Ltd. Publisheers. Printed in Bulgaria.

Piringer M. and Baumann-Stanzer K. Selected results of a model validation exercise. *Advances in Science and Research.* 2009; **3**: 13–16.

Piringer M. and Baumann K.U. Modifications of a valley wind system by an urban area – Experimental results. *Meteorology and Atmospheric Physics.* 1999; **71**: 117–125.

Piringer M., Joffre S., Baklanov A. *et al.* The surface energy balance and the mixing height in urban areas – Activities and recommendations of COST-Action 715. *Boundary-Layer Meteorology.* 2007; **124**: 3–24.

Piringer M., Baumann K. and Langer M. Summertime mixing heights at Vienna, Austria, estimated from vertical soundings and by a numerical model. *Boundary-Layer Meteorology.* 1998; **89**: 25–45.

Piringer M., Petz E., Groehn I. and Schauberger G. A sensitivity study of separation distances calculated with the Austrian Odour Dispersion Model (AODM). *Atmospheric Environment.* 2007; **41**: 1725–1735.

Reuter H.U. Die Ausbreitungsbedingungen von Luftverunreinigungen in Abhängigkeit von meteorologischen Parametern. *Arch. Met. Geoph. Biokl. A.* 1970; **19**: 173–186.

Rühling A. and Lohmeyer A. *Modellierung des Ausbreitungsverhaltens von luftfremden Schadstoffen/Gerüchen bei niedrigen Quellen im Nahbereich* (1998) FuE Vorhaben. Sächsisches Landesamt für Umwelt und Geologie, Radebeul.

Santos J.M., Griffiths R.F., Reis Jr N.C. and Mavroidis I. Experimental investigation of averaging time effects on building influenced atmospheric dispersion under different meteorological stability conditions. *Building and Environment.* 2009; **44**: 1295–1305.

Sarkar U. and Hobbs S.E. Odour from municipal solid waste (MSW) landfills: A study on the analysis of perception. *Environment International.* 2002; **27**: 655–662.

Sarkar U., Longhurst P.J. and Hobbs S.E. Community modelling: A tool for correlating estimates of exposure with perception of odour from municipal solid waste (MSW) landfills. *Journal of Environmental Management.* 2003; **68**: 133–40.

Schauberger G., Piringer M. and Petz E. Weighting of odour sensation by the time of the day and time of the year to improve the reliability of the calculated separation distances. Wasser – Abwasser – Umwelt, Schriftenreihe des Fachgebiets Siedlungswasserwirtschaft der Univ. Kassel. 2009; **31**: 160–168.

Schauberger G., Piringer M. and Petz E. Odour episodes in the vicinity of livestock buildings: A qualitative comparison of odour complaint statistics with model calculations. *Agriculture, Ecosystems and Environment.* 2006; **114**: 185–194.

Schauberger G., Piringer M., Jovanovic O. and Petz E. A new empirical model to calculate separation distances between livestock buildings and residential areas applied to the Austrian guideline to avoid odour nuisance. *Atmospheric Environment.* 2012; **47**: 341–347.

Schatzmann M., Olesen H.R. and Franke J. COST 732 model evaluation case studies: approach and results (2010) COST Office, Brussels.

Schauberger G., Piringer M. and Petz E. Diurnal and annual variation of the sensation distance of odour emitted by livestock buildings calculated by the Austrian odour dispersion model (AODM). *Atmospheric Environment.* 2000; **34**: 4839–4851.

Schauberger G., Piringer M. and Petz E. Calculating direction-dependent separation distance by a dispersion model to avoid livestock odour annoyance. *Biosystems Engineering.* 2002; **82**: 25–37.

Schauberger G., Piringer M., Schmitzer R., *et al.* Concept to assess the human perception of odour by estimating short-time peak concentrations from one-hour mean values. Reply to a comment by Müller *et al. Atmospheric Environment.* 2012; **54**: 624–628.

Schauberger G., Schmitzer R., Kamp M., *et al.* Empirical model derived from dispersion calculations to determine separation distances between livestock buildings and residential areas to avoid odour nuisance. *Atmospheric Environment.* 2012; **46**: 508–515.

Schauberger G. and Piringer M. Assessment of the odour concentration in the near-field of small sources. *VDI-Berichte,* **1850**: 2004; 343–352 + 576.

Seibert P., Beyrich F., Gryning S.E. *et al.* Review and intercomparison of operational methods for the determination of the mixing height. *Atmospheric Environment.* 2000; **34**: 1001–1027.

Seinfeld, J.H. and Pandis, S.N. (1998) *Atmospheric Chemistry and Physics.* John Wiley & Sons, Inc., New York.

Sheridan B.A., Hayes E.T., Curran T.P. and Dodd V.A. A dispersion modelling approach to determining the odour impact of intensive pig production units in Ireland. *Bioresource Technology*. 2004; **91**: 145–152.

Smith M.E. *Recommended Guide for the Prediction of the Dispersion of Airborne Effluents* (1973) ASME, New York.

Sucker K., Both R., Bischoff M. *et al.* Odor frequency and odor annoyance Part II: Dose – Response associations and their modification by hedonic tone. *International Archives of Occupational and Environmental Health*. 2008; **81**: 683–694.

TA Luft Erste Allgemeine Verwaltungsvorschrift zum Bundes–Immissionss-chutzgesetz. (Technische Anleitung zur Reinhaltung der Luft – TA Luft) (2002).

van Langenhove H. and De Bruyn G. Development of a procedure to determine odour emissions from animal farming for regulatory purposes in Flanders. *Water Science and Technology*. 2001; **44**: 205–210.

van Langenhove H. and van Broeck G. Applicabiliby of sniffing team observations: Experience of field measurements. *Water Science and Technology*. 2001; **44**: 65–70.

VDI 3782 Part 1. Environmental meteorology – Atmospheric dispersion models – Gaussain plume model for air quality management (2009).

VDI 3940 Part 1. *Measurement of odour impact by field inspection – Measurement of the impact frequency of recognizable odours* – Grid measurement (2006) Beuth Verlag, Berlin.

VDI 3940 Part 2. *Measurement of odour impact by field inspection – Measurement of the impact frequency of recognizable odours* – Plum measurement (2006) Beuth Verlag, Berlin.

Wallis I. and Oma R. Integrated Odour Management at Western Australian Wastewater Treatment Plants. *Water Practice & Technology*. 2009; **4** (2): 1–14.

Wallis I. and Cadee K.U. Odour exposure criteria and odour modelling in Western Australia. *Water*. 2008; **35**: 144–148.

Watts P. and Sweeten J. Toward a better regulatory model for odour. Feedlot Waste Management Conference. (1995).

Yang G. and Hobson J. Odour nuisance – advantages and disadvantages of a qualitative approach. *Water Science and Technology*. 2000; **41**: 97–106.

Yu Z., Guo H., Xing Y. and Lagué C. Setting acceptable odour criteria using steady-state and annual hourly weather data. *Biosystems Engineering*. 2009; **103**: 329–337.

Zhang Q., Feddes J.J.R., Edeogu I.K. and Zhou X.J. Correlation between odour intensity assessed by human assessors and odour concentration measured with olfactometers. *Canadian Biosystems Engineering/Le Genie des biosystems au Canada*. 2002; **44**, 6.27–26.32.

Zhang Q., Zhou X.J., Guo H., Li Y.X. and Cicek N. Odour and greenhouse gas emissions from hog operations. Final Report submitted to the Manitoba Livestock Manure Management Initiative Inc. Project MLMMI. (2005).

Part 6

Odour Regulation and Policies

S. Sironi, L. Capelli, L. Dentoni and R. Del Rosso

Politecnico di Milano, Department of Chemistry, Material and Chemical Engineering, Italy

6.1 Introduction

The regulation of industrial odour emissions has been an issue in industrialized nations for several decades now, particularly since rapidly increasing population density in certain regions resulted in either the building of new residential areas alongside existing industrial facilities or, vice versa, new industrial plants in already settled areas.

Odour is a pollutant that can have a significant negative impact on both the quality of life and economic activity, by, for instance, affecting property values or tourism revenues. Overall, however, it poses little threat to human health in the generally-accepted meaning of the term. This is why industrial odour emissions in the United States have only ever been locally rather than federally regulated. With the exception of Germany, Australia and the United States, it is only in recent years that olfactometry or the measurement of ambient odour concentration has been nationally and internationally standardized. The introduction and standardization of olfactometry has served to overcome the subjective nature of odour perception, thus permitting the issuing of specific control regulations.

There have been various different legislative responses to the need to protect air quality from being affected by industrial odour emissions in industrialized nations. Independent of the measurement methods used, the approaches that can be adopted to regulate odour-related problems can be classified into four different categories (Frechen, 2001).

- *Regulations based on air quality standards and limit values:* Different levels of action can be adopted. The first level includes regulations focusing on protecting general air quality without any specific provision for odour emissions. These are the oldest kind of regulations and impose a qualitative limit on odour emissions even though they do not

Odour Impact Assessment Handbook, First Edition. Edited by Vincenzo Belgiorno, Vincenzo Naddeo and Tiziano Zarra.
© 2013 John Wiley & Sons, Ltd. Published 2013 by John Wiley & Sons, Ltd.

set any objective criterion for establishing when odour annoyance from industrial sources becomes unacceptable. Despite this vagueness, these regulations can still give the (generally local) authorities the power to close a facility because of the problems its emissions are causing. The second level of action includes regulations aimed at defining minimum distance criteria or Minimum Distance Standards (MDS). The latter specify the minimum distance from the closest inhabited area that possible odour-producing industrial or agricultural facilities (grouped according to sector) can be located. The mDS are based on a purely empirical 'prediction of olfactory annoyance' for each type of industrial process. The public authority may only use the mDS to ascertain that a facility will not cause any olfactory annoyance at the design stage. Lastly, the third and most specific level of action includes directives aimed at defining Maximum Emission Standards (MES) which set out the emissions limits for the odour sources expressed in odour concentration, as an odour emissions rate or a combination of both. Compliance with these standards must be verified at both the design (e.g. granting of licences/permissions) and operational phases, through periodic emissions analysis. The extent to which the MESs can be enforced depends on the presence of recognised standards for emissions sampling and analysis methods (CEN 13725, 2003; VDI 3880, 2011). Verification of MES throughout the authorization procedure means that the public authority must investigate the processes and equipment used to contain emissions.

- *Regulations on direct exposure assessment:* These regulations define maximum impact criteria (Maximum Impact Standards, MIS) and contain provisions aimed at the overall protection of air quality through the definition of odour impact limits at sensitive receptors. The impact limits may vary according to the type of area being impacted; for instance, urban or artisanal, densely or sparsely populated. The MIS provide the public authority with more detailed insight into source-receptor interaction and allow it to set more restrictive limits where very special air quality protection is required (for instance, in tourist or densely populated areas). In most cases, these directives are based on the application of suitable atmospheric dispersion models which allow for the evaluation of ground level odour concentrations. Thus, in such cases, the applicability of the MIS is dependent not only on the existence of olfactometry standards, but also the agreed reliability of the dispersion models as well as the availability of appropriate meteorological data, at both the design and operational stages. Alternatively, the MIS can be applied on the basis of suitable direct odour measurement (field inspections).

- *Regulations based on no-annoyance:* These regulations aim to guarantee maximum satisfaction to the population potentially affected by emissions (whether odour-related, noise-related and caused by other agents deemed damaging to air quality) originating from one or more facilities. Specifically, these directives define the Maximum Annoyance Standards (MAS). With regard solely to odour, MAS are based, at least concerning authorization for a plant, not only on the ability to calculate the exposure of the population to the odour released, but also on the relationship between exposure and annoyance, which is difficult to ascertain with any certainty. When the facility is operational, however, the MAS can be verified by using psychometric methods (Longhurst, 2001). These methods will involve, for instance, the distribution of questionnaires to people living in the vicinity of the facility with the aim of establishing the level of annoyance the odour-related emissions cause (Sucker *et al.*, 2009). These methods need to be subject to standards, but thus far only Germany has made provisions in this regard.

- *Regulations based on best practice:* These directives aim to define best practice for processes and protection. Legislation has been put in force in many nations that has resulted in the establishment of the best practices for process techniques and protection in relation to deodorizing exhaust gases. Conformity of the plant to established parameters is often essential to the granting of integrated environmental authorization by the various national and regional authorities.

6.2   Regulation Based on Air Quality Standards and Limit Values

Regulations aimed at protecting overall air quality standards contain provisions that apply to all plants generating odours of any kind that alter normal air quality, but do not set any specific limit on odour emissions. In this sense, olfactory pollution is deemed to be a modification of air quality that in turn alters health or environment.

Various standards, all aimed at containing emissions from existing facilities, have been drawn up in various countries and emissions limits for certain specific substances (ammonia, hydrogen sulfide, particulate matter, VOCs, etc.) and certain types of facilities (for instance, waste and wastewater treatment plants).

Directives aimed at defining minimum distance standards were drafted in the 1980s and 1990s when various European nations (Austria, Belgium, Germany, Netherlands, Switzerland) adopted mDS to reduce olfactory annoyance caused by intensive animal farming facilities. In certain cases, the minimum distances were set and tabulated, by taking into account the use (e.g. residential or agricultural area) or the residential density of the area in which the facility was located (JORF, 2005; VROM, 1996). In other cases, the minimum separation distance was not tabulated but calculated using specific mathematical expressions, for instance power functions, the coefficients and exponents of which depend on the number of animals and empirical parameters regarding prevailing meteorological conditions, animal housing (e.g. forced or natural ventilation) and other characteristics of the facility and the site (Schauberger *et al.*, 2012; Piringer and Schauberger, 1999; Grimm *et al.*, 1999).

Lastly, regulations aimed at defining Maximum Emissions Standards (MES) provided a major leap forward in the odour and air quality legislation field. The prerequisite for the issuing of a MES is, of course, the choice or the definition of an olfactometry standard. In this regard, huge impetus has come from the publication of American (ASTM E679-04, 2011) and European (CEN 13725, 2003) standards as well as the recent German guidelines (VDI 3880, 2011).

One sector in which the public authorities have been quick to intervene with MES has been that of industrial composting plants because, even though their emissions are generally neither toxic nor noxious, they can result in olfactory annoyance and be the source of complaints from the population living in their vicinity. For this reason, monitoring of these facilities has focused on odour emissions rather than individual polluting compounds. Furthermore, because the odour sources in these facilities are either point sources or area sources with outward flow, sampling for olfactometric analysis is simpler and less scientifically controversial than at other plants that typically cause olfactory annoyance (such as landfills or wastewater treatment plants). Lastly, composting plants use relative simple technologies that do not vary much from one to another. Thus, thanks to the data accumulated in the past through the monitoring of other plants, it is reasonably possible to predict what configuration will be required in order to guarantee that a new plant design will comply with a established exhaust air emissions limits.

Examples of MES applied to composing plants include the S 2205-1 standard in Austria (ONORM S 2205-1, 1997) and the Region of Lombardy's 'Guidelines for the construction and operation of compost-producing plants' (DGR Lombardia n.7/12764, 2003) in Italy. Both set the upper limit of odour concentration in emissions from treatment systems at 300 ou_E/m^3. The Austrian standard also stipulates a limit to the overall odour emission rate at 5000 ou_E/s; any facilities exceeding the aforementioned limit must present an expert appraisal that demonstrates compliance with the MIS.

Another MES-type approach that is quite different to those outlined above comes from Switzerland where the federal authorities have set particularly strict limits on around 150 individual compounds to simultaneously reduce toxic and olfactory annoyance effects (Frechen, 2001).

6.3   Regulation Based on Direct Exposure Assessment

The MIS picture worldwide is quite complex. MIS are based on limits fixed at receptors. Compliance with these odour impact limits can be verified using two different approaches: the simulation of odour impact using suitable atmospheric dispersion models or direct exposure assessment through field inspections.

With regard to MIS based on dispersion model application, the acceptability standards are usually expressed in terms of the frequency with which a given odour concentration is exceeded. *Integrated Pollution Prevention and Control (IPPC) – Horizontal Guidance for Odour Part 1 – Regulation and Permitting* published by the Environmental Agency of the United Kingdom (PC H4 IP/Part 1, 2002) is a good example of this. The approach it takes is to establish exposure criteria in terms of ground level odour concentration at the 98th percentile, that is, the maximum odour concentration that may only be exceeded by 2% of the hours in a year. The limits set by the guidelines are expressed in terms of hourly average odour concentration values at the 98th percentile, and are differentiated on the basis of the level of potential olfactory annoyance ('low', 'medium' or 'high') associated with the industrial category under consideration (Table 6.1).

Table 6.1   *Exposure criteria in terms of ground level odour concentration as a 98th percentile, in the United Kingdom.*

Relative 'offensiveness' of odour	Indicative Criterion
HIGH	1.5 ou_E/m^3
(e.g. activities involving putriscible waste, processes involving animal or fish remains, wastewater treatment, oil refining)	98th percentile
MEDIUM	3.0 ou_E/m^3
(e.g. intensive livestock rearing, fat frying, sugar beet processing)	98th percentile
LOW	6.0 ou_E/m^3
(e.g. chocolate manufacture, brewery, fragrance and flavourings, coffee roasting, bakery)	98th percentile

In this specific case, these limits are set at 1.5 ou_E/m^3, 3.0 ou_E/m^3 or 6.0 ou_E/m^3 (at the 98th percentile), in the case of industrial categories that are, respectively, considered to have a 'high', 'medium' or 'low' annoyance potential, on the basis of the concentration and quality of the odour emitted.

The Australian state of New South Wales has drawn up very detailed guidelines with regard to the containment of odour nuisance (NSW Framework, 2006; NSW Notes, 2006). In relation to maximum impact standards, they specify that the 99th percentile of ground level concentration must not exceed a threshold that varies between 2–7 ou_E/m^3 as population density of the area affected decreases. New Zealand, which is broadly in line with New South Wales with regard to the olfactometric and impact estimation techniques, has also set the maximum odour exposure limit at 5 ou_E/m^3 at the 99.5th percentile (Freeman *et al.*, 2000). In Belgium, no uniform odour-related standards have been established to protect air quality, but many licences for the construction of potentially odour-producing facilities do specify that the 98th percentile of the ground level concentrations must not exceed a threshold level of 1 ou_E/m^3 at sensitive receptors outside the facilities.

In Italy, on the other hand, guidelines are only at the publication stage at regional level (Regione Lombardia, 2012). The Region of Lombardy guidelines for the characterization and authorization of atmospheric gaseous emissions from activities with an odour impact specify that impact maps must be drawn up indicating annual peak odour concentration values at the 98th percentile, as resulting from atmospheric emission dispersion simulation at 1, 3 and 5 ou_E/m^3.

In certain cases, the legislation in force is not horizontal that is, not applicable to all types of plants, even though specific regulations do exist for, for instance, composting or bio-stabilization plants (France) or intensive livestock facilities (Netherlands). With regard to France, article 26 of the document that sets the technical regulations for composting and aerobic bio-stabilization plants (JORF, 2008) specifies that the concentration of odour detectable at the plant, assessed via an impact survey in the human-occupation zones within a 3000 m radius of the plant, must not exceed 5 ou_E/m^3 for more than 175 hours per year which is the same as 2% of the hours of the year (i.e. 98th percentile).

In the Netherlands, for categories of animals for which odour emission factors are established in ministerial decrees (e.g. pigs and poultry), the limits are set on the basis of odour emissions assessments. Specifically, maximum emissions levels depend on the density of such plants in the area under consideration ('concentrated area' or 'non-concentrated area') as well as whether the zone in which the emissions occur is inside or outside a built-up area (Bongers, 2009; VROM, 2008). Emissions limits are, once again, expressed as odour concentration at the 98th percentile, and are listed in Table 6.2.

In other cases, MIS exist that set not the concentration that may be exceeded only for a given number of hours in the year but the maximum odour concentration that a source may cause outside the confines of the facility itself. This is the case with the State of Massachusetts (US), where odour concentration beyond the confines of effluent treatment facilities must never exceed 5 D/T, where D/T is the dilution factor at the perception threshold in line with ASTM E 679-91 (Frechen, 2001).

With regard to MIS based on odour assessment carried out not through mathematical modelling but actual field inspections, reference can be made to the German guidelines on odour emissions, dated 13 May, 1998 *GIRL – Geruchsimmissions-Richtlinie* (GIRL, 1998). These guidelines set the maximum number of 'odour hours' that may be perceived by the

Table 6.2   Odour emissions limits for livestock facilities in the
Netherlands (VROM 2006/2008).

	Odour concentration [ou_E/m^3 as a 98th percentile]	
	Built-up area	Outside built-up area
Concentration area	3	14
Outside concentration area	2	8

population surrounding an industrial area, at 15% per year. The limit for industrial plants in residential or mixed areas, however, is 10%. This standard is based on field inspections for which a panel of specialists are asked to rate their olfactory sensations every 10 s for a minimum time of 10 min at the location being tested. One odour hour is defined as an hour during which odour was detected for more than 10% of its duration, that is, for more than 6 min.

   As happened in the case of EN 13725 (CEN 13725, 2003), that is, the European Dynamic Olfactometry Standard, which was drawn up to closely mirror the German VDI 3881 guidelines (VDI3881, 1986), the aforementioned German guidelines (VDI3940/1, 2006; VDI3940/2, 2006) on field inspection acted as a precursor to the draft of a European field inspection standard. In fact, CEN (Comité Européen de Normalisation) will soon be publishing a standard that will regulate field inspections to determine the olfactory impact of odour emissions sources on surrounding areas. The standard currently being debated regulates the execution of on-site Plume Inspections using panellists to determine the exteny of odour plumes relating to specific odour-producing emissions.

6.4   Regulation Based on 'No Annoyance'

As there is still very little available literature that investigates the relationship between exposure and annoyance, the MAS issued thus far deal only with exposure from the viewpoint of the receptor, that is, their aim is to reduce the number of complaints from the population. Analogous and alternative objectives include reducing the percentage of the population that feel the annoyance level is serious, or increasing the ratio between the number of people who, having made complaints in the past, now feel satisfied, and the total number of people who make complaints (Longhurst, 2001). To achieve these objectives, 'annoyances indexes' need to be defined and demoscopic (or psychometric) surveys of the population need to be carried out in order to quantify annoyance impact. It is also important to identify the sources of the annoyance and, in the case of industrial odour emissions, make any necessary modifications to the equipment and variables involved in the odour-generating process.

   In Switzerland, the annoyance index is measured on a scale of 1–10 as a function of the percentage of people claiming that the annoyance level is serious (Frechen, 2001). The public authorities in the Netherlands set a goal of reducing the percentage of people who feel they are subject to industrial-related annoyance, to 12% and 3% the percentage of people who consider the annoyance serious (Frechen, 2001). These percentages are calculated, year after year, as follows: a panel made up of people residing in the vicinity of the facility in question is formed. These individuals are invited to sniff the air in the area at a specific time on a specific day and then report whether they can detect any odour or not. If they do detect odour, they must then rate it on a scale ranging from 'no annoyance' to 'extreme annoyance'.

6.5   Regulation Based on Application of Best Practice

Various directives (BAT, IPPC, BREFs, NeR) have been published in recent years which have forced industrial facilities to equip themselves with systems that will contain emissions as much as possible.

The goal of the European directive on integrated pollution prevention, Directive 96/61/CE (Council Directive 96/61/EC, 1996), is to reduce emissions to all areas of the environment (air, water, soil, waste) and sets out the criteria for the issuing of integrated authorization, for the various kinds of emissions, for new and existing plants as well as the operating criteria for the latter. The directive applies to specific categories of facilities which are listed in an annex of the norm. One of the conditions of authorization is that Best Available Techniques (BAT) for pollution prevention must be applied. Predictably, for at least some production sectors, measures aimed at preventing and/or reducing odour-related emissions will be taken into consideration when drawing up the aforementioned guidelines.

The facilities targeted by the directive include, in fact, those of dimensions larger than a specific threshold and the following types which may produce odorous substances:

- Chemical plants;
- Hazardous waste recovery and disposal facilities;
- Landfills;
- Paper mills;
- Tanneries;
- Slaughterhouses;
- Food industries;
- Animal by-product collection and disposal facilities (rendering plants);
- Poultry and pig livestock facilities; and
- Surface treatment plants (that use organic solvents).

In accordance with the IPPC Directive (Council Directive 96/61/EC, 1996), permit conditions must include emission limit values (ELV) for pollutants and waste minimization measures, in relation to results achievement by applying BAT.

Another type of best practice regulation is based on setting a limit in terms of minimum abatement efficiency of the installed odour control systems. An example of this is the *Resoluçao Sema N° 054* in the state of Paraná in Brazil (SEMA n.054, 2006).

Article 12 of the aforementioned directive declares that activities that generate odorous substances with odour emission rates exceeding $5 \cdot 10^6$ ou$_E$/h (equivalent to approx. 1400 ou$_E$/s), must install odour capture and removal systems that will reduce odour by over 85% (based on olfactometric analysis). The type of system to be installed will depend on

Table 6.3   *Regulations and laws covering industry-related odour emission control in European and non-European countries.*

Country	Regulation	Contents
Germany	BImSchG-5/90 Bundesimmisionschutzgesetz TA Luft Technische Anleitung zur Reinhaltung der Luft (GIRL) (GIRL, 1998)	Limits to odour emissions and technical guidelines for air pollution prevention. Indications on methods for odour measurement and impact assessment.
	VDI 3880 (VDI 3880, 2011)	Procedures for static sampling
	VDI 3882 – Part 1 (VDI3882/1, 2008)	Procedure for the measurement of odour intensity
	VDI 3882 – Part 2 (VDI3882/2, 2008)	Procedures for the measurement of odour hedonic tone
	VDI 3477 (VDI3477, 2004)	Criteria for designing and running biofilters for the treatment of odorous air flows
	VDI 3478 – Part 1 (VDI3478/1, 2011)	State of the art of applications and criteria for the design of bioscrubbers as odour abatement systems
	VDI 3478 – Part 2 (VDI3478/2, 2008)	State of the art of applications and criteria for the design of biological trickle bed reactors for effluent treatment
	VDI 3883 – Part 1 (VDI3883/1, 2003)	Determination of annoyance parameters by questioning
	VDI 3883 – Part 2 (VDI3883/2, 2008)	Determination of Annoyance Parameters by Questioning Repeated Brief Questioning of Neighbour Panellists
	VDI 3940 – Part 1 (VDI3940/1, 2006)	Measurement of odour impact by field inspection – Measurement of the impact frequency of recognizable odours – Grid measurement
	VDI 3940 – Part 2 (VDI3940/2, 2006)	Measurement of odour impact by field inspection – Measurement of the impact frequency of recognizable odours – Plume measurement
France	Code Permanent Environnement et Nuisances (CPEN, 1987)	Limits for odour pollution (in terms of concentrations and fluxes)
	jorf 2005 (JORF, 2005)	Measures for reducing odour emissions from specific livestock operations
	jorf 2008 (JORF, 2008)	Reduction of odour emissions from composting plants. Composting plants odour impact should be evaluated by application of a dispersion model and comparison with specific limits. Periodic controls should be run in order to verify the plant actual odour impact.

Table 6.3 (Continued)

Country	Regulation	Contents
	NF X 43-101 (NF X 43-101, 1986)	Best practice for odour measurement of a gaseous flow and determination of the dilution factor at the odour detection threshold
	NF X 43-103 (NF X 43-103, 1996)	Best practices for odour measurement of a gaseous flow by means of supraliminary methods
	NF X 43-104 (NF X 43-104, 1984)	Best practices for odour emission sampling
USA	ASTM E1432 – 04 (E1432-04, 2011)	Standard Practice for Defining and Calculating Individual and Group Sensory Thresholds from Forced-Choice Data Sets of Intermediate Size
	ASTM E679 – 04 (ASTM E679-04, 2011)	Standard Practice for Determination of Odor and Taste Thresholds By a Forced-Choice Ascending Concentration Series Method of Limits
	ASTM E544 – 10 (E544-10, 2010)	Standard Practices for Referencing Suprathreshold Odor Intensity
Austria	S 2205-1 (ONORM S 2205-1, 1997)	Technical requirements for composting plants. Limit of 300 ou_E/m^3 at biofilter outlets. Limit of 5000 ou_E/s for plants with airflows up to 60 000 m^3/h. For plants with higher airflows, limits should be fixed case by case.

local dispersion conditions, the proximity of residential areas as well as the quantity of odour-causing substances emitted (quantified using olfactometry and expressed in odour units per hour) (SEMA n.054, 2006).

6.6 Comparison of Different Regulatory Approaches

The regulations and laws covering industry-related odour emission control currently in force in European and non-European countries are listed in Table 6.3.

References

AFNOR, Normalisation Francaise NF X 43-104, *Qualité de l'air. Atmosphères odorantes. Méthodes de prélèvement*, Paris (1984).

AFNOR, Normalisation Francaise NF X 43-101, *Qualité de l'air. Mesurage de l'odeur d'un effluent gazeux. Détermination du facteur de dilution au seuil de perception*, Paris (1986).

AFNOR, Normalisation Francaise NF X 43-103, *Qualité de l'air. Mesurage de l'odeur d'un effluent gazeux. Méthode supraliminaire*, Paris (1996).

ASTM Standard E544 – 10, Standard Practices for Referencing Suprathreshold Odor Intensity, *ASTM International*, West Conshohocken, PA (2010).

ASTM Standard E679 – 04, 2011, *Standard Practice for Determination of Odor and Taste Thresholds By a Forced-Choice Ascending Concentration Series Method of Limits*, ASTM International, West Conshohocken, PA (2011).

ASTM Standard E1432 – 04, 2011, Standard Practice for Defining and Calculating Individual and Group Sensory Thresholds from Forced-Choice Data Sets of Intermediate Size, *ASTM International*, West Conshohocken, PA (2011).

Austrian Standards ONORM S 2205-1:1997 04 01, *Technische Anforderungen an Kompostierungsanlagen zur Verarbeitung von mehr als 3000 t pro Jahr – Bioabfall aus Haushalten*, Austrian Standards plus GmbH, Wien, Austria (1997).

Bongers M.E., Recent developments in odour nuisance policy for livestock farming in The Netherlands, in *Odour and VOCs: Measurement, Regulation and Control*, Frechen F.B. (ed.), Kassel University Press, Kassel, Germany (2009).

CEN, EN 13725, Air quality – *Determination of odour concentration measurement by dynamic olfactometry*, Comité Européen de Normalization, Brussels (2003).

Code Permanent Environnement et Nuisances, *Editions Législatives et Administratives*, Paris, France (1987).

Council of the European Union, Council Directive 96/61/EC of 24 September 1996 concerning integrated pollution prevention and control, *Official Journal L 257*, 0026-0040 (1996).

Frechen F.B., Regulations and policies, in R. Stuetz and F.B. Frechen (Eds.), *Odours in Wastewater Treatment: Measurement, Modelling and Control*, IWA Publishing, London (2001).

Freeman T., Needham C. and Schulz T., *Analysis of Options for Odour evaluation for industrial or trade processes*, Auckland Regional Council, CH2M BECA Ltd, Auckland (2000).

Geruchsimmissionen-Richtlinie (GIRL), BImSchG-5/90 Bundesimmisionschutzgesetz TA Luft Technische Anleitung zur Reinhaltung der Luft. Feststellung und Beurteilung von Geruchsimmissionen, Berlin, Germany (1998).

Grimm E., Kypke J., Martin I. and Krause K.H., German Regulations on Air Pollution Control in Animal Production, in *Proceedings of the Regulation of Animal Production In Europe International Congress*, KTBL, Wiesbaden, Germany, pp. 234–242 (1999).

Journal Officiel de la République Française (JORF), Arrêté du 22 Avril 2008 fixant les règles techniques auxquelles doivent satisfaire les installations de compostage ou de stabilisation biologique aérobie soumises à autorisation en application du titre Ier du livre V du code de l'environnement, JORF (2008).

Journal Officiel de la République Française (JORF), Arrêté du 7 Février 2005 fixant les règles techniques auxquelles doivent satisfaire les élevages de bovins, de volailles et/ou de gibier à plumes et de procs soumis à déclaration au titre du livre V du code de l'environnement, JORF (2005).

Longhurst P., Monitoring nuisance and odour modelling, in Stuetz R. and Frechen F.B. (eds.), *Odours in Wastewater Treatment: Measurement, Modelling and Control*, IWA Publishing, London (2001).

NSW Environment Protection Authority, *Technical Framework – Assessment and Management of Odour from Stationary Sources in NSW*, Sydney (2006).

NSW Environment Protection Authority, *Technical Notes – Assessment and Management of Odour from Stationary Sources in NSW*, Sydney (2006).

Piringer M. and Schauberger G., Comparison of a Gaussian diffusion model with guidelines for calculating the separation distance between livestock farming and residential areas to avoid odour annoyance, *Atmospheric Environment*, **33**, 2219–2228 (1999).

Regione Lombardia, *Linee guida relative alla costruzione e all'esercizio di impianti di produzione di compost* (Guidance for the construction and operation of compost-producing plants), DGR 16 April, 2003 n. 7/12764, Milan (2003).

Regione Lombardia, Linea guida per la caratterizzazione e l'autorizzazione delle emissioni gassose in atmosfera, http://www.reti.regione.lombardia.it/cs/Satellite?c=Redazionale_P&childpagename=DG_Reti%2FDetail&cid=1213355419039&pack edargs=NoSlotForSitePlan%3Dtrue%26menu-to-render%3D1213355318948&pag ename=DG_RSSWrapper (accessed 6 February, 2012) (2012).

Schauberger G., Schmitzer R., Kamp M., *et al.* Empirical model derived from dispersion calculations to determine separation distances between livestock buildings and residential areas to avoid odour nuisances, *Atmospheric Environment*, **46**, 508–515 (2012).

Secretaria de Estado do Meio Ambiente e Recursos Hídricos (SEMA), Resoluçao Sema N° 054, de 22 de dezembro de 2006, *Estabelece padrões de emissões atmodféricas*, Curitiba, Paranà, Brazil (2006).

Sucker K., Both R. and Winneke G., Review of adverse health effects of odours in field studies, *Water Science and Technology*, **59** (7), 1281–1289 (2009).

Technical Division Environmental Protection Technologies – Verein Deutscher Ingenieure, VDI 3477, *Biological waste gas purification – Biofilters*, Verein Deutscher Ingenieure e.V. VDI Guidelines Department, Dusseldorf, Germany (2004).

Technical Division Environmental Protection Technologies – Verein Deutscher Ingenieure, VDI 3478 – Blatt 2, *Biological waste gas purification – Biological trickle bed-reactors*, Verein Deutscher Ingenieure e.V. VDI Guidelines Department, Dusseldorf, Germany (2008).

Technical Division Environmental Protection Technologies – Verein Deutscher Ingenieure, VDI 3478 – Blatt 1, *Biological waste gas purification – Bioscrubbers*, Verein Deutscher Ingenieure e.V. VDI Guidelines Department, Dusseldorf, Germany (2011).

Technical Division Environmental Quality – Verein Deutscher Ingenieure, VDI 3881, *Olfactometry – Odour threshold determination*, Verein Deutscher Ingenieure e.V. VDI Guidelines Department, Dusseldorf, Germany (1986).

Technical Division Environmental Quality – Verein Deutscher Ingenieure, VDI 3883 – Blatt 1, *Effects and assessment of odours – Psychometric assessment of odour annoyance – Questionnaires*, Verein Deutscher Ingenieure e.V. VDI Guidelines Department, Dusseldorf, Germany (2003).

Technical Division Environmental Quality – Verein Deutscher Ingenieure, VDI 3940 – Blatt 1, *Measurement of odour impact by field inspection – Measurement of the impact frequency of recognizable odours – Grid measurement*, Verein Deutscher Ingenieure e.V. VDI Guidelines Department, Dusseldorf, Germany (2006).

Technical Division Environmental Quality – Verein Deutscher Ingenieure, VDI 3940 – Blatt 2, *Measurement of Odour Impact by Field Inspection – Measurement of the Impact Frequency of Recognizable Odours – Plume measurement*, Verein Deutscher Ingenieure e.V. VDI Guidelines Department, Dusseldorf, Germany (2006).

Technical Division Environmental Quality – Verein Deutscher Ingenieure, VDI 3882 – Blatt 1, *Olfactometry; determination of odour intensity*, Verein Deutscher Ingenieure e.V. VDI Guidelines Department, Dusseldorf, Germany (2008).

Technical Division Environmental Quality – Verein Deutscher Ingenieure, VDI 3882 – Blatt 2, *Olfactometry; determination of hedonic odour tone*, Verein Deutscher Ingenieure e.V. VDI Guidelines Department, Dusseldorf, Germany (2008).

Technical Division Environmental Quality – Verein Deutscher Ingenieure, VDI 3883 – Blatt 2, *Effects and assessment of odours; determination of annoyance parameters by questioning; repeated questioning of neighbour panellists*, Verein Deutscher Ingenieure e.V. VDI Guidelines Department, Dusseldorf, Germany (2008).

Technical Division Environmental Quality – Verein Deutscher Ingenieure, VDI 3880, *Olfactometry. Static Sampling*, Verein Deutscher Ingenieure e.V. VDI Guidelines Department, Dusseldorf, Germany (2011).

UK Environmental Agency, Technical Guidance PC H4 IP Integrated Pollution Prevention and Control (IPPC), *Horizontal Guidance for Odour Part 1 – Regulation and Permitting*, Bristol (2002).

VROM, Richtlijn Veehouderij en Stankhinder (Regulation on livestock farming and stench nuisance) (1996).

VROM, *Wet geurhinder en veehouderij* (Odour nuisance and livestock farming act), Ministry of Public Planning and the environment, The Hague, The Netherlands (2006/2008).

Part 7

Procedures for Odour Impact Assessment

V. Naddeo, V. Belgiorno and T. Zarra
Sanitary Environmental Engineering Division (SEED), Department of Civil Engineering,
University of Salerno, Italy

7.1   Introduction

Odour impact is defined as the alteration of the air quality in terms of odours that cause nuisances. The factors contributing to nuisance and then odour impacts are multiple and in depth discussed in the following sections (Nicell, 2009).

Odour impacts can be measured or assessed respectively by a monitoring plan at the receptors or by dispersion modelling from sources. Measurement of odour exposure can obviously only be done for existing activities, in this case we talk of odour impact assessment from exposure measurement. The evaluation of odour impact could also be carried out through the interpretation of results by dispersion modelling; in this case we generally talk of odour impact assessment from sources. In this section, we explore both approaches with relative tools; and for each one the advantages, disadvantages, situations where it is useful and where it is inappropriate.

Odour mitigation measures as well as the general guidelines for monitoring plan are also discussed in the following in terms of tools for support odour impact studies.

7.2   Factors Contributing to Odour Impact

Odour impact depends on a number of factors. There is no single method of reliably measuring or assessing odour impacts, and any conclusion is best based on a number of

Odour Impact Assessment Handbook, First Edition. Edited by Vincenzo Belgiorno, Vincenzo Naddeo and Tiziano Zarra.
© 2013 John Wiley & Sons, Ltd. Published 2013 by John Wiley & Sons, Ltd.

pieces of evidence. The FIDOL acronym is a useful reminder of the main factors that will determine the significance of odour impacts:

- Frequency of detection (F);
- Intensity as perceived (I);
- Duration of exposure (D);
- Offensiveness (O);
- Location (L).

The FIDOL factors encompass the pattern of odour impacts and the receiving environment where these occur (Nicell, 2009). These are the main factors that influence the extent to which odours adversely affect individuals, and this information can be utilised to assist with odour investigation and assessment.

All these factors should be considered concurrently. Odours may occur in frequent short bursts or for longer less-frequent periods. However, all of these odour patterns can cause a significant adverse effect; although an odour of high intensity (concentration) occurring for a short period of time is likely to cause a different type of adverse effect to a low-intensity odour occurring almost constantly.

In addition to the FIDOL factors, other cultural issues, perception, and background odours even a person's mental and physical state may need to be considered in assessing the degree of adverse effects.

In the following subsections we will give a concise description of these factors.

7.2.1   Frequency of Detection

Frequency of odour occurrence refers to how often an individual is exposed to odour in the ambient environment (Sucker *et al.*, 2008a). Frequency is influenced by the odour emission source and characteristics, the prevailing wind conditions, the location of the source in relation to the individual affected, and the topography of the area (Sucker *et al.*, 2008b). The frequency of odour exposure is generally greatest in areas that are most often downwind of an odour source, especially under stable conditions with low wind speeds (provided that the odour is not emitted at a significant height above the ground).

7.2.2   Intensity and Duration of Odour

Intensity of odour refers to an individual's perception of its strength. This is different from the odour's character, or quality (Both *et al.*, 2004). The relationship between the perceived strength (or intensity) of an odour and the overall mass concentration of the combined chemical compounds (mg/m^3) was discussed in Section 2.1.3 of this book. Duration of exposure to the odour is related to the type of odour source, the local meteorology and the location of the odour source.

7.2.3   Offensiveness

Offensiveness, or 'hedonic tone', is the subjective rating of pleasantness or unpleasantness of an odour. This double meaning can be confusing: on the one hand offensiveness is one of the FIDOL factors and can be used in relation to a pleasant odour; on the other hand we use the general definition of 'offensive or objectionable odours', where offensive means unpleasant. It may help to remember that this general definition has a much broader meaning, which encapsulates the combined effect of all the FIDOL factors.

The hedonic tone should be considered separately from intensity, and to be a useful parameter it needs to indicate the inherent character of the odour (Gostelow *et al.*, 2001). When assessing the extent of adverse effects, the physical effects of the odour on the affected people should be considered here (as described in Section 2.1.6). The odour source and its associated characteristics (such as dispersion, and impregnation onto surfaces such as skin or washing) also influence how offensive an odour may be to an individual.

7.2.4   Location

Location is an essential factor when assessing the likelihood of adverse effects from odours. It accounts for the type of area in which a potentially affected person environment. These factors determine the likelihood of a person being adversely affected to the point where they find an odour to be offensive or objectionable. The absence or presence of background odours also has a significant effect.

The sensitivity of the receiving environment can generally be categorised according to land use (Van Harreveld, 2001). For example, odours occurring in an industrial area are less likely to be considered offensive by people working on neighboring industrial premises, and if the odours occur at night they are unlikely to have any effect at all unless people are working night shifts. In addition, people who are more likely to be sensitive to odour effects may not be present in such environments.

Conversely, odours occurring in the evenings in residential areas (particularly in summer) could have a significant adverse effect on residents using their backyards for outdoor dining and entertaining, or trying to sleep with windows open (Suffet and Rosenfeld, 2007). People coming to a place for, say, recreation who are not familiar with the odours in the area are also more likely to be adversely affected.

People living in and visiting rural areas generally have a high tolerance for rural type odours, such as from silage or decomposing cow manure, which are acceptable to most rural people and fit the description of a rural odour in a rural area (Littarru, 2007). Some types of odour are quite different to the normally expected rural odours, however, and much less acceptable; for example, odours from rendering plants, wastewater treatment and large-scale intensive factory farms.

Under these aspects, receptor sensitivity (Location) needs to be considered carefully. Some receptors are more sensitive than others. Domestic residences, or a pub with a beer garden are more likely to be sensitive than an industrial complex or passers-by.

7.2.5   Other Factors

In some cases, such as with human sewage treatment, it is the knowledge of the activity generating the odour that causes an offensive reaction in the people smelling the odour. If the people know it is there, and they find the activity itself to be offensive, they are more likely to find the odour itself offensive even if the FIDOL factors would indicate otherwise. In same way, strong odours early in the morning coming from a bakery will be much more acceptable to an Italian rather than strong odours coming from a Chinese restaurant. All these aspects are related to cultural issues.

With regards to perception, if a person who is exposed to an odour associates it with a natural occurrence, such as mudflats, seaweed, or some rural activities, they often do not consider the odour to be offensive or objectionable, so a significant adverse effect is less likely to occur even if the true source of the odour is an industrial activity (Henshaw *et al.*, 2006). Conversely, if the odour is associated with an activity such as wastewater treatment, landfill, composting, industrial production or factory farming, the same odour is more likely to cause an adverse response in people.

In addition we can easily imagine that some individuals will be extremely tolerant of odours at high intensities while others will be unable to tolerate an odour as soon as they identify it. Evidence that, for example, only one person finds the odour unacceptable whereas most others, similarly exposed, find it acceptable in that context (e.g. in a rural village) would be relevant to the assessment of the degree of pollution. In this way, there are a very small number of people, who have conditions which put them well outside the normal range of sensitivities and make them able to detect very low concentrations of odour. We would not expect an operator to design a system to satisfy those individuals, and at same way we would not consider these people in the odour impact assessment procedure.

As already discussed, the background odours are also very important in the assessment of odour impacts. Some environments have a high level of background odours contributing to the overall ambient air; for example, silage and cow manure in rural areas, hydrogen sulfide from volcanic activities, and commercial smells from traffic and restaurants. If a person lives in such an environment, they can become desensitised to these odours, and the addition of other similar odours may not be noticed (unless the odour is strong enough to dominate the background odours).

Similarly, in a community dominated by one or two industries, and where most of the community is employed by that industry, the people who live there are likely to find the odours more tolerable than individuals whose livelihood is not directly or indirectly related to the local industry (Henshaw *et al.*, 2006). Problems can then arise in these situations when visitors or new residents come into an area and they are not conditioned to the background odours, or when a small community begins to grow and become populated by residents with no association with the dominant industry.

7.3   Odour Impact Assessment from Exposure Measurement

Odour impact assessment from exposure measurement could be used only when the odorous activities exist and working causing adverse effects in terms of odour. In this type of application the applicant seeks consent to continue with their current activity in the same manner, without any changes to the ways in which odours are generated and discharged. Evaluation techniques that rely on assessing the degree of significant adverse effect experienced by people occupying land near the activity and a systematic field observation programme is more applicable.

When investigating existing activities that may be causing adverse effects, it is important to determine the type of adverse effect that is most likely to be occurring: whether they are chronic or acute odour effects, due to normal and controlled or uncontrolled releases of emissions to air, respectively. The correct identification of the type of adverse odour effects will help in the selection of the appropriate odour assessment tool. In this category the main assessment tools are:

- sensorial field investigations that include sniff testing (which gives a judgement of intensity and offensiveness) like field inspection method (VDI 3940 Part 1) or plum measurement (VDI 3940 Part II);
- complaints and odour diaries;
- odour annoyance surveys;
- continues monitoring by e-noses.

Each one of these assessment tools is discussed in Part 3 of the book. Deciding which of these evaluation techniques is most appropriate will depend on the type of application and the particular activity and location. Often a number of techniques will be used in combination. An overview of the various assessment tools, their advantages, disadvantages and applications is given in below. In any case using every one of the above tools we should take account of the following points:

- it is often difficult for investigators to witness odour incidents that are episodic and short-lived;
- emissions are greatly diluted from their point of release, and are often below detection limits of instruments but can still be detected by people;
- peaks in exposure may be due to changing dispersion conditions (wind direction, turbulence) or variable emissions (doors opened);
- emissions from elevated stacks may reach the ground beyond the monitoring point;
- it can be difficult to work out where an emission comes from or to distinguish it from other sources;
- the variable nature of many odour exposure scenarios and the short term of some sampling methods mean that it is much easier to demonstrate exposure than to conclude that no exposure has taken place.

7.3.1 Field Sniff Testing

In these categories there are different assessment tools where the main are:

- the field inspection method (VDI 3940 Part I);
- the plum measurement (VDI 3940 Part II).

Between these two tools, we retain the most useful for odour impact assessment is the field inspection according to VDI 3940 Part I (see Section 3.7.2). Other methods don't consider all FIDOL factor for the impact assessment. On one hand, the field inspection method implements a system of observations of odours versus time and versus space. It can be used to collate a data base of odour intensity versus time patterns for different weather conditions and provides detailed FIDOL information over relatively short time periods. It does not require emotive judgements of odour character or effects and results can be linked to existing intensity versus odour concentration relationships, and used to infer odour exposure levels and peak-to-mean ratios in the environment. On the other hand, the field inspection method, as with odour diaries, obtains information that indicates the exposure pattern and is not a direct measure of adverse effects being caused. It requires detailed procedures and training of assessors to ensure consistent results and relative high costs of implementation.

In any case this tool is able to confirm the character of existing impacts of odour, and is useful for investigating odour complaints and self-monitoring of odour impacts beyond the site boundary. It is also useful for providing further supporting information to odour diary or survey programmes.

7.3.2   Complaints and Odour Diaries

These tools provide a method of obtaining information from the community on odour impact patterns (see Section 3.7.3). They give exposure pattern information, such as frequency of strong, moderate or weak odour impacts at various locations, over a defined period of time. Diarists can record adverse effects as they happen (e.g. 'had to shut windows, felt nauseous').

On other hand, there are difficulties in implementing it in a way that provides useful data unless the diarists are well motivated; considerable diligence and effort are required from all parties to sustain diary entries over a long period. Overall, this tool has proved less straightforward in evaluating the significance of results compared to annoyance survey data, as the information is not a direct measure of adverse effects being caused.

In any case this tool is useful for rural or isolated areas where population density is low and when the major type of effect is either chronic (frequent, low intensity impacts) or due to short-term acute impacts. It is also useful for measuring odour occurrence patterns (e.g. frequency, duration) and to confirming whether a particular industrial or trade site is causing occasional odour impacts. It can be used in most situations, but is less effective than surveys for establishing effects in densely populated communities.

7.3.3   Community Surveys

The main advantages of this assessment tool are related to the fact it accounts for real effects and interactions of physical and social factors. In addition it is a simple and cost-effective approach for assessing the relative extent of nuisance being caused within a community. This assessment tool can also help to rank different industrial facilities according to their contribution to the overall cumulative stress or annoyance within a specific community (see Section 3.7.2). On the other hand, it can be implemented for urban or semi-urban population densities to ensure statistical significance and requires specialist design for each case to ensure results relate to a particular level of impact.

Then surveys are a useful tool for the measurement of degree of adverse effect in terms of annoyance, in urban or semi-urban areas and where odour is being contributed by a number of sources of varying character. In any case it is a useful tool for quantifying the zone of influence from a specific odour source and It is less useful in rural isolated areas where population density is low.

7.3.4   Continuous Monitoring by E-Noses

This is a new assessment tool proposed by international researchers. E-noses provide continuous and daily feedback on the relative level of odour present at receptors. Information is obtained in a formal report that is able to identify different types of odours and in this way a relative source for each type of odour (see Section 3.9.2).

E-noses continuously provide feedback to indicate the likely level of odour on a particular day, and trends over the longer term (Di Francesco *et al.*, 2001). They are used as evidence of either significant or minor adverse odour effects caused in a community. The main weaknesses are related to the calibration phase of the e-nose by dynamic olfactometry. While emissions monitoring by e-nose are usefully documented, the odour exposure monitoring by e-nose it is still under testing.

7.4   Odour Impact Assessment from Sources

Odour impact could be also assessed from sources by dispersion modelling (see Part 5). This approach is of course the only way for new proposals or projects. The evaluation of new proposal could be the most complex type of application to prepare or evaluate. In this case there is no history of complaints, nor plant performance from which to determine the existing odour effects and the regulator must rely heavily on dispersion model results or past experience with similar activities and proposed controls. The evaluation can be complicated if there is little information on which to select odour emission rates, and often a conservative approach to evaluation will be required.

In any case the assessment should be made of the sensitivity of existing and likely future receptors, and the main steps of this approach are:

- identification and characterization of odour sources;
- estimation or measurement of odour emission rate;
- characterization of meteorological conditions;
- characterization of topography of a possible exposed area;
- identification of receptors and their sensitivity;
- evaluation of the exposure levels by modelling results;
- assessment of odour impacts.

These approaches could be useful:

- to predict the impact of a new proposal, comparing with benchmarks or maximum exposure levels;
- as a tool to assist in the investigation of the cause of odour complaints from existing facilities and the influence of changing weather conditions on odour dispersion;
- to compare the cost effectiveness of odour mitigation options;
- to work out emission limits for point source emissions, either mg/m^3 for a single odorous substance or ou_E/m^3 for mixtures of substances;
- to indicate how much improvement is needed or size abatement equipment;
- to calculate a suitable chimney height to provide an acceptable exposure at receptors.

7.4.1   Odour Sources

Identification and characterization of odour sources is the first step of this approach and can be carried out with a comparative approach of similar activities and for new project and with an accurate surveying of the existing process. Understanding the nature and extent of the stock of odorous materials held on site is key to recognizing odour sources and exploiting control opportunities. Management of these materials is related to potential odour sources and their total quantity. Assessment of holding conditions is related to a material's odour potential.

Holding times or conditions for feedstock materials before they arrive at the site are frequently very important sources in waste management activities.

Finding fugitive (including diffuse) emissions can sometimes be quite straightforward, be it open doors on a waste plant or the spreading of manure on farmland. But it is important that we don't focus solely on sources that are easy to identify and measure. Looking for fugitive emissions in a complex process requires a detailed knowledge of valves, flanges and vents, what processes are taking place and what substances are where.

When available, management plans must include an inventory, with descriptions and quantities, of all potentially odorous solid, liquid and gaseous materials held on site across the full range of operating conditions. From the consultation of a management plan or project it is also possible to identify air releases including reducing evaporation and, if needed, containment and abatement. These could help in the identification of main odour sources.

For each emission source the characterization must include the type of odorous gasses released, the height of release through a stack and/or the timing of releases through management of activities. All these aspects can influence the emission rate and the dispersion before there is an impact on people.

7.4.2   Odour Emission Rate

Odorous emissions can be characterized in terms of frequency, intensity, duration and offensiveness, flux and concentration. This could be carried out on a new plant according to scientific literature or a technical report for similar activities, while for existing activities this step must be supported by a measurement of real emissions by a monitoring plan. For the main methods of sampling and measurement of odour emission and the estimation of odour emission rate, see Section 3.6.

Particularly in cases where emissions are released through one or more vents or stacks, it is often appropriate to specify performance criteria for any abatement equipment (Sironi *et al.*, 2010). This may be in the form of odour units through dilution olfactometry (taking volumes into account) or, where available, suitable surrogate measurements which can be more easily monitored. Generally several Environment Agencies in EU will seek to incorporate these performance criteria into the environmental permit in the form of Emissions Limit Values; accordingly this could be very useful this aspect to include in the odour impact study, especially for a new plant were is mandatory verify the acceptability of any limitation to emissions in terms of costs and impacts.

The measurement of odour emission rates and subsequent prediction of ambient odour levels using dispersion models is more complicated than for many other air contaminants (Burlingame *et al.*, 2004). Monitoring odour sources commonly produces emission rates that are quite variable. For example, an odorous pond may have emission rates that vary significantly across the surface, from season to season, and at different times of the day. Emission rates can be several orders of magnitude higher during upset events than under normal operation.

For simple dispersion model scenarios with only one or two sources, the maximum measured emission rate from the source is typically used for dispersion calculations. However, because of the intermittent nature of odours, use of worst case emission rates assumed to occur continuously may result in overly conservative and unrealistic results. Infrequent bursts of strong odour may actually be considered to be less of a nuisance than long exposures to lower intensity odours. The choice of these aspects is strongly related to the use of the dispersion model and boundary conditions; these criteria are described in Part 5.

In any case the starting point for existing activities is from the results of a monitoring plan that could be very helpful for the estimation of the main emission rate. Monitored releases must provide good evidence of real emissions during the duration and on efficiency of any control measures if adopted.

7.4.3   Meteorological Conditions

For proposed new activities, as well as existing activities, the assessment of the local micro-meteorology in both qualitative and quantitative terms is an essential step that is often underestimated. The occurrence of unexpected adverse odour effects from newly established activities often results from the local micrometeorology being too simplistically or wrongly characterized.

To represent conditions for an 'average year' hourly meteorological data for a period of at least three, preferably five years should be used. Data can be sourced from a representative meteorological station. If such a station is not available or the site has specific local features that are likely to influence dispersion significantly, consideration should be given to the use of site specific predictive meteorological datasets derived from analysis of synoptic data. Data of sufficient quality for use in steady state and nonsteady state models is available commercially from a number of sources. In simple studies, on-site meteorological data for a minimum period of a year is also required to allow for an acceptable prediction of percentile odour concentration values.

This simplicity is often driven by the relatively low requirements of steady-state Gaussian dispersion models or by the availability of data for only short periods and / or from station situated very far from the study area. In practice, a robust odour impact assessment generally requires an in-depth analysis of potential odour plume behaviour (see Section 5.7). A high level of expertise in micro-meteorology is therefore necessary to at least produce a qualitative assessment of the significance of cold-air drainage flows, high terrain and/or sea breeze patterns at a particular site.

The analysis of site micro-meteorology is also desirable when importing experiences from other sites. This allows for the use of buffer distances from other similar activities to be assessed for the new site. For example, an adequate buffer distance cannot necessarily be confirmed from the experiences of other sites when potential odour effects at these sites occurred during different types of atmospheric conditions. This is also an argument against the use of fixed buffer distances as a robust odour assessment tool (see Part 6).

7.4.4   Receptors and Sensitivity

Receptors must be selected in the possible impact area and their sensitivity is often regulated by guidelines and legislation (see Part 6). Possible odour impact is not generalizable and is strongly related to the dispersion of the odours in atmosphere. For existing activities it is possible to use different methods to identify the area of investigation (see Part 3). For new activities this step is often related to the preliminary result of the dispersions modelling to identify the area of investigation.

The sensitivity of the receptors is generally linked to the land use, distance from the activity and maximum exposure levels; these values are reported in Part 6. On other hand, the mechanisms that occur during odour exposure, people's responses and sensitivity of receptors are described in the Section 2.5.

In any case for proposed new activities, as well as existing activities, a preliminary monitoring plan to assess the background odours and/or real exposure level at receptors is strongly suggested. The technique to be implemented depends of the objectives of the study, according to the advantages and disadvantages of each of the implemented tools.

7.4.5   Dispersion Modelling

Dispersion modelling is a useful source of predictive information to assess the likely impact of odour. Dispersed odour concentrations predicted by modelling are the results of complex statistical calculations which take account of variations in odour perception over hourly time periods. Through the use of a dispersion model it is possible to assess odour exposure, expressed as frequency of 'odour hours'. This method offers the option of assessing the odour exposure levels around a site over the long-term and is therefore very useful in terms of providing a definitive answer on odour impact.

It is important to give evidence from all predictive methods appropriate weight depending upon their relevance and reliability in the circumstances of a particular site. A detailed discussion on different dispersion models, on their implementation is given is at Section 5.3.

Each model has its own advantages and limitations that must be taken into account when considering an effective assessment strategy. For example, some of these techniques are predictive, while some tools may be able to draw inferences from historical events. Some techniques are qualitative, whilst others give quantitative, numerical data. In practice, it can sometimes be difficult to predict when a situation will lead to a statutory nuisance. Many tools involving prediction for example, modelling, are less effective for the endpoint of statutory nuisance than they are for a planning impact assessment 'no significant loss of amenity'.

However, real-time tools such as direct sensory assessments in the field by 'sniff test' and retrospective techniques such as complaints monitoring are more likely to be effective or useful for the first model calibration. It is important not to consider these tools/techniques in isolation. Such assessments work best when brought together with other assessment techniques and confidence in the conclusions reached can generally be improved by using multiple assessment tools.

7.4.6   Odour Impact Assessment

Dispersion models commonly used asses the odour exposures at receptors in terms of a percentile (generally ranged from 95–98th percentiles) of hourly average odour concentrations over a year. The results are expressed as odour unit contours on a map by isopleths. Unacceptable levels of odour or better odour impact can be found against exposure benchmarks. When the results are presented and interpreted, they must take uncertainty into account, especially in terms of emissions and weather data.

As a matter of clarification, it should be noted that the odour impact standards used in impact assessments do not relate directly to receptor experience because of the statistical methods used in dispersion modelling. Where planning applications concern developments which have not been built, then odour emission rate model inputs may have to be based on measurements made on other similar plants elsewhere. Alternatively, the modelling might more simply be used to define maximum permissible emissions limits to achieve a required level of odour impact protection.

For example, when impact criteria used in interpreting modelling exercises is expressed as the 98th percentile, this represents a small proportion of time (around 14 h per month) during which odour concentration might reach or exceed the specified concentration. The use of percentiles is also consistent with criterion used to assess environmental impacts of other air 'contaminants'.

It is necessary to appreciate that the percentile hourly mean odour concentrations used to interpret dispersion modelling predictions are not directly analogous to odour concentrations measured in an odour laboratory. In free field conditions, the odour concentrations will vary and the actual concentration within one inhalation may be orders of magnitude higher or lower than the hourly average, as described by the 'peak to mean' ratio of the variations within that hour (see Section 5.4).

Dispersion modelling results are normally interpreted using either some kind of site specific analysis of dose-response relationships in the community or area around an odour source, or more commonly by 'custom and practice' benchmarks. A sensitivity analysis, to enable the overall uncertainties to be understood, should also be provided including:

- likely uncertainties in the source term, including a consideration of fugitive emissions;
- the degree to which the emissions are likely to be steady or fluctuating and the impact of this on the model chosen;
- likely uncertainties associated with the meteorological data;
- plausible worst case scenarios;

These uncertainties should be acknowledged in consideration of the isopleths. Once built, the model should be run for different design/what if options in order to show that BAT/appropriate measures are being proposed and to test the uncertainties.

7.5   Mitigation of Odour Impact

Any study on odour impact assessment will need to consider the measures to control odour. This section provides a general explanation of how it is possible tackle odour issues and describes types of control measures and plant that should be considered to prevent or abate pollution. However, it does not consider the detail of plant design, operation or maintenance.

A conventional way to detect mitigation tools or responses in an Environmental Impact Assessment (EIA) context is to apply a DPSIR model. DPSIR is a causal framework for describing the interactions between environment and socio-economic activities society. This framework is an extension of the pressure-state-response (PSR) model developed by the OECD and adopted by the European Environment Agency (EEA), were DPSIR stands for:

- Driving forces (D);
- Pressures (P);

- States (S);
- Impacts (I);
- Responses (R).

Under this approach, for detecting appropriate responses (R), it is necessary considering each of the following headings:

- at driving forces level (D): localization of the plant (or sources) and managing inventory;
- at pressures level (P): controlling emissions and containment and abatement systems;
- at state level (S): Dispersion techniques;
- at impact level (I): techniques to reducing annoyances of the odours at exposures.

The most effective strategies may or may not involve large capital investment, but most measures will need careful management.

7.5.1   Responses to Driving Forces

The main aspects that must be analysed are related to the type of plant, and its localization. These aspects are related to the environmental impact assessment of new project and all alternatives must be evaluated using the approach of odour impact assessment from sources.

For an existing plant, it is also possible reduce the causes of possible emissions managing odourises materials. Many feedstock materials, particularly putrescible wastes or animal by-products, can become very odorous before they arrive at the site-plant. It is recommended to contact suppliers of raw materials (Stuetz *et al.*, 1998). For example, for waste management facilities, contracts may need to specify:

- which types of waste the processing plant or local authority collection teams will receive, and which they will reject;
- how long the waste can be held before it is delivered;
- storage and treatment conditions;
- any appropriate pre-treatment before the waste is dispatched;
- transport conditions (refrigeration, for example);
- the need to divert wastes if you have operational difficulties or you have exceeded your capacity.

In any case, the following will be necessary:

- treat odorous materials promptly in a way which reduces their odour potential;
- keep odorous materials on site to a minimum, rotating stock where appropriate;
- generate as little extra odorous chemicals as possible by, for example, minimizing temperatures or maintaining aerobic conditions;
- consider a housekeeping regime and select building materials which can be easily cleaned;

If this is not enough, then it is necessary to have procedures in place so that you can identify and reject highly odorous wastes. Some sites will be specifically designed to manage odorous feedstock materials, or materials over which they have more limited control. These sites will require much more robust management controls.

7.5.2 Responses to Pressures

The control of odour emissions (pressures) can be made by considering the following aspects:

- control of odour emission rate;
- control of odour sources;
- end of pipe treatment.

It is possible control many odorous chemicals (at least partly) by reducing their rate of emission. The methods to do so can be either chemical or physical, for example:

- lower the temperature by avoiding direct sunlight or otherwise reducing the water evaporation rate and the release of dissolved odorous chemicals;
- increase humidity in the immediate environment to reduce evaporation, as above;
- reduce airflow over the surface of odour-releasing materials to reduce the rate of evaporation;
- control the acidity/alkalinity of a material to make specific smelly chemicals much more soluble in water and less likely to evaporate; for example, acidic conditions (low pH) can suppress the evaporation of alkaline chemicals such as ammonia. Conversely, alkaline conditions (high pH) can suppress odorous acidic chemicals such as propionic acid or acetic acid;
- introduce temporary surface treatments to lower the surface temperature or create a chemical barrier. Plain water is the simplest and is often helpful; these treatments can also contain pH buffers as above or other chemicals to make odorous chemicals more soluble;
- reduce the surface area of an odorous material; this will cut the emission rate;
- avoid disruptive activities such as shredding or screening, which dramatically increase exposed surface area and emissions, unless adequate containment is provided.

Other important point is the control of odour sources. If we cannot avoid producing significant levels of odorous air, we will need to contain the emissions before treating them. Under these points the main aspects to consider could be:

- choose containment and treatment methods together so that will be possible coordinate the most appropriate treatment with management of ventilation rates;
- localized containment lowers the volume of air required to be treated. It will normally be much more cost effective than if you rely entirely on a large building for primary containment;
- where the control of continuous odour relies on s containment we should maintain effective airflow by pressure control within the process plant or within process buildings. 'Air-lock' entry and exit doors will enable the integrity to be maintained. Complex air management systems which are affected by thermal lofting or complex ducting arrangements will need to be designed by competent engineers;
- keep windows and doors on buildings used for containment shut. All doors should be self-closing;
- be aware that two or more openings on either side of a building can create a through-draft and carry odours out;

- consider all of the conventional techniques for minimizing VOC emissions from tanks and pipework;
- check pipes, valves and tanks regularly for leaks and damage.

Last but not the least, the aspects to control odour emissions are the abetment systems or end of pipe treatments. There are many ways to treat air from contained sources and are discussed in-depth in Part 3. In any case, briefly they are, in general, the same techniques used for chemical abatement:

- adsorption using activated carbon, zeolite, alumina (disposable or with regeneration);
- dry chemical scrubbing; solid phase impregnated with chemical agents such as pH modifiers, chlorine dioxide or permanganate;
- biological treatment; trickling biofilters, soil bed biofilters, non-soil biofilters (peat, heather, wood bark, compost), bioscrubbers;
- absorption (scrubbing); spray and packed towers, plate absorbers (single pass or recirculating);
- thermal treatment; existing boiler plant, thermal or catalytic oxidation;
- other techniques; odour treatment chemicals, condensation, plasma technology (ozone), catalytic iron filters and UV.

It is very common to use hybrid or combined methods. For example, many activated carbon products are impregnated with dry chemical scrubbing reagents and ozone methods sometimes work best after excess moisture have been reduced by condensation.

The main characteristics of abatement systems and their relative strengths and weaknesses are discusses in Part 4.

7.5.3   Responses to State

At the state level, we could image different solutions to control the dispersion of odorous substances in the atmosphere, that can imply both structural and management aspects. High stacks and/or high trees downwind may be used to allow odorous emissions to disperse before they reach the ground. Similarly, where it is possible to increase the physical distance to receptors, this can also reduce exposure. Dispersion modelling, discussed in the Part 5, can be used to assess the benefits of these measures. Where feasible, some dispersion benefits may be realised through arranging emissions points at locations which are further away from nearby receptors.

To avoid peak impacts it strongly recommended to schedule operations in the allotted time. For example, suspending operations when there are inversion or cold drainage flow conditions or when the wind direction is towards nearby residents, or by undertaking activities at a time of day when residents are not present or are likely to be indoors. Where this is part of the control strategy, the monitoring of weather and forecasts can suggest the adequate actions to avoid impact.

7.5.4   Responses to Impact

It is really important to engage with the people who may be affected by activities. Many operators do this as a matter of course and have well-established procedures for interfacing with the general public. However, some operators overlook this essential step.

When neighbours are likely to perceive odours from some identified or not-identified sources, several responses must be taken in consideration starting from driving forces (D), then to pressures (P) and on to status (S). When these tools are not sufficient, additional responses could be applied directly to receptors to increase the acceptability of the odour impacts. It is important to take in to consideration for exposed people (neighbours of odorous plant), odours may:

- cause annoyance;
- reduce enjoyment of home and gardens;
- reduce property values;
- raise concerns about exposure to harmful emissions (e.g. bio aerosols);

Engaging with, and becoming an active member of, the local community may enable operators to mitigate the impact of their activities and increase tolerance of odours, particularly where those odours are relatively transient. Engagement can include a wide variety of activities, but communication is always a key aspect. This means being a reliable source of information to the community and being available to hear what they have to say. Exactly how you establish channels of communication depends upon what you and the community are comfortable with.

Active participation in the community not only helps people to get to know you and your staff, but also helps people to understand what you do and, possibly, even view you as an asset. Some of your employees may live in the surrounding community and can be important as ambassadors for your business. In addition to active participation in the community is possible foresee a spray of odorous substances in the flow enzymatic or near sensitive receptors. This solution does not reduce the pressure, and it increases the concentration of chemicals in the atmosphere, but will cover unpleasant odours making them more acceptable impacts.

7.6 Odour Monitoring

In any study of odour impact assessment is the inclusion a monitoring activities and a monitoring plan. Of course monitoring activities could be necessary in specific situation for the mensuration of background odours or the odour impacts, the monitoring plan will be used to verify assessment results and/or the efficacy of some odour control solutions (Van Harreveld, 2004).

According to their main objectives there are different types of monitoring:

- assess impact (using complaints, community questionnaires, interviews and field sniff testing);
- assess exposure (using field surveys, field inspection, e-noses, surrogate monitoring);
- investigate sources and pathways (using fence line monitoring, meteorological monitoring);
- measure releases (using dynamic dilution olfactometry, surrogate monitoring, assessment against emission limit values);
- control processes (using temperature, oxygen levels, pH, moisture, etc.).

The monitoring of odour emission and generally of the processes is generally part of the odour management plan of the source plant. With this it is possible to assess odorous substances emitted and how effective control measures are. On the other hand, the measurement of odour exposures is directly related to the measurement of odour impacts on the local community, this could be related to specific plant or to municipality and/or environmental agency action to control the acceptability of air quality in terms of odour. In this section there is a brief overview of available monitoring approaches and their applicability; for the techniques of odour sampling and measurement, see Part 3.

7.6.1   Monitoring Plan

We must be clear about reasons for monitoring in order to identify how best to carry it out; in this way the monitoring plan can take several different forms:

- field investigations that include sniff testing (to check ambient air on or off site);
- meteorological monitoring: very simple, low risk, sites may get away with indirect (e.g. local airfield met data) or observation methods; most though, will require appropriately configured on-site data-logging instruments;
- complaints (direct complaints, as well as those made to the Environment Agency or a third party such as a local authority);
- odour diaries;
- surrogate chemicals or process parameters (e.g. H_2S, ammonia, odourless methane as an indicator of odorous landfill gas, etc., pH and flow in a scrubber);
- emissions monitoring if there is a point of discharge;
- community surveys.

In any case a monitoring plan should include:

- why and how monitoring will take place, for example:
  - steady state monitoring to confirm that odour is under control: regular sniff tests and if appropriate, continuous monitors or process surrogates;
  - assessment against any emissions limits in permit or odour management plan;
  - if an odour problem arises, the monitoring will carry out to establish what needs to be done;
  - if a solution in place was found, the monitoring will confirm that problem was really resolved.
- how to interpret the results including, whenever feasible, trigger values for further monitoring or remedial action;
- if the terrain is complex, or if odours come from many places, how monitoring will handle this;
- record-keeping and reporting.

7.6.2   Monitoring Report

Whatever the monitoring plan, the records need to include enough information about the emissions measurement for you to use that data in your analysis. For example, results for a sample analysed by laboratory-based olfactometry must, include:

- date, time and details of emissions point sampled, and why you chose them;
- how samples were preserved (condensation, holding time and conditions);

- method of sampling (e.g. stack sampling through a 3 m stainless sampling tube);
- the laboratory where the results were analysed, and any certification status;
- any laboratory observations that might affect how results were interpreted;
- process parameters;
- weather conditions.

References

Both, R., Sucker, K., Winneke, G. and Koch E. (2004) Odour intensity and hedonic tone–important parameters to describe odour annoyance to residents?, *Water Sci. Technol.*, **50** (4), 83–92 .

Burlingame, G.A., Suffet, I.H., Khiari, D. and Bruchet A.L. (2004) Development of an odor wheel classification scheme for wastewater, *Water Sci. Technol.*, **49** (9), 201–209.

Di Francesco, F., Lazzerini, B., Marcelloni, F. and. Pioggia, G. (2001) An electronic nose for odour annoyance assessment, *Atmospheric Environment*, **35** (7), 1225–1234.

Gostelow, P., Parsons, S.A. and Stuetz, R.M. (2001) Odour measurements for sewage treatment works, *Water Research*, **35** (3), 579–597.

Henshaw, P., Nicell, J. and Sikdar, A. (2006) Parameters for the assessment of odour impacts on communities, *Atmospheric Environment*, **40** (6), 1016–1029.

Littarru, P. (2007) Environmental odours assessment from waste treatment plants: Dynamic olfactometry in combination with sensorial analysers 'electronic noses', *Waste Management*, **27** (2), 302–309.

Nicell, J.A. (2009) Assessment and regulation of odour impacts, *Atmospheric Environment*, **43** (1), 196–206.

Sironi, S., Capelli, L., Céntola, P. *et al.* (2010) Odour impact assessment by means of dynamic olfactometry, dispersion modelling and social participation, *Atmospheric Environment*, **44** (3), 354–360.

Stuetz, R.M., Engin, G. and Fenner, R.A. (1998) Sewage odour measurements using a sensory panel and an electronic nose, *Water Science and Technology*, **38** (3), 331–335.

Sucker, K., Both, R., Bischoff, M. *et al.* (2008a) Odor frequency and odor annoyance Part II: dose–response associations and their modification by hedonic tone, *International Archives of Occupational and Environmental Health*, **81** (6), 683–694.

Sucker, K., Both, R., Bischoff, M. *et al.* (2008b) Odor frequency and odor annoyance. Part I: assessment of frequency, intensity and hedonic tone of environmental odors in the field, *International Archives of Occupational and Environmental Health*, **81** (6), 671–682.

Suffet, I.H. and Rosenfeld, P. (2007) The anatomy of odour wheels for odours of drinking water, wastewater, compost and the urban environment, *Water Sci. Technol.*, **55** (5), 335–344.

Van Harreveld, A.P. (2001) From odorant formation to odour nuisance: new definitions for discussing a complex process, *Water Sci. Technol.*, **44** (9), 9–15.

Van Harreveld, A.P. (2004) Odour management tools: filling the gaps, *Water Sci. Technol.*, **50** (4), 1–8.

Part 8

Case Studies for Assessment, Control and Prediction of Odour Impact

8.1 Urban Wastewater Treatment Plant

J. Lehtinen
Department of Biological and Environmental Sciences, University of Jyväskylä, Jyväskylä,
Finland

8.1.1 Motivation for the Study

Wastewater treatment plants are common sources of odours and VOC (volatile organic compound) emissions all over the world. The location of the plants is often very close to residential areas. In many cases, the plant was built beforehand, with the residential area and other communal services being built in the direct vicinity of the plant afterwards. Therefore, the situation can be very delicate and difficult concerning odour emissions. When it comes to odours, people are very keen to defend their properties from the devaluation caused by odour annoyance and the vicinity of the wastewater treatment plant.

Moreover, the annoyance is not the only reason for odour complaints. The unpleasant nature of the wastewater treatment plant odour can be considered a health hazard. Although in most cases, it is not an actual threat, especially for the people living in the proximity of the plant. The reason for concern is that odorous air can be interpreted as unhealthy among people (Rosenkranz and Cunningham, 2003; Zarra *et al.*, 2008).

Usually, odour problems can be very complicated due to the nature of the wastewater treatment plant design. The tanks are large and are often placed outdoors. In addition, the odour originates from several different odour emission sources in the plant area and the actual total emission to the surroundings is very difficult to measure or even estimate. In many cases, the severe odour emissions are consequences of some kind of disturbance in the process conditions, which cannot be predicted beforehand. Maintenance operations such as sludge transporting or washing the tanks are usually predictable odour emission sources but they are quite hard to measure or estimate with contemporary methods.

Odour Impact Assessment Handbook, First Edition. Edited by Vincenzo Belgiorno, Vincenzo Naddeo and Tiziano Zarra.
© 2013 John Wiley & Sons, Ltd. Published 2013 by John Wiley & Sons, Ltd.

Furthermore, odour itself is a complicated subject. In wastewater treatment, the odour is a combination of several volatile organic compounds that are released from different treatment processes. The overall odour type varies depending on the source at the wastewater treatment plant (Lehtinen and Veijanen, 2011b). Some of the odorous compounds are present in the incoming wastewater, while others may be generated during the transport or treatment processes (van Durme *et al.*, 1998). Different chemical and biological processes and processing conditions like pH, temperature, retention time and so on, may have a great effect on the odour characteristics (van Durme *et al.*, 1998; Bonanni, 1998). It is known that the long retention times of effluents and anaerobic conditions or even low oxygen levels in the sewers favour the generation of malodorous sulfur compounds and carboxylic acids. In addition, people's odour perceptions and tolerances to annoying odours are highly subjective, meaning that different people find different odours offensive and at different concentrations. Odour thresholds that people have for different odorous compounds vary widely due to the chemical nature of the compounds and between persons depending on age, gender and state of health. This makes odour measurement and determination very challenging (Stuetz *et al.*, 2001).

Due to the complicated nature of odour and odour emissions, the goal in this study was to define the odour concentrations from different emission points in the plant in order to achieve a general view of the magnitude of odour concentrations. In the study, odour causing compounds and their possible health threatening nature were determined. Odour impact to the surroundings of the plant and the residents' opinions about the odour emissions from the plant were also examined. These investigations were made by measuring the odour concentration by olfactometry, by measuring the single odour components and organizing a survey about people's reactions and experiences of odour annoyance from the wastewater treatment plant.

8.1.2   Description of the Situation

In Finland, wastewater treatment is required and regulated by EU directives and Finnish environmental laws by plant specific licenses. Wastewater processing methods and parameters vary by plant but primary and secondary wastewater treatment is usually used (Finland's Environmental Administration, 2010).

The wastewater treatment plant in the study is one of the largest in Finland, treating effluents of 140 000 citizens and industrial plants that are connected to the municipal sewer system. The daily effluent intake of the sewage treatment plant was 42 648 m³/day in average during the period of 2005–2007, and the processes are based on the biological-chemical treatment method (see Figure 8.1.1). In the environmental licence, the plant is obligated to keep the odour levels so low that it does not cause odour annoyance to the neighbourhood, but no numerical emission limits are given in the licence.

Bar screens, grit removal, sludge thickening and dewatering as well as bio-filter units are placed indoors where ventilation is in use. Air from the bar screens and grit removal tanks is collected and conducted to the bio-filter treatment. From the sludge treatment, the air is conducted outdoors through ozonation.

The bio-filter has a surface area of 60 m² and the filter material consists of sludge compost and woodchips. The outdoor units include three round, primary clarifying tanks of 20 m in diameter with a combined volume of 5700 m³, biological treatment tanks of combined volume of 12 000 m³ and three secondary clarifying tanks with a combined

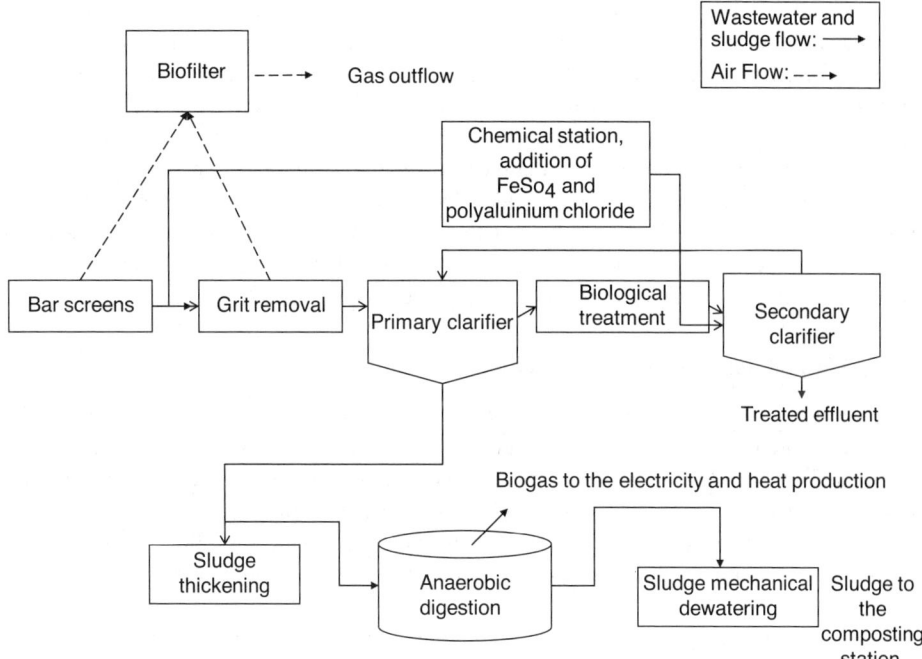

Figure 8.1.1   *Simplified scheme of the treatment process in the plant.*

volume of 13 480 m^3. The excess sludge is thickened and digested anaerobically and thereafter dewatered and transported to a tunnel composting plant.

The studied wastewater treatment plant is located on the lakeshore Päijänne and is surrounded by a nice residential area of family houses, town houses and blocks of flats. There is also a daycare centre and a bakery 400 m from the plant. The closest residences are 200–250 m from the plant. The most parts of the residential area have been built afterwards in the vicinity of the wastewater treatment plant. The area between the residences and the plant is mainly forest (birches, pines, spruces). On the western side, there is a rock area about 10 m high that prevents the odour emission spreading to the western residential area. On the eastern side of the plant, there is Lake Päijänne. The prevailing wind direction in plant area is from the south-southwest, but seasonal variation is great. In spring, the prevailing wind direction is from the north-west, in summer from the east.

Considering the delicate location of the plant, it has been impossible to avoid odour complaints. Due to this situation, the plant was under enormous public pressure and was forced to make a lot of improvements in its processes. When considering the odour emissions, many critical process sites have been built indoors and the biofilter was installed in 2005 in order to reduce the odours emitting from the primary treatment processes.

8.1.3   Specific Objectives of the Study

The plant location in the middle of residential area is very delicate, with the plant and environmental authorities having received odour complaints in the early 2000s. Therefore, it was considered important to characterize the odour and find out the most odorous sites and also find out the magnitude of the odour concentrations at the plant. In addition, odour causing compounds were determined in order to assess the possible severe odour sources as well as determine whether the emissions contained some health threatening components. Research was carried out by measuring single odour components (VOCs, ammonia, hydrogen sulfide) and odour concentration at the same sites.

Since odour is a very subjective matter, the people's personal experiences of odour emission from the plant were investigated. Therefore, the residents of the nearby area surrounding the plant were inquired about the odour annoyance by organizing a detailed survey. The survey was sent for the residents within the distance of 800 m from the plant. Odour emission measurements were taken in the years 2005–2007.

8.1.4 Methodology and Data Collection

When dealing with odour measurements, there is a wide variety of methods that can be used to estimate the odour and odour annoyance. Odour measurement methodologies are presented elsewhere in this book. In this particular case, the overall odour concentration was measured by olfactometry. The single odour causing compounds were determined by GC-MS-Sniff-technique. Since it was important to estimate the odour impact to the surroundings, a detailed odour survey was arranged in 298 households within a distance of 200–800 m of the plant.

The sampling sites in the wastewater treatment plant were primary clarifier, sludge thickening, sludge dewatering and biofilter, incoming gas and outflow gas (incoming gas includes the bar screens and grit removal exhaust gases).

8.1.4.1   *VOC Measurement*

For the VOC analyses, duplicate air samples were collected in-situ to the adsorption tubes (Supelco 20920-U) filled with Tenax GR adsorption resin (200 mg) by using an air pump (Gilian pump, adjusted flow rate 0.1 l per min). The average sample volume was 4–10 l depending on the expected the VOC concentrations. On measurement days, the process conditions at the plant were normal, which is recommended for this type of sampling (Albrecht *et al.*, 2008). Duplicate samples were collected every season once or twice from several sites at the wastewater treatment plant. The samples were collected during the years 2005–2007.

The laboratory analyses of the VOC samples were performed using a thermal desorption or purge and trap-thermal desorption/gas chromatograph/mass spectrometer device (PT-TD-GC-MS) (Tekmar 3000/Agilent 6890 + /5973N MSD spectrometer), connected with simultaneous sniffing done by a professional conversant with odours and odorous compounds. Helium (Aga, 99.9996% purity) was used as the carrier and purge gas. (Lehtinen and Veijanen, 2011a, 2011b).

8.1.4.2   *Hydrogen Sulfide Measurement*

Hydrogen sulfide (H_2S) was analysed in-situ by a portable infrared gas analyser GA 94 which has an external electrochemical cell for hydrogen sulfide analysis. The measuring range for H_2S was 1.45–290 mg/m^3. A portable gas chromatography (Perkin Elmer Photovac Voyager) was also used.

8.1.4.3   *Ammonia and Reduced Sulfur Compound Measurements*

Ammonia was measured at every sampling site in-situ using detection tubes or diffusion tubes (Rae Systems and Dräger, range 0.86–25.8 mg/m³). A Perkin Elmer Portable gas chromatograph (Perkin Elmer Photovac Voyager) equipped with a photo ionization detector (GC/PID) was used for in-situ measurements of reduced sulfur compounds such as H_2S, methyl mercaptan (MeSH), dimethyl sulfide (DMS) and dimethyl disulfide (DMDS). The measuring range for MeSH, DMS and DMDS was 0.13–25.8 mg/m³ and for H_2S 0.15–14.5 mg/m³.

8.1.4.4   *Odour Concentration Measurements*

Samples for the olfactometric analyses were taken from the same sites as the VOC samples. Sample air was collected in Nalophan® sampling bags (7 l) by Ecoma Vacuum Sampling device. The olfactometric analyses were conducted in the Olfactometric Laboratory of Jyväskylä University Department of Biological and Environmental Sciences with an Ecoma TO7 olfactometer device. The odour analysis method was compatible to the European Standard EN 13725 (2003). The panellists were selected and their sense of smell was tested according to the guidelines presented in the European Standard EN 13725 (2003) (Lehtinen and Veijanen, 2011b).

8.1.4.5   *Survey to the Residential Area after the Improvements in the Plant*

The survey on the residential area surrounding the wastewater treatment plant was carried out in 2006 following the installation of the biofilter. The survey form was sent to 298 households in a range of 200–800 m from the plant at different compass points. In the survey, plant specific and peoples attitude reflecting questions about plant odour and odour annoyance were presented. The questions were formulated in order to adhere the instructions given in a guidebook about Finnish odour annoyance study (Arnold, 1995). The main questions of the survey were as follows:

1. Is the quality of the outdoor air on your property:
   1. fresh,
   2. slightly polluted,
   3. moderately polluted, or
   4. polluted?
2. Is there plant odour on your property? Yes/No
3. How strong is the effect of the plant odour on the quality of outdoor air on your property?
   1. no effect,
   2. slight effect,
   3. moderate effect,
   4. clear effect,
   5. very strong.

The opinion reflecting questions were:

4. In your opinion, is the plant odour a comfort diminishing factor?
5. In your opinion, is the plant odour an environmental hazard?
6. In your opinion, is the plant odour a health hazard?

In the survey, people were also asked to evaluate the level of annoyance the plant odour causes. The answering categories were:

1. no annoyance
2. very little annoyance
3. moderately annoying
4. very annoying
5. extremely annoying.

People were also asked about the effects or symptoms they are having, if any, due to the odour. The alternatives were headaches, nausea, eye and respiratory problems, fear of becoming ill, comfort diminishing outdoors, and comfort diminishing indoors. The answering scale was 0–4, where 0 meant never and 4 meant very often.

8.1.5   Results and Discussion

The analysed single odorants from the most odorous sites at the plant during the years 2005–2007 as well as the odour concentration results from the same sites are presented in Table 8.1.1.

Compounds that exceeded their odour threshold concentrations multiple times being therefore the dominant odour components were DMS, DMDS, toluene, heptanal, alpha-pinene, DMTS, octanal and limonene (Table 8.1.1) (Lehtinen and Veijanen, 2011a). The concentration of sulfur compounds is higher in the sludge processing units as well as in the gas coming from the pre-treatment units (bar screens and grit removal), as assumed. The origin of the organic sulfur compounds and aldehydes is most likely the result of the anaerobic degradation of the organic matter; alpha-pinene and limonene are possibly originated from household discharges, because they are very common odorants in cosmetics and cleaning agents (Lehtinen and Veijanen, 2011b).

The thresholds for all of the VOCs are not present in current scientific literature, and therefore it is possible that other compounds may have an effect on the overall odour. These compounds include many sulfur compounds such as DMTS and allyl methyl sulfide and terpenes, which caused odour perception during the sniffing analysis. In this case, the studied wastewater treatment plant did not have high H_2S concentrations. In fact, H_2S was found only once in detection limit exceeding concentration twice in biofilter incoming gas. In studies considering wastewater treatment and odours, H_2S is mentioned to be one of the major odorants (Stuetz *et al.*, 1999; Hvitved-Jacobsen and Vollertsen, 2001; Hobson and Yang, 2001). Currently, treatments like oxidation and the precipitation of sulfides by iron have reduced hydrogen sulfide concentrations and it cannot be considered as the major odour component in this particular wastewater treatment plant.

The VOCs analysed in the biofilter incoming and outflowing gas are presented in Table 8.1.2. The highest reduction regarding single compounds was achieved by 2-butanone, allyl methyl sulfide and hexanal. Limonene, methylene chloride and chloroform were not reduced very well. This is possibly due to the chemical nature of these compounds as well as the capacity of the microorganisms to use different substances as nutritional substrates. The biodegradation processes depend on the molecular structure of the pollutant: aliphatic compounds are easily degraded, then the aromatics, chlorinated compounds and after that the cyclic hydrocarbons (Le Cloirec *et al.*, 2005). Limonene and other terpenes may have been released into the exhaust gas from the woodchips in the biofilter material.

Table 8.1.1 VOC concentrations (arithmetic average in µg/m³) and odour concentrations (OU/m³) at the sewage treatment. (Lehtinen and Veijanen, 2011a), in parenthesis min-max values. (n = 4, bdl: below detection limit, nd: not detected). Odour threshold concentrations (in µg/m³) and odour descriptions (O'Neill and Phillips, 1992; Ruth, 1986), analysers' own descriptions.

Compound	Primary clarifier	Sludge thickening	Sludge dewatering	Odour threshold concentration, µg/m³	Odour description
DMS	0.9 (bdl-2.9)	56.3 (51.6-69.0)	26.4 (3.9-103.8)	0.3	decayed cabbage
allyl methylsulfide	nd	8.2 (bdl-22.7)	nd	–	garlic, unpleasant
DMDS	1.6 (bdl-3.6)	83.9 (34.5-162.0)	22.5 (5.0-32.5)	0.1	decayed cabbage
DMTS	nd	2.7 (bdl-10.8)	1.9 (bdl-9.6)	–	garlic, decayed
hexanal	7.6 (3.9-15.9)	4.9 (bdl-8.6)	1.5 (bdl-4.4)	28	green, grass
heptanal	3.2 (bdl-6.5)	6.7 (bdl-15.3)	1.5 (bdl-7.4)	6.0	greasy, pungent
octanal	16.5 (bdl-43.5)	16.1 (bdl-37.6)	nd	–	greasy
toluene	10.2 (5.7-16.7)	154.5 (11.7-450.6)	6.9 (17.9-219.4)	80	paint, solvent
alfa-pinene	2.5 (0.8-4.7)	39.4 (3.6-109.3)	13.8 (1.0-13.7)	16	woody, coniferous
3-carene	nd	nd	28.44 (3.3-61.4)	–	lemon
limonene	4.0 (2.9-5.4)	41.2 (17.6-95.5)	59.8 (13.1-128.5)	10	lemon
diethyl ether	30.5 (bdl-110.8)	272.9 (66.4-536.6)	14.0 (0.6-19.4)	900-28 000	ether
OU/m3	157 (57–380)*	4330 (1100–7100)	3530 (1500–7600)		

Table 8.1.2 *VOC concentrations at the biofilter (arithmetic average in μg/m³), Odour concentrations (in OU/m³) and the average biofilter removal efficiency for single volatile organic compounds at the biofilter unit, min-max values in parenthesis (bdl: below detection limit, n: number of measurements) (results according to (Lehtinen and Veijanen, 2011b)).*

Compound	Incoming waste gas to biofilter	Exhaustion gas from biofilter	Biofilter removal efficiency (%)
DMS	43.4 (6.3–116.7)	17.8 (8.3–27.4)	59.0
DMDS	69.6 (4.1–215.5)	32.1 (6.2–67.2)	53.9
allyl methylsulfide	5.1 (bdl-18.7)	nd	100.0
DMTS	16.1 (bdl-33.1)	9.9 (2.9-25.5)	38.5
methylene chloride	3.3 (bdl-10.6)	4.0 (bdl-8.9)	−30.0
chloroform	23.1 (bdl-92.5)	19.5 (bdl-76.0)	15.6
toluene	375.4 (64.1-879.2)	132.7 (47.4–303.2)	64.7
hexanal	36.3 (bdl-141.5)	2.8 (bdl-11.1)	78.7
heptanal	12.6 (bdl-50.2)	4.9 (bdl-11.3)	61.1
octanal	25.6 (bdl-102.3)	8.8 (bdl-35.2)	65.6
alfa-pinene	37.2 (21.0-59.4)	22.5 (5.6–38.7)	39.5
limonene	93.4 (13.1–194.4)	90.6 (20.8–149.3)	3.0
diethyl ether	272.3 (47.3–648.3)	153.5 (20.5–268.2)	43.6
2-butanone	11.4 (4.1–26.9)	nd	100.0
H2S	0.75	nd	100.0
OU/m3	11630 (1400–30 000)	4230 (1300–8000)	67.7

For single compounds, the total reduction percentage in average was 58.2% (Table 8.1.2). The biofilter reduction potential for odour concentration (OU/m³) was 67.7% in average (Table 8.1.2). The highest odour concentration values were measured at the biofilter incoming gas, as expected, because it is a combination of air from the bar screen and grit removal rooms.

As the results show, the odorous compound concentrations are not especially high and nor are the odour concentrations in the normal process conditions. However, the variation in results is highly significant. The highest odour concentrations were measured in the pre-treatment section where the gas was conducted to the biofilter. The emission will be the highest from the biofilter due to the air velocity of the ventilation. This causes the total emission and odour burden to the environment to be strongest in this section. In order to estimate the total odour burden to the environment, the emission and flow velocities should be measured. Due to the nature of odour emissions in the plant, where there are several dispersed emission points, the total emission rates have been quite hard to measure and estimate at the time of the study. Therefore, the odour impact was estimated through the traditional survey method.

8.1.5.1   Odour Impact Assessment by Means of the Survey

The percentage answering the survey was moderately low, only 44.3% returned the survey form. The survey revealed that 66.7% of all the respondents have noticed the plant odour in their property or at home. 45.5% rated the surrounding air in their residential area as fresh (Figure 8.1.2b). 38.6% evaluated the air to be slightly polluted. 2% of the respondents judged the air to be polluted. When dividing the answers by compass points, the southern, south-western and northern regions seemed to suffer more from the odour than the western region. This is also logical when considering the prevailing wind direction on the area. Most of the time, the odour is carried to the northern area by the southern or south-western wind, or to the southern area by the north-western wind, especially in spring time. 47.9% of the northern region respondents evaluated the air to be slightly polluted rather than fresh. 36.4% of the southern area respondents evaluated the air to be slightly polluted and 42.4% evaluated the air to be fresh although 81% of them have noticed the odour from the plant on their property. The answers from the survey imply that most of the people living at a distance of 500 m or more from the plant did not suffer from the odour originating from the plant. The strongest annoyance of odour was found to be in the streets nearer to the plant, naturally.

When asking the odour frequency on the people's properties, the most usual answers were: less than once in a month or 2–3 times per month. In the northern area, 29% perceived odour less than once a month and 27% 2–3 times a month. In the southern section, 36.4% perceived the odour emission less than once a month and 21.0% 2–3 times per month. In the western area, 47.8% suffered the odour less than once in a month and in the southwest region, 28.6% of people never experienced the odour on their property. 42.9% of south-western residents answered that the odour occurs less than once a month on their properties. According to all the answers, the odour occurred in the residential area occasionally at every season. In most of the areas, the odour was mentioned to occur when the wind direction is from the plant to the residential area, of course, and also on whether the moisture content is high and depression occurs. According to previous results about the odour frequency, it is very likely, that the odour in the neighbourhood occurs mostly on days when some maintenance operations, for example sludge transporting, are being carried out.

When investigating the inhabitants' opinions about the effects of odour annoyance, 57.7% of the respondents felt that odour is a comfort diminishing factor and 33.3% thought that it does not affect the comfort of their living. The distribution of the odour annoyance level was that 30.3% felt that the odour is not an annoying at all: 16.7% felt the odour very slightly annoying and 28.8% thought the odour to be moderately annoying: 14.5% considered the odour clearly annoying and 4.5% thought it to be very annoying. Extremely annoying odour was experienced by 1.5% of the respondents (Figure 8.1.3a).

The odour was considered an environmental hazard by 42.4% of the people. When asking the possibility of odour as a health hazard, 32.6% of the respondents considered the odour to be a health threatening factor (Figure 8.1.3b). On the other hand, when asking about the effects or reactions the odour is causing, the effects were very minor: 1.7% very rarely suffered from headaches and 1.0% sometimes. While, 1.0% very often suffered from headache. Other health effects that people had experienced were eye and respiratory track symptoms; 2% suffered from them very rarely and 1.0% was having these symptoms sometimes. 1.7% was having eye and respiratory symptoms very often (Table 8.1.3).

In relation to the question about the wastewater treatment plant's influence on the quality of air in their property, 18.9% thought the plant does not have any effect on the quality of air: 36.4% felt that the plant has a slight effect and 27.3% of the respondents thought that

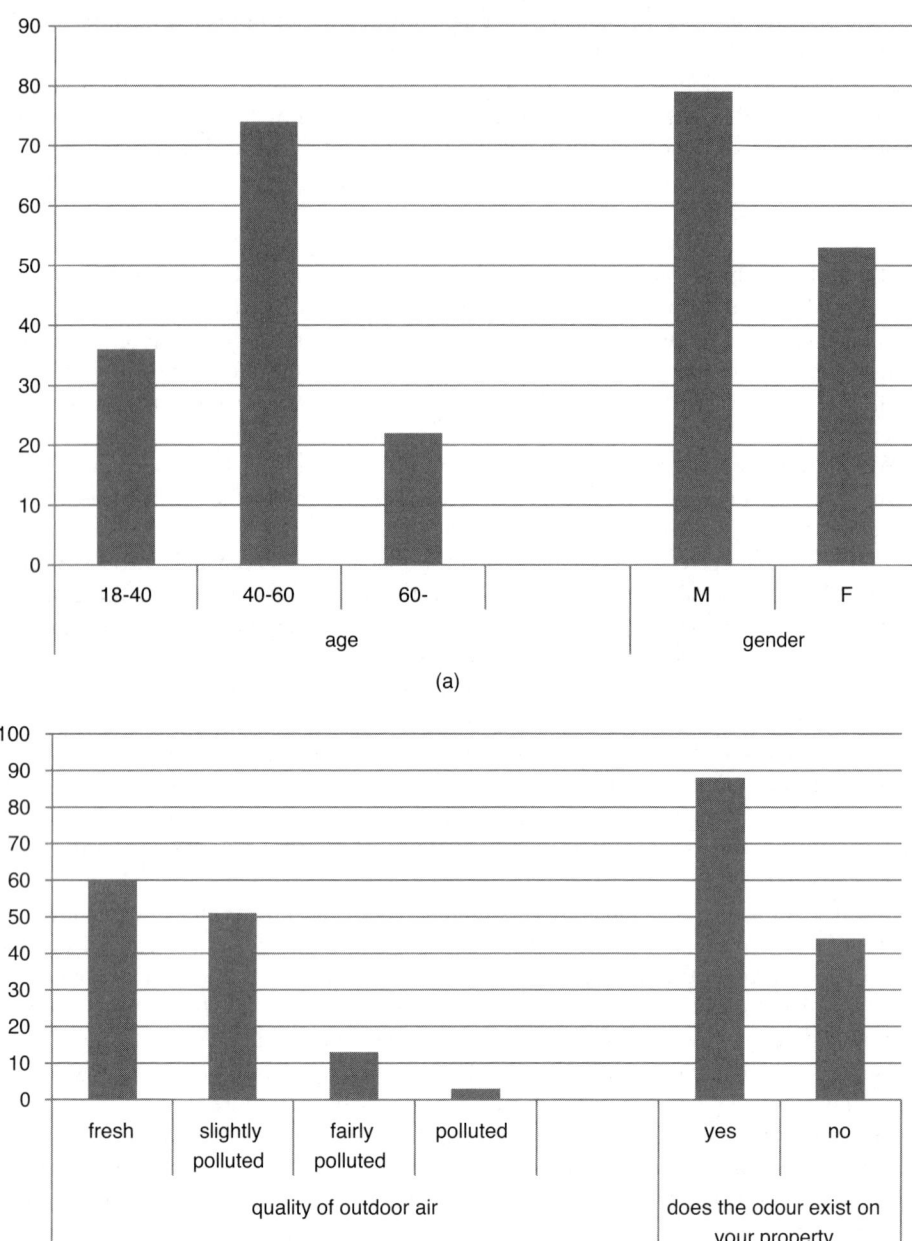

(a)

(b)

Figure 8.1.2 *(a) Distribution of age and gender of the survey participants and (b) their evaluation of the prevailing quality of air on their property.*

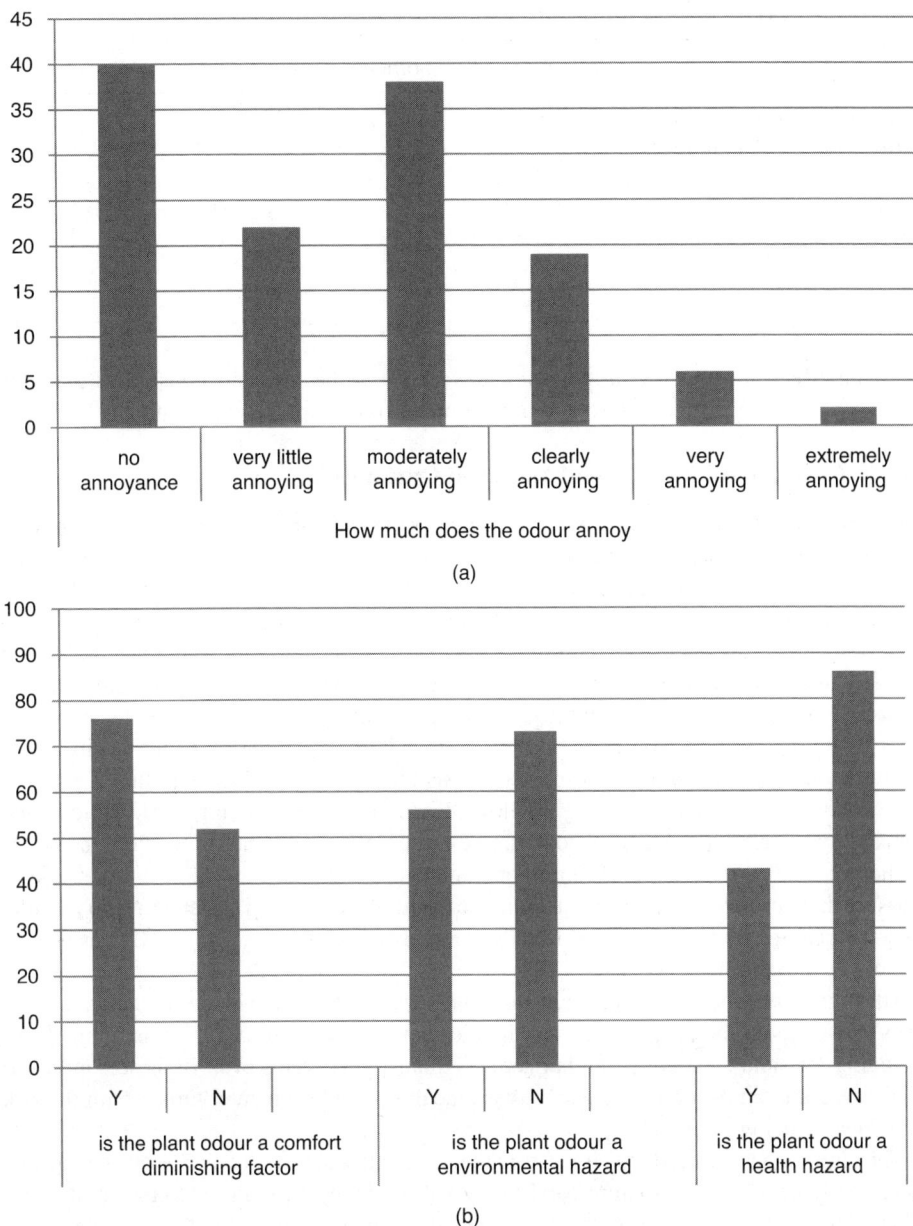

(a)

(b)

Figure 8.1.3 *(a) Distribution of the respondents' opinions about the odour annoyance level and (b) the opinions about the odour as a comfort diminishing factor or health or environmental hazard. Distribution of age and gender of the survey participants and (b) their evaluation of the prevailing quality of air on their property.*

Table 8.1.3 *Distribution of answers (in %) about the effects or symptoms odour is causing.*

Symptom/effect	Very rarely (%)	Sometimes (%)	Often (%)	Very often (%)
comfort diminishing outdoors	8.7	11.4	6.0	2.3
comfort diminishing indoors	9.1	5.7	1.3	1.3
annoyance	4.0	4.0	1.7	1.7
headache	1.7	1.0	0	1.0
nausea	3.7	1.7	0	0
fear of getting illness	2.3	1.0	1.0	0

the plant effects on the air quality moderately: 4.5% of the residents considered the plant odour to have a very strong effect on the air quality of their properties (Figure 8.1.4).

8.1.6   Conclusions, Recommendations and Outcomes

The odour causing compounds determined here were typical for wastewater; sulfur compounds, some terpenes and aldehydes. The odour concentrations in the units of incoming gas to the biofilter and sludge processing were the highest. Under normal conditions, when the measurements were taken, the VOCs and odour concentrations were fairly low. However, despite the low odour concentrations, there are numerous point sources in the area of the plant and therefore the odour burden to the surroundings is created as the sum of all the emission sources, with it not being possible to ignore the effect of odours to the surroundings. If the precise emission rates and total odour burden estimation to the environment are to be defined, an extensive set of very demanding emission flow measurements need to be conducted. When considering the flow measurements, they can be very difficult and demanding to carry out in open odour sources where there is no active air flow in use (as in the pre-treatment tanks). The emission flow can be so small that it is not possible to measure them properly. Therefore, it would be desirable to develop easier and field friendly measurement methods or analysers in order to improve the accuracy of analyses as well as reduce the measurement costs.

The highest odorous compound concentrations were determined in the sludge processing units as well as in the incoming gas to the biofilter consisting of pre-treatment unit gases including the sand separation and bar screen exhaust gases. However, in these concentrations, the compounds are not considered as a health or environmental threat, with the risk for exposure being minimal.

The biofilter has reduced the odour from the pre-treatment units by 67.7% on average and there is still the possibility to improve the removal efficiency. In addition, today the stronger odour emissions are restricted to the days when maintenance operations, for example sludge transporting, are being carried out.

According to the survey, odour emission is considered to be quite a minor problem in the residential area. The strongest odour annoyance was determined among the residents living at a distance of 200–300 m from the plant to the south and north, naturally. Despite the good odour abatement at the plant, these opinions cannot be ignored and therefore the plant is still obligated to reduce odour emissions as well as, for example, carry out odour emission modelling.

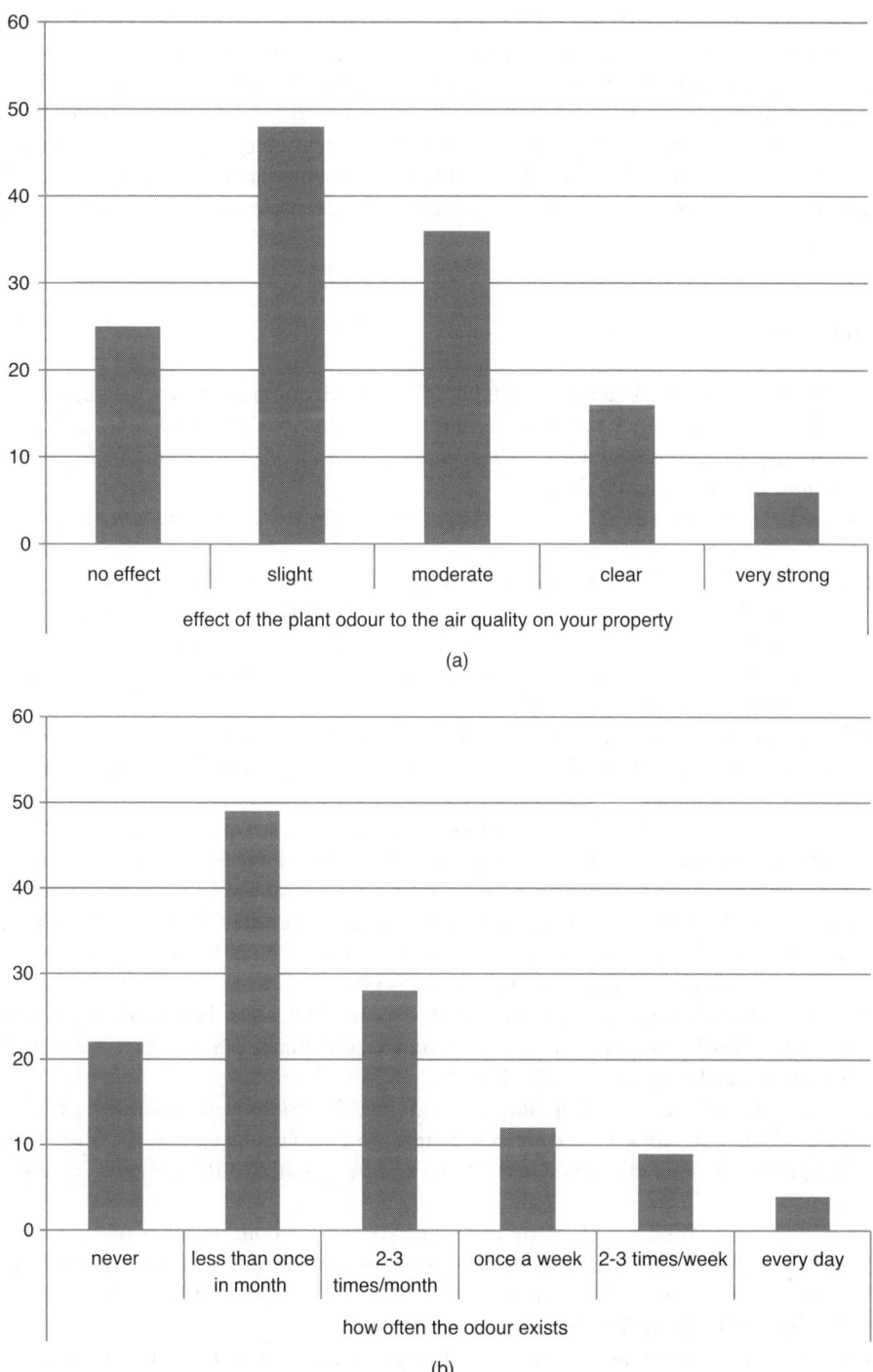

Figure 8.1.4 *(a) Distribution of answers about the effect of odour to the overall air quality and (b) distribution of answers showing the frequency of odour existing in answerers' property.*

The odour emission from the plant is not very strong under normal process conditions. The odour emission from the pre-treatment unit has decreased by 67.7%, following the installation of the biofilter unit in 2004. According to the CEO of the plant, no complaints about the odour have been since 2008, so the situation with odour management is very good. Despite the good odour abatement at the plant, occasional stronger emissions occur during the transporting of sludges. Therefore, environmental authorities have obliged the plant to reduce odour emissions further as well as, for example, carry out odour emission modelling

References

Albrecht, A., Fischer, G., Brunemann-Stubbe, G., *et al.*, Recommendations for study design and sampling strategies for airborne micro-organisms. MVOC and odours in the surrounding of composting facilities, *International Journal of Hygiene and Environmental Health*, **211**, 121–131 (2008).

Arnold, M., (1995) *Hajuohjearvojen perusteet (Principles for odour emission guidelines)*, VTT research notes: 1711.

Bonanni, E.A., The addition of chemicals to liquid to control odors, in H. J. Rafson (ed.), *Odor and VOC Control Handbook,* pp. 8.2–8.17, McGraw-Hill, US (1998).

Finland's Environmental Administration *Jätevesien Puhdistus (Wastewater Treatment)*. Available at: http://www.ymparisto.fi/default.asp?contentid=317538&lan=FI (accessed 21 January, 2010) (2010).

Hobson, J. and Yang, G., Odour mapping using H_2S measurements, in R. Stuetz and F.B. Frechen (eds), *Odours in Wastewater Treatment,* pp. 214–231, IWA Publishing, UK (2001).

Hvitved-Jacobsen, T. and Vollertsen, J., Odour formation in sewer networks, in R. Stuetz and F.B. Frechen (eds), *Odours in Wastewater Treatment*, pp. 33–65, IWA Publishing, UK (2001).

Lehtinen, J. and Veijanen, A., Odour monitoring by combined TD-GC-MS-sniff technique and dynamic olfactometry at the wastewater treatment plant of low H_2S concentration, *Water, Air and Soil Pollution*, **218** (1–4), 185–196 (2011a).

Lehtinen, J. and Veijanen, A., Determination of odorous VOCs and the risk of occupational exposure to airborne compounds at the wastewater treatment plants, *Water Science and Technology*, **63** (10), 2183–2192 (2011b).

Le Cloirec, P., Andrés, Y., Gérente, C., and Pré, P., Biological treatment of waste gases containing volatile organic compounds, in Z. Shareefdeen and A. Singh (eds), *Biotechnology for Odor and Air Pollution Control*, pp. 280–302, Springer, Germany (2005).

O'Neill, D.H. and Phillips, V.R., A review of the control of odour nuisance from livestock buildings: Part 3, properties of the odorous substances which have been identified in livestock wastes or in the air around them, *Journal of Agricultural Engineering and Research*, **53**, 23–50 (1992).

Rosenkranz, H.S. and Cunningham, A. R., Environmental odors and health hazards, *The Science of the Total Environment*, **313** (1–3), 15–24 (2003).

Ruth, J.H., Odour thresholds and irritation levels of several chemical substances: A review. *American Industrial Hygiene Association Journal*, **47** (1986).

Stuetz, R.M., Fenner, R.A. and Engin, G., Assessment of odours from wastewater treatment works by an alectronic nose, H2S analysis and olfactometry, *Water Research*, **33** (2), 453–461 (1999).

Stuetz, R.M., Gostelow, P. and Burgess, J.E., Odour perception, in R. Stuetz and F.B. Frechen (eds), *Odours in Wastewater Treatment: Measurement, Modelling and Control*, pp. 3–15, IWA Publishing, UK (2001).

van Durme, G.P., Wastewater, in H.J. Rafson (ed.), *Odor and VOC Control Handbook*, pp. 6.36–6.55, McGraw-Hill, US (1998).

Zarra, T., Naddeo, V., Belgiorno, V., *et al.*, Odour monitoring of small wastewater treatment plant located in sensitive environment, *Water Science and Technology*, **58** (1), 89–94 (2008).

8.2 Composting Plant

S. Giuliani[1], T. Zarra[1], M. Reiser[2], V. Naddeo[1], M. Kranert[2] and V. Belgiorno[1]

[1]*Sanitary Environmental Engineering Division (SEED), Department of Civil Engineering, University of Salerno, Fisciano, Italy*

[2]*ISWA (Institut für Siedlungswasserbau, Wassergüte- und Abfallwirtschaft) University of Stuttgart, Stuttgart (Büsnau), Germany*

8.2.1 Motivation for the Study

The current state of the art of solid waste disposal is based on the concept of an integrated management system. As a result, the composting process of the municipal solid waste organic fraction allows for a recovery of biomass as well as a reduction of biowaste for landfilling, according to European Council Directive on the Landfill 1999/31/EC (Belgiorno *et al.*, 2003).

In all the composting processes, the aerobic and/or anaerobic breakdown of solid organic matter by microorganisms is a crucial step (Derikx *et al.*, 1990). Generally, aerobic processes are used to convert biowaste into compost. In these plants, the biowaste is aerated during several weeks, and possibly up to several months by forced suction or blowing, in order to remove moisture and heat as well as create an optimal environment for the aerobic mesophilic and thermophilic microorganisms performing the biodegradation (Haug, 1986).

A relevant environmental impact of operating aerobic composting plants is the odour pollution due to the emission of volatile compounds (Krauss *et al.*, 1992). Emission of volatiles already starts upon arrival of the fresh biowaste to the composting plant. According to Eitzer (1995), most volatile organic compounds (VOC) in aerobic composting plants are emitted at the early processing stages, that is, at the tipping floors, the shredder and the initial active composting region. Pöhle and Kliche (1996) classified the aerobic composting process (ACP) in an acid start stage, a thermophilic stage and a cooling stage, with the production of specific odorants in each stage. Anaerobic conditions in composting piles due to incomplete or insufficient aeration will produce sulfur compounds with an intense smell, while incomplete aerobic degradation processes result in the emission of alcohols, ketones, esters and organic acids (Homans and Fischer, 1992). Zarra *et al.* (2009) identified limonene, 2-butanone and α-pinene as key VOC compounds at a composting facility to evaluate odour emissions and their impact on the surrounding area.

Odorous emissions from composting plants are essentially caused by the presence of reduced catabolic exhaust gases in the air (Favoino, 2002). In composting plants, different diffusive and non-diffusive odour sources can be identified (Zarra, 2007). The composition of the exhaust air is indicative of the composting process (Day *et al.*, 1999) as well as the quality of the compost. For instance, under anaerobic conditions, specific compounds are released, such as carboxylic acids (Brinton, 1997) and ammonia (Beck-Friis *et al.*, 2001). The compounds emitted depend on the type of waste material (for the early stages of composting), age of the pile, temperature, O_2 level, humidity and the pH (Tchobanoglous *et al.*, 1993). There are typically three origins of the released substances: compounds of the waste (their emission is a function of their vapour pressure that depends on the temperature, like the hydrocarbons), biogenic components (aerobic and anaerobic degradation by micro-organisms) and abiogenic substances (released by purely chemical reactions) (Day *et al.*, 1999; Brinton, 1997; Komilis *et al.*, 2004).

Odours from a composting plant are usually dispersed several kilometres into the surroundings from the emission sources, depending on the weather and topographical conditions. However, the dispersed odours are generally only judged as significant within 500 m off site (Zarra *et al.*, 2009). Legislation is, consequently, oriented to set minimum distances from the facility to the nearest housing (Schlegelmilch *et al.*, 2005) to reduce the potential impact.

In this study, dispersion modelling and field inspections were used to assess the odour impact of a composting plant.

8.2.2   Material and Methods

8.2.2.1   *Location, Local Conditions and Relevant Receptors*

The study was carried out at a composting plant located in Teora (AV) (Campania Region, Southern Italy) (see Figure 8.2.1).

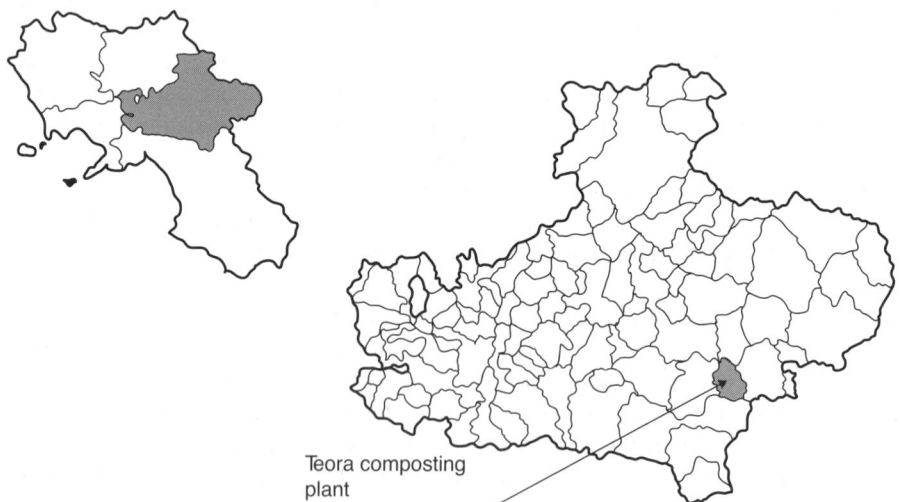

Teora composting
plant

Figure 8.2.1   *Localization of Teora (AV) composting plant in the Campania Region, Avellino, Italy.*

Table 8.2.1   *Characteristics of the Teora composting plant.*

Parameter	Data
Design capacity	6000 t/y
Treated quantity	2500 t/y
Input fractions of SSOF	55%
Input fractions of vegetables	45%
Input fractions of sludges	0%

The design capacity, treated quantity and input materials, in terms of SSOF (Source Separated Organic Fraction), vegetables and sludges from biological wastewater treatment plants are summarized in Table 8.2.1.

The Teora composting plant has a total annual capacity of 6000 t of VFG-waste (Vegetable, Fruit and Garden waste). The process takes place in under-pressure sheds. Under these conditions, odours emitted by the process can be collected and treated.

The composting system was characterized by a set of 12 biocells divided in two sets of six active units respectively. The flow chart of the treatment units of the composting plant is shown in Figure 8.2.2.

Active composting time in biocells was 15 days, with the temperature and moisture content being automatically controlled. Air was blown into every biocell at a flow rate

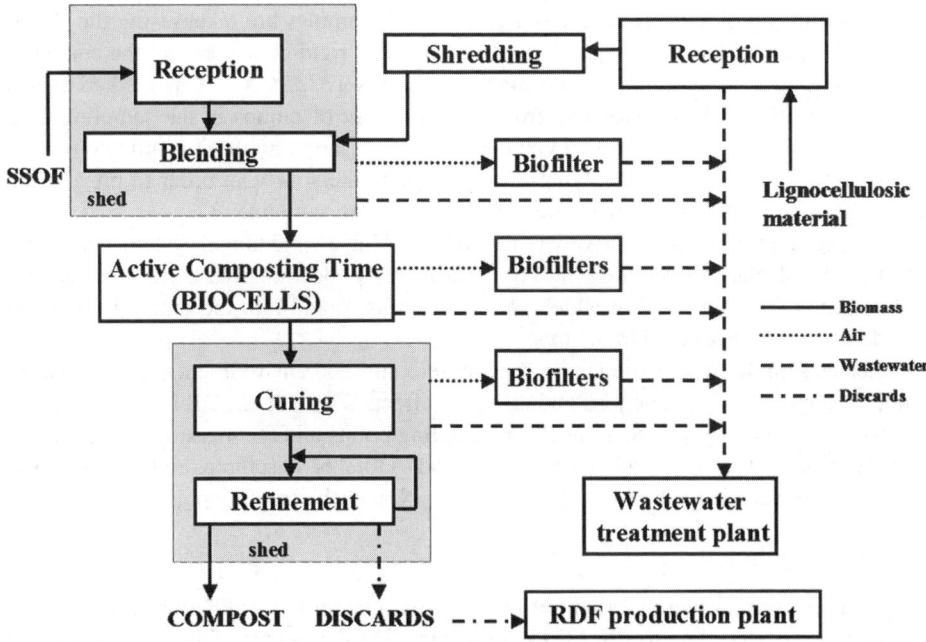

Figure 8.2.2   *Flow chart of the treatment units of the composting plant at Teora (AV, Italy).*

Table 8.2.2   *Characteristics of the odour abatement systems in Teora.*

Characteristics	Static biofilter	Mobile biofilter
Size	1.45 × 9.8 × 2,5	6.0 × 2.2 × 2.1
Filling height	1.45 m	1.45 m
Contact time	0.52 s	0.46 s
Gaseous affluent velocity	0.87 m/s	1.12 m/s
Load loss	16 mm	14 mm

of 600 m^3/h, while an air flow rate of 720 m^3/h was sucked, so as to set commensurate depression conditions.

The curing phase was carried out in an enclosed shed by means of periodically turned windrows system (two or three times per week), for a period of 40–50 days. Moreover, the plant was equipped with odours treatment facilities: six mobile biofilters (two for the two sets of biocells and four relative to the curing shed) in containers like biocells and one static biofilter (for the shed of SSOF reception, blending and biocell loading). Table 8.2.2 reports the main characteristics of the odour abatement systems present at the Teora plant.

During the monitoring activities, the composting plant was managed under the following constant condition: input quantity of treated waste reached 40% (8 t/d) of the daily potentiality (20 t/d).

8.2.2.2   Sampling Program

A complete characterization of the odour emission of the composting plant was carried out by the monitoring of six different odour sources. Air samples are taken using the 'lung' technique, whereby the sampling bag is placed inside a rigid container and the container evacuated using a vacuum pump in accordance with EN 13725:2003. This method avoids any contamination, which may arise from the direct use of pumps in the sampling line. Nalophan® sampling bags with a 7 l volume are used for the sampling. Sampling on active area sources was carried out applying a total cover to the source in order to prevent the sample from being influenced from the ambient air during sampling.

Sampling on passive area sources was carried out using a wind tunnel system, consisting of a PET hood placed over the emitting surface. Air collection was carried out after a contact time of 10 min, during which an air flow was forced into the tunnel in order to provide a flux velocity equal to 0.3 m/s.

Collected samples were stored in a temperature-controlled environment at 20°C, so as to minimize phytochemical reactions and diffusion effects (Zarra *et al.*, 2011).

Table 8.2.3 describes the position of the sampling points and the measurement program carried out during the two month monitoring period. A total of 48 samples and concentration evaluation were carried out with a weekly frequency at each sampling point.

8.2.2.3   Analysis

Collected air samples were carried out by dynamic olfactometry. Olfactometric analyses were conducted according to EN 13725:2003 at the SEED (Sanitary Environmental Engineering Division) Olfactometric Laboratory of University of Salerno using an olfactometer

Table 8.2.3 *Sampling points and measurements program.*

ID	No. of analyses	Location
P1	8	wastewater treatment plant
P2	8	shed 1 (blending) – crude gas
P3	8	shed 2 (curing)
P4	8	static biofilter – clean gas
P5	8	biocells
P6	8	pile of compost

model TO8 (ECOMA GmbH, D), based on the 'yes/no' method. All the measurements were carried out within 30 h after sampling, relying on a panel composed of four trained panellists. The odour concentration was calculated as the geometric mean of the odour threshold values of each panellist, multiplied by $\sqrt{2}$.

The Odour Emission Rate (OER, OU/s) associated with each emission was calculated as the product of the Specific Odour Emission Rate (SOER, OU/s m^2), determined according to the equation proposed by Jiang *et al.* (1995), and the emitting surface (m^2) of the considered area source. Air flow velocity (m/s) from the source was measured through a Kestrel® Pocket Wind Meter (Nielsen-Kellerman, PA, US) anemometer.

8.2.2.4 Dispersion Model

A Calpuff model system (Earth Tech. Inc., 2000) was used for the simulation of the odours emission dispersion into the surrounding area. In the model, a spatial domain of 2000 m × 2000 m, with a square grid of receptors every 200 m was set. The characterization of the 'terrain following' was carried out by seven vertical layers. Coefficients of used land were selected according to the Scire *et al.* (2000) proposition. Hourly average data of meteorological parameters (wind direction, wind speed, pressure, temperature, precipitation) were collected for a period of 12 months at the meteorological station localized in the composting plant on the office building. The level of clouds and cloud cover were considered constant and equal respectively to 1500 m and 5/10 (Scire *et al.*, 2000).

A long term simulation was carried out, referring to one year of data and with a calculation step of 1 h, implemented with an odour emission rate determined through olfactometric measurements. For the dispersion evaluation of each unit, modelled as area sources, the odour emission rate values calculated considering the highest values of specific odour emission rate (SOER) detected out over all analysis period were used.

According to the German GIRL guidelines (Geruchsimmission–Richtlinie), since it was an industrial area, odour impact was evaluated using the methodology of peak to mean ratio (P/M), checking the maximum frequency of the odours and with this being understood to mean the relative frequency of times when odours are clearly perceptible, fixed at 1 OU/m^3, not exceeding 15% odour hours.

8.2.2.5 Field Inspection

Field inspection was implemented according to VDI 3940 Part II: 2006. The investigated area covered a surface of 1.5 km^2 and was divided into a squared grid, with the sides of

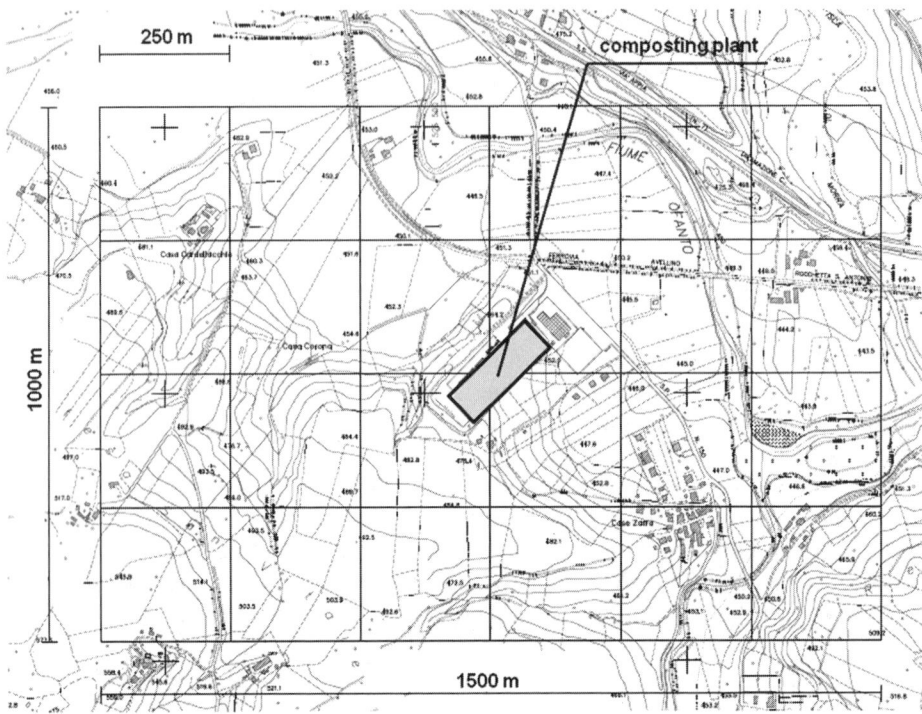

Figure 8.2.3   *Squared grid of the investigated area.*

the mesh being fixed as 250 m (Figure 8.2.3). The theoretical grid was adapted to the real world area by moving the intersection points when necessary to an accessible point of the ground.

Odour monitoring was carried out at all the corner points identified from the grid of the investigated area, twice a week for a three month period. The field panels consist of 6–8 trained, qualified panel members selected using the same criteria as for the odour laboratory, according to EN13725:2003. A total of 26 measurements for each corner point were carried out. For each measurement, the odour hours (n_h) for each corner point as well as the odour load (OL) for each mesh according to VDI3940 were calculated. Meteorological measurements (wind velocity and direction, temperature, humidity) were carried out during the odour field inspection observations.

For the field inspection, the odour impact was carried out according to the German GIRL guidelines (Geruchsimmission–Richtlinie) checking the maximum frequency of the odours and with this being understood to mean the relative frequency of times when odours are clearly perceptible, not exceeding 15% odour hours, for industrial areas.

In this research, the results of the field-inspection analyses, combined with the meteorological conditions during the field observations, were used to validate the dispersion modelling results according to 'reverse dispersion modelling' (RDM).

8.2.3 Results and Discussion

8.2.3.1 Odour Emission Characterization

Table 8.2.4 shows the detected odour emission concentrations by dynamic olfactometry over the investigated period at the sampling points (P_i) of the composting plant.

Table 8.2.4 *Detected odour emission concentration by dynamic olfactometry.*

Sampling week	Odour concentration (OU/m³)					
	P1	P2	P3	P4	P5	P6
1	6235	2521	1460	285	345	169
2	4103	3215	2477	153	386	174
3	5236	3121	2703	278	451	106
4	4978	2698	951	105	398	109
5	3265	1589	2106	314	258	154
6	1488	4005	589	197	356	175
7	2364	1893	1112	209	223	201
8	1028	3332	1861	258	254	184

The results show that the highest odour concentration was detected at the wastewater treatment plant (6235 OU/m³) while the lowest for the compost pile (106 OU/m³).

Table 8.2.5 shows the values of maximum specific odour emission rate ($SOER_{max(i=1-8)}$), calculated out over all the analysis period at the investigated sources and the corresponding OER values.

The results show that the wastewater treatment plant has the maximum SOER value (90.3 OU/m² s). Furthermore, the static biofilter has the highest OER value (3175.5 OU/s).

8.2.3.2 Field Assessment

Figure 8.2.4 shows the odour load (OL) calculated for the investigated area for each mesh.

The results show that the odours emissions of the composting plant cause elevate nuisance in their immediate surroundings. While a decrease of the odour perception is detected with

Table 8.2.5 *SOER Values and corresponding OER values.*

ID	Area (m²)	$SOER_{max(i=1-8)}$ (OU/m² s)	OER (OU/s)
P1	15	90.3	1354.5
P2	76	11.5	874.0
P3	58	9.4	545.2
P4	145	21.9	3175.5
P5	88	7.2	633.6
P6	90	0.8	72.0

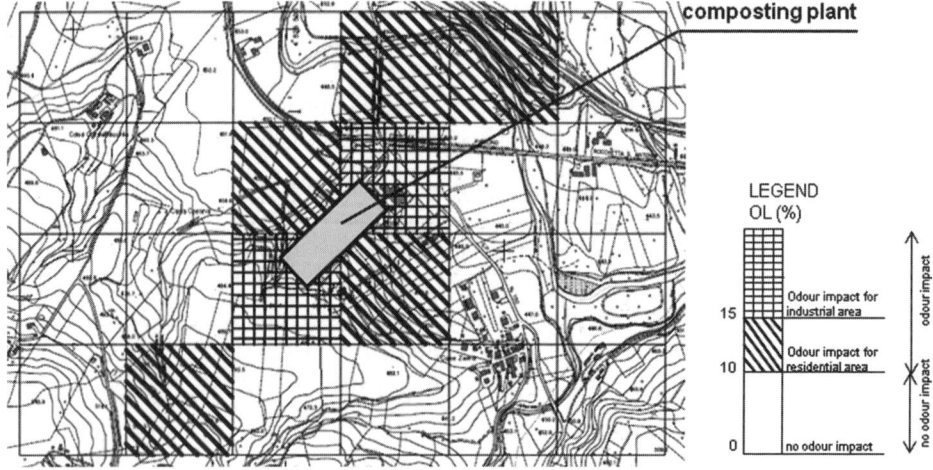

Figure 8.2.4   Odour load for investigated area according to VDI3940.

an increasing distance from the plant. Seven meshes out of 24 are impacted by odour of which two have a strong odour impact.

8.2.3.3   Odour Impact Assessment

The results of the odour dispersion simulation using the Calpuff Model System and the OER values as data input, determined by dynamic olfactometry, are shown in Figure 8.2.5.

The results show that the maximum distance at the composting plant where the odour impact, calculated according to the German guidelines, is observed at 550 m in a south-south-west direction.

Figure 8.2.6 reports the overlap of the odour impacts results obtained by applying the dispersion models and the field inspections, according to the 'reverse dispersion modelling' validation.

The results show a good correspondence between the dispersion model elaborations and the field inspection assessment for the evaluation of the odour impact area. All the meshes characterized with an odour impact (OL >10%) by field inspection are located into the odour impact area determined by the dispersion modelling according to the German guidelines. This confirms the validity of the adopted modelling procedure, without need, in this specific case study, of their calibration phase.

8.2.4   Conclusions, Recommendations and Outcome

The odour impact assessment of a composting plant was presented. In this study, the evaluation was carried out through the use of dispersion models, starting from the characterization of the emission using olfactometric analyses according to EN13725:2003. While a field assessment study was used to validate the obtained results, according to 'reverse dispersion modelling'. For both approaches, the odour impact assessment was performed based on the

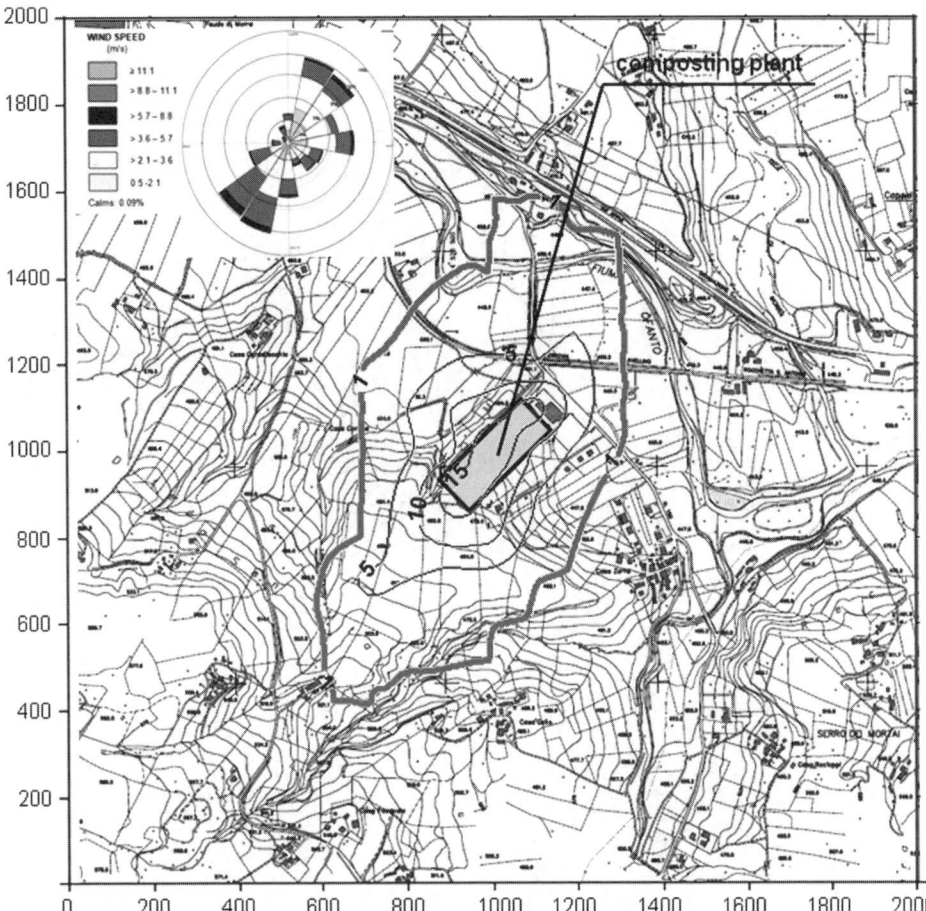

Figure 8.2.5   *Relevant isopleths to the 85th percentile according to German guidelines and odour annoyance level.*

German GIRL guidelines, where it was evaluated in terms of frequency of exceeding of the concentration on the basis of 'odour hour'. Specifically, a Calpuff model system for a long term application was used for the odour dispersion evaluation.

The results show that for a composting plant, one of the major odour emission sources is related to the treatment of the produced wastewater. The work shows that dispersion modelling and field measurements provide a suitable method for determining the odour impact from an odour annoyance emission site. These two distinct approaches can be used independently, but it is always recommended to apply them in an integrated manner. An integrated procedure makes it possible to validate the odour impact area as well as increase the obtained results and confidence of the work done against the social partners. Field assessment, through field inspection with trained observers or sociological questionnaires

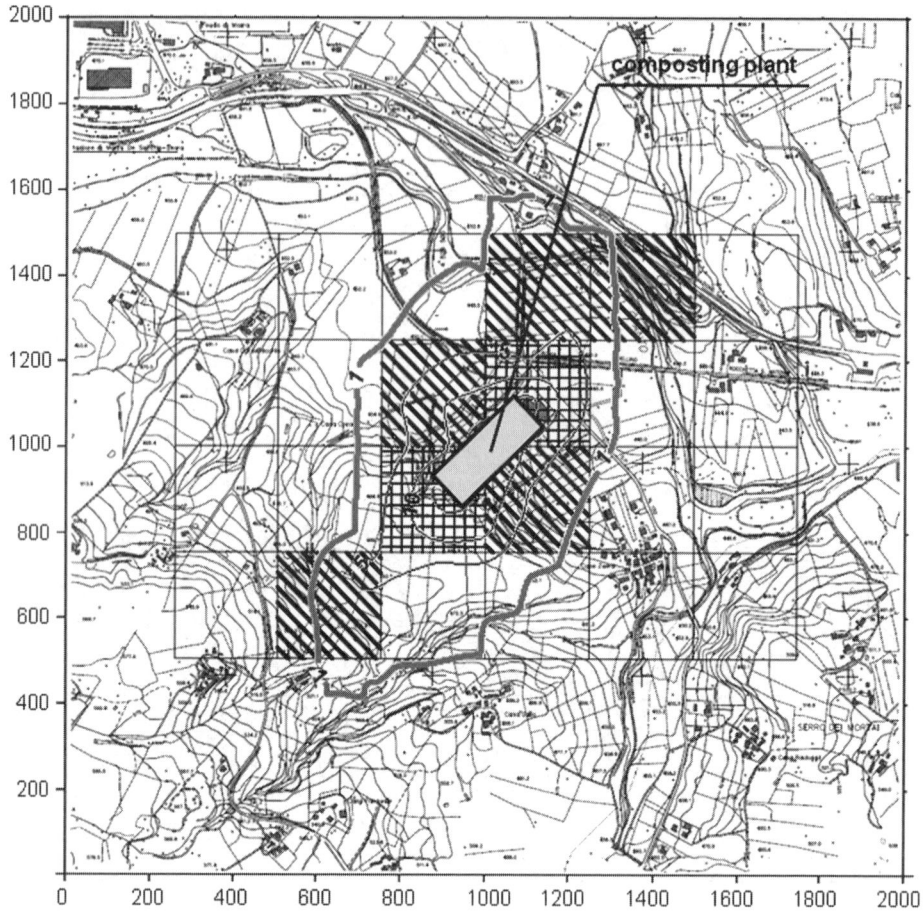

Figure 8.2.6 *Correspondence between dispersion model and field inspection.*

given to the surrounding population, are suggested to obtain the involvement of the annoyed population and their confidence in problem analysis and resolution.

References

Beck-Friis, B., Smars, S., Jonsson, H. and Kirchmann H. (2001) Gaseous emission of carbon dioxide, ammonia and nitrous oxide from organic household waste in a compost reactor under different temperature regime. *J. Agric. Eng. Res.*, **78** (4), 423–430.

Belgiorno, V., De Feo, G., Panza, D. and Napoli, R.M.A. (2003) An 'alternative' disposal for Refuse Derived Fuel. *Proceedings Sardinia 2003, Ninth International Waste Management and Landfill Symposium*, S. Margherita di Pula, Cagliari, Italy, 6–10 October.

Brinton, W.F. (1997) Compost volatile organic acids- Production and odorant aspects, *Proceedings of the Sixth Annual Conference on Composting*, Beltsville, MD, 11–13 October.

Day, M., Shaw, K. and Krzymien, M. (1999) Composting odours: what can chemistry tell us? *Proceedings of the International Composting Symposium*, Halifax/Dartmouth Nova Scotia, Canada, 19–23 September, 1999.

Derikx, P.J.L., Op Den Camp. H.J.M., van der Drift, C. *et al.* (1990) Odorous sulfur compounds emitted during production of compost used as a substrate in mushroom cultivation. *Appl. Environ. Microbiol.*, **56**: 176–180.

Eitzer, B.D. (1995) Emissions of volatile organic chemicals from municipal solid waste composting facilities. *Environ. Sci. Technol.*, **29** (4), pp. 896–902

EN 13725 (2003) *Air quality – determination of odour concentration by dynamic olfactometry*, Comitè Europè en de Normalisation, Brussels, pp. 1–70.

Favoino, E. (2002) Gli odori negli impianti di compostaggio. La prevenzione e la gestione del problema, Emissioni: Gas e odori nel trattamento dei rifiuti, Rapporti GSISR.

Haug, R.T. (1986) Composting process design criteria, part 1: feed conditioning. *BioCycle*, **27**, 8, 36–43.

Homans, W.J. and Fischer, K. (1992) A composting plant as an odour source, compost as an odour killer. *Acta Horticulturae*, **302**, 37–44.

Jiang K., Bliss P.J. and Schulz T.J. (1995) The development of a sampling system for determining odor emission rates from areal surfaces: part I. aerodynamic performance. *Journal of the Air & Waste Management Association*, **45**, 917–922.

Komilis, D.P., Ham, R.K. and Park J.K. (2004) Emission of volatile organic compounds during composting of municipal solid wastes, *Water Res.*, **38** (7), 1707–1714.

Krauss, P., Krauss, T., Mayer, J. and Wallenhorst, T. (1992) Examination of odour formation and odour reduction in composting plants. *Staub Reinhalt. Luft.*, **52**, 245–250.

Pöhle, H. and Kliche, R. (1996) Emission of odors from composting of biological waste. *Zentralbl. Hyg. Umweltmed.*, **199**, 38–50 (in German).

Schlegelmilch, M., Streese, J., Biedermann, W. *et al.* (2005) Odour control at biowaste composting facilities. *Waste Management*, **25** (9), 917–927.

Scire, J.S., Strimaitis, D.G. and Yamartino, R.J. (2000) *A User's Guide for the CALPUFF Disperion Model*. Earth Tech, Inc.

Tchobanoglous, G., Theisen, H.V. and Samuel, A. (1993) *Integrated Solid Waste Management: Engineering Principles and Management Issues*. McGraw-Hill.

VDI 3940 (2006) Measurement of odour impact by field inspection – Measurement of the impact frequency of recognizable odours – Grid measurement.

Zarra, T. (2007) *Procedures for detection and modelling of odours impact from sanitary environmental engineering plants*. PhD Thesis, University of Salerno, Italy.

Zarra, T., Naddeo, V. and Belgiorno, V. (2009) A novel tool for estimating the odour emissions of composting plants in air pollution management. *GLOBAL NEST Journal*, **11** (4), 477–486.

Zarra, T., Reiser, M., Naddeo, V. *et al.* (2011) A comparative and critical evaluation of different sampling materials and methods in the measurement of odour concentration by dynamic olfactometry. *Proceedings of NOSE2012 Conference*, 23–26 September 2012, Palermo (Italy).

8.3   Landfill of Solid Waste

A.C. Romain and J. Nicolas

Department of Environmental Sciences and Management, Arlon Campus Environment, Faculty of Sciences, University of Liége (ULg)

8.3.1   Motivation for the Study

There are two main motivations for this study. The first is linked to the decision made by a regional authority. Indeed, in 1998, the environmental authorities of Wallonia decided to organise a monitoring network of sanitary landfills. The objectives were the improvement of the standards in the landfill process and, in particular, monitoring. This environmental management tools were put in place with a slight development in the philosophy of the EU landfill Directive (EEC/1999/31/EC). The study covers the monitoring of liquid emissions in order to control their impact on ground waters and the monitoring of gaseous emissions as well as the assessment of their influence on ambient air quality. An interdisciplinary air survey was carried out around 12 municipal solid waste (MSW) landfills in Wallonia (Belgium). The surveying campaigns included four axes of investigations: landfill gas (LFG) surface emissions detection, ambient air quality control, odours annoyance assessment and measurements of the exhaust fumes from the LFG valorisation units. In this context, ISSEP (Public Scientific Institute) was designated to manage this network and selected the ULg research team to carry out the odour investigation. Among the 12 landfills under investigation, the 'Habay, Les Coeuvins' (HLC) landfill was presented.

The second motivation is that various studies have been conducted by the research team and its 'odometric' spin-off on this MSW. This particular landfill is fully accessible to the research team and considered to be a laboratory in which various odour studies can be realized.

8.3.2   Description of the Situation

8.3.2.1   Process and Installations

The landfill site has received municipal solid wastes and non-hazardous, inert industrial wastes since 1979. HLC is a waste treatment plant with different installations and sorting systems (see Figure 8.3.1).

The non-hazardous and non-recyclable inert industrial wastes as well as the bulky household wastes are deposited by trucks in the exploitation cell of the landfill. The oldest cell of the landfill was restored in 2005 and the second one is currently in the rehabilitation phase. Landfill gas (LFG) collection systems were installed in 2005 on the first cell, in 2006 on the second and, in 2009 on the operating cell. In mid-2006, a torch was installed for the LFG collected from the two old cells. Recently, since 2008, a motor has been used to produce electricity with the LFG collected from the whole landfill area.

The non-organic fractions of the domestic garbage are dumped in a cavity with a capacity of $1000\,m^3$. The metallic parts are sorted and the non-organic wastes are crushed. Previously, the organic part of the domestic solid waste was transformed in compost. However, since 2009, the organic matter has been transported to another site to be transformed into biogas for energy production. The non-metallic inorganic residues (plastics, food packaging . . .) of these municipal wastes are deposited into cubic compartments inside the shelter where

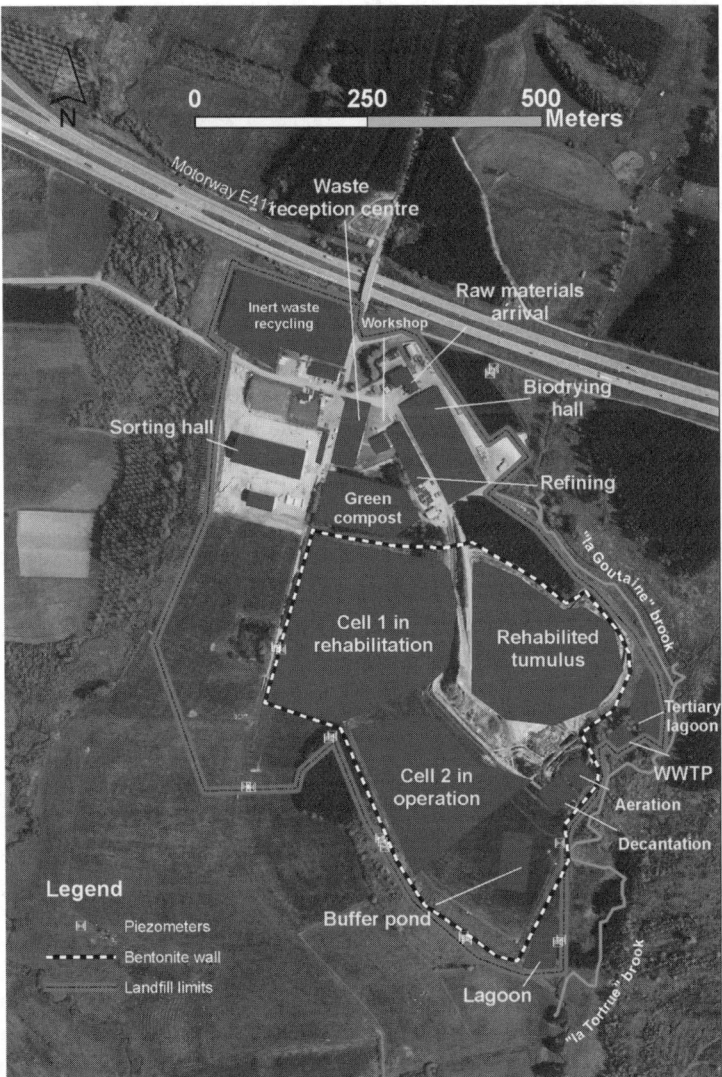

Figure 8.3.1   *Installations of the considered landfill.*

they are dried by air, pulsed from small holes in the concrete slab. This drying hall was previously used for the composting process. The dried residues are then exported to another plant to be either transformed into combustibles for industrial sites like cement factories, paper mills or dropped on the landfill. The site also receives vegetables waste like branches, leaves, yard waste, and Christmas trees. They are composted in open air windrows that are frequently turned.

The annual average load is about 60 000 tons for dried fraction, 20 000 tons for vegetables waste, 20 000 tons for inert industrial waste and 20 000 tons for bulky household waste. Household papers, plastics bottles, soda cans, and clothes are sorted manually by several employees in a sorting hall. A wastewater treatment plant (WWTP) is located downstream the site to clean the leachates before their release into the nearby river.

The landfill site covers an area of 45.5 ha, included sorting, treatment, deposal, technical and administrative zones. The waste deposal area of the landfill is 15.5 km^2 with a filling capacity of 2.1×10^6 m^3. The wastes are deposited in tumulus which has created an elevation change of the landfill area.

The evolution of this site and its activities has modified the odour sources. Indeed, two years ago, the municipal solid waste was either composted or dumped on the exploitation cell. These activities were the most significant sources of odour complaints. Today, due to the development of regulations as well as the positive inclination towards recycling and producing energy with wastes, the household waste is transformed into biogas for the production of electricity on another site.

The four potential sources of odour annoyance for the HLC landfill are therefore the storage of raw waste, the waste drying hall (also called bio-drying hall), the LFG leakages and the windrows of green compost. The odour generated by the wastewater treatment plant is weak and not dispersed into the habitat area.

8.3.2.2   Location, Local Conditions and Relevant Receptors

The HLC landfill is located in the south of Belgium (in Wallonia), in a rural region, characterised by a temperate climate, with a prevailing wind coming from the south-west. The topography around the site is quite unvaried.

To the north and east of the site, the areas are forest zones. To the south and south-west, there is an agricultural zone. There are a few companies in a small economic park near the site and the closest habitat area is located more than one kilometre to the north of the landfill. A farm (the nearest house) is located about 900 meters to the south of the site. Within a radius of 2 km, the residents of several villages can perceive the odours generated by the MSW landfill (see Figure 8.3.2). The site is along a highway and the motorists can smell the odours. Even if this perception is brief, it gives a bad image of the site.

8.3.3   Specific Objectives of the Study

There are several objectives of the study. The ultimate goal is to assess the odour impact of the landfill according to the authorities' decision. In order to reach this objective, the development of a methodology is solicited. The use and optimisation of different techniques to measure landfill odours is the first step to define and harmonize a methodology to assess the odour impact of a MSW landfill. To validate the results obtained by the different approaches, relationships with the odour perception in the neighbourhood have to be established. This study has to initiate this validation procedure for the Wallonia landfills.

Another objective is to compare the odours of the site in different periods, for instance after process modifications. For the HLC landfill, two campaigns were carried out, one in 2005 and the second in 2009 after several process optimisations.

The ULg team has been working on the development of electronic noses to assess the odours in the environment (Romain *et al.*, 1997; Nicolas *et al.*, 2006b) for several years However, while detailed research in e-nose technology has resulted in significant progress in the domain of continuous odour monitoring in the field, more successful long term case-studies are still needed in order to overcome the early overoptimistic performance expectations. To go ahead with this objective, a network of e-noses has been installed

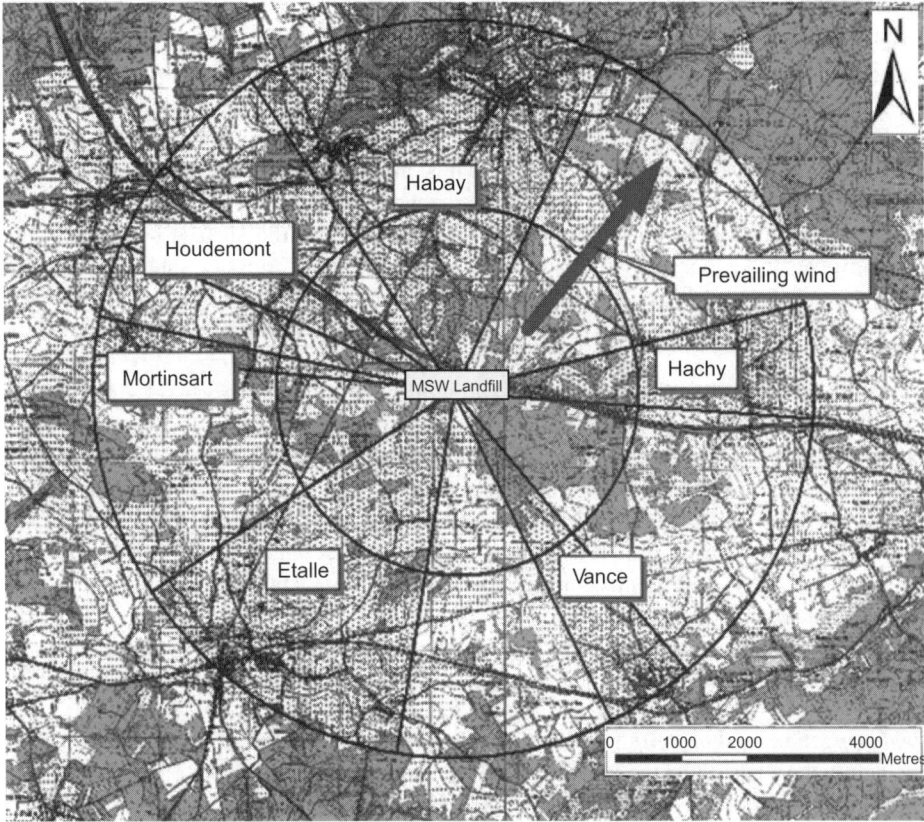

Figure 8.3.2   *The surroundings of the landfill.*

around this landfill to improve the outputs of this technology in the prediction of odour impact (Muñoz *et al.*, 2010).

8.3.4   Methodology

Several methods have been used, compared and improved. Depending on the specific objective, one method can be preferred to another. These methods used sensorial approaches (odour sampling and dynamic olfactometry, field perception and dispersion modelling, questionnaire, resident diaries, 'watchmen'), chemical analyses and E-nose technology. The interests and disadvantages of these complementary techniques are discussed in (Romain *et al.*, 2008).

8.3.4.1   Sensorial Techniques

Dynamic olfactometry and sampling.   Dynamic olfactometry (DO) has been performed in the laboratory, according to the European standard EN13725 with the Odile olfactometer (Odotech Inc., Canada). A panel of six members judges samples of odour gas. A decreasing step sequence in the geometric series of a factor of 1.58 and the triangular choice are used. A 'No odour' response is allowed. Only answers 'with certainty' are considered as correct.

DO is used to obtain the European odour concentration of samples, in ou_E/m^3. This analysis is carried out after odour gas sampling in the field. Odour gas is collected in a Tedlar bag placed in a sealed-barrel maintained under negative pressure by a vacuum pump.

To assess the odour impact in the field, the odour emission rate has to be known. A landfill site is typically an area source with various odours, for which the effluent flow rate cannot be measured like in a chimney. Thus, an evaluation of the flow rate has to be performed using for instance a flux chamber or a wind tunnel.

Multiplying the odour concentration by the flow rate for the surface covered by the chamber gives the odour emission rate ($ou_E/s\ m^2$). The global emission rate for the whole emitted area is then evaluated by multiplying the odour emission rate by the total emission area (ou_E/s). This value can be introduced into a dispersion model to calculate the odour percentiles.

However, a MSW landfill is characterised by different dynamic processes and different odour sources, with it not being possible to evaluate the flow rate with a flux chamber. How can the flow be evaluated during the windrows aeration by turning, during passages of trucks filled with odorous waste or emerging from diffuse sources and on a large heterogeneous area? A flux chamber and spot samples collection are really not appropriate in determining a global emission rate for a landfill. The exactitude of the value obtained would be very weak. To evaluate a global emission rate for a landfill, an adaptation of the sniffing team campaigns described in German standards (VDI, 2003) is preferred.

For this case study, odour gas sampling with dynamic olfactometry was only performed to quantify the various sources in order to classify them as well as calibrate the e-noses.

Field inspection method coupled with reverse modelling and percentile evaluation.   This original method is an adaptation of the sniffing team campaigns for the particular case of landfill (Nicolas *et al.*, 2006a; Van Langenhove and Van Broeck, 2001). The global odour emission rate of the site is determined with an evaluation of the odour concentration in the field.

In a first step, the assessors (or the 'sniffing team': a team of at least three tested persons, whose perception has been checked against n-butanol) delineate the region in which the odour is perceptible. They build the perimeter by odour detection downwind around the plume axis. The 'odour'-'non-odour' perceptions are recorded and located on a map thanks to a GPS device. A description of the odour source (biogas, fresh waste, compost . . .) and the intensity complete the observations. A portable weather station simultaneously gives the odour measurement: the wind direction, the wind speed, the temperature and the solar radiation. The typical duration of a sniffing field inspection is from 20–60 min. After this measurement, reverse modelling is applied. The emission rate entered into the models is adjusted until the simulated average isopleths for 1 OU/m^3 (without 'E' subscript, valid only for olfactometry according to the European standard) at about 1.5 m height fits the observed perimeter and the maximum perception distance. The field inspection is repeated for various climatic conditions and different emission characteristics. Finally, the mean emission rate is introduced into the same dispersion model for the typical reference year of the region to calculate the percentiles. A long term dispersion area can then be used to delineate a zone of potential nuisance.

Typically, for this study case, about 10 campaigns were organised. The field inspection was carried out between 10 am–2 pm. A bi-Gaussian model (Tropos from Odotech) was used. If the relief had been less flat around this landfill, a 3D model would have been applied.

Questionnaire survey and residence diaries ('watchmen').   This approach has become very popular in Wallonia. The implication of the residents and the creation of a discussion group where all the actors are implied (the residents, the landfill manager team, the local authorities, the 'odour expert team') has shown its efficiency (Nicolas *et al.*, 2010). Social participation and community involvement in detecting odour events offers significant public relations benefits for the concerned stakeholders. This approach is the most relevant in assessing the odour impact perceived by the residents.

If the residents are trained with regular contact with the odour team, the percentages of answers are often more than 30% and with a high reliability level. To increase the objectivity of the study, a statistical treatment of the data has to be applied to identify the false responses. This approach was used to assess the impact in the neighbourhood as well as compare the resident's perception with the results of the dispersion model. Identification of the most frequent odour source(s) was also an issue of this approach. Moreover, the results are exploited to improve the information given by the e-noses.

Questionnaire.   A preliminary survey was conducted among the population to describe the general situation of their living environment concerning the odour and noise. For this study, the proposed questions (see Figure 8.3.3) were adapted from VDI 3883 Part 1 guideline (VDI, 1997).

In the mail sent to the population to present the survey, it was also asked if they would accept to participate in a regular recording of their perception over several months.

Resident diary.   After analysing the questionnaires, a form adapted to this case study was distributed to the volunteers. The perception of these persons was checked. The test consists of measuring the odour intensity of a series of n-butanol solutions in different concentrations, according to the French standard NFX 43-103 ('Méthodes supraliminaires'). The persons who passed the training test are included in the panel.

These panel members have to regularly record their odour perception, respecting some requirements, if possible:

- At least three days per week (Tuesday, Thursday and Sunday) and two times a day (two set periods: 7–9 am and 6–8 pm), but any particular observation outside those ranges can be mentioned,
- Sniffing near their home, but outside,
- Estimation of the odour rating on a six-level scale (0 = no odour, 1 = scarcely perceptible, 2 = weak, 3 = sharp, 4 = strong, 5 = very strong, 6 = unbearable),
- Selection of a unique odour descriptor among a list of specific terms,
  Data are encoded and processed by the odour research team and its odometric spin-off.

8.3.4.2   Chemical Analyses

Air above specific odour sources are sampled on Tenax cartridges and analysed by TD-GC-MS (from Markes and Thermo). An exhaustive analysis was performed to characterize the different chemical profiles of the sources. This information is useful for the manager in order to better understand the process as well as improve the odour control tools. The

Survey to assess the environmental quality around the site of HLC

Address *(to locate your residence in regard to the landfill))*

....

Phone number *(for complementary information))*

....

Your age range

| < 20 years | from 20 to 40 years | from 40 to 60 years | > 60 years |

What is the type of location?

| Your residence | You workplace |

Since when do you live or do you work in this quarter?

....

Do you think, generally, that the neighbourhood is polluted? | Yes | No |

If you have to rate the nuisance on a scale of 0-10, what level would be reached (0 = no discomfort, 10 = unbearable discomfort) ?

For the noise	0	1	2	3	4	5	6	7	8	9	10
For the odours	0	1	2	3	4	5	6	7	8	9	10
Other pollutions *(identify: ....)*	0	1	2	3	4	5	6	7	8	9	10

Especially regarding the smell, how do you perceive when you're outside of the building?

| No odor | scarcely perceptible | weak | sharp | strong | Very strong | unbearable |

How often do you perceive an odour when you're outside the building?

| never | once a month | two, three times per month | once a week | two, three times per month | about each day |

Usually, how is your sensitivity to the odours? Do you have a "good nose"?

| No responsive | Weak reaction | Moderately responsive | Very responsive | Extremely responsive |

How do you characterize the perceived odours?

| garbage | gas | sewer | chemistry | vegetal |
| livestock | essence | engine exhaust gas | Other:................. | |

Are there situations, periods in the year or days in the week during which you feel particularly the odour?
Specify these periods: | yes | no |

....

Are there any hours during which you perceive a particular odor nuisance | yes | no |
Specify these hours:

....

How odors have evolved since you perceive them?

....

Other comments:

Figure 8.3.3   *Questionnaire distributed to the residents.*

Figure 8.3.4   *Location of the five E-noses (N1 to N5) around the site.*

different chemical patterns are also used to develop an e-noses network adapted to this landfill (Romain *et al.*, 2005).

ISSEP has also conducted two campaigns of ambient air quality monitoring (ISSEP, 2009). Mobiles laboratories including various analysers and meteorological tools carry out continuous analyses of some tracers in ambient air around the landfill (CH_4, H_2S, NO_x, SO_2, BTEX, limonene and pinene, particulate matter). The laboratories are positioned simultaneously upstream and downstream from the site regarding the direction of the prevailing winds. In normal dominant wind conditions, the first station measures the background ambient air pollution. If there is a change in the wind direction, the role of the laboratories is inverted or a new position is selected.

E-noses.   An E-nose network called FIDOR®, providing in real time the five components of the odour annoyance has been operational around the landfill site since 2009. *F*requency and *D*uration are provided through the detection of odour events emerging above a given threshold. *I*ntensity of the odour (the dose level) is usually estimated by a regression model calibrated with dynamic olfactometry measurements compared to the sensor signals. *O*ffensiveness is obtained by identifying the odour type on the site and the *R*eceptor exposure is translated in terms of downwind perception distance.

The implemented FIDOR network consists of five home-made electronic noses, each one comprising of six metal oxide sensors and of three weather stations. The instruments are placed around the site at distances between 20–25 m from the sources (see Figure 8.3.4).

Through a user-friendly graphical interface, raw signals are processed by a central computer to supply in real time the various FIDOR dimensions on background maps.

Different categories of models are used, either to recognise the more worrying odour emission types, estimate the global odour level released by the facility or assess the direction and size of the odour plume in the neighbourhood. Particular attention was given to the field validation of the information given by the e-noses as well as the testing of the robustness of the methods used.

8.3.5   Data Collection

Two campaigns were performed in 2005 and 2009 for the Wallonia Ministry of Environment. In addition to these campaigns, various other studies were carried out for the research activities of the team. For this purpose, dynamic olfactometric measurements were realised, after point sampling above specific odour sources as well as chemical analyses to establish chemical profiles.

The 2005 study was organized between 25 July and 25 November. The sniffing team realised 12 field inspections corresponding to twelve different days during the same period of the day (between 10 am–2 pm). Due to the large distances between the different odour sources, each source was considered in the dispersion model (with its area and location) to evaluate the global emission rate. During this period, the wind was mainly coming from the East or South-Southwest. The weather conditions were variable (Augustus was rainy, September and October were sunny and the end of October and November were colder) and typical of the local conditions, excepted for September and October that were warmer than usual. At this time, the composting of the organic domestic waste was still active. Mobile laboratories continuously monitored some tracer compounds.

In 2009, a second assessment was organised with the sniffing team method between 30 March and 2 July. The winds were coming from the North-East or South-West with quite a high average speed of 3.5 m/s. The temperature was higher than in 2005.

In addition to this field inspection, a resident diary was planned. Prior to the organisation of this survey, 440 questionnaires were distributed in seven spatial clusters to villages downwind, others upwind and around the landfill area (in the small economic park): Hachy, Habay-la-Neuve, Habay-la-Vieille, Houdemont, Mortinsart, Nantimont and Etalle.

Only 11.6%, corresponding to 51 persons, filled in the questionnaire. Among those 51 persons, 21 accepted to participate in a resident diary and record their perception, at least two days per week and twice a day. This low participation in 2009 is explained by an important reduction of the global odour rate (as explained in Section 8.3.6).

This resident diary is in progress since December 2010. Recordings for December have been received from all 21 panel members. This survey should last at least one year in order

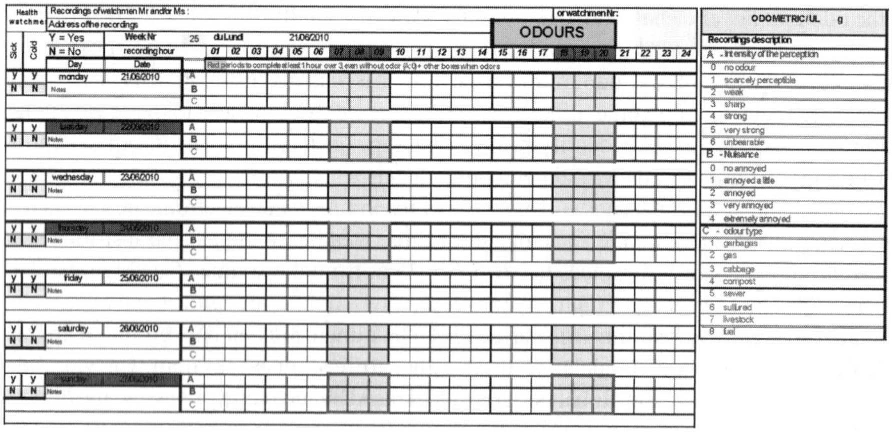

Figure 8.3.5   *Framework example of a resident diary.*

to have enough data for the statistical treatment as well as to cover different situations. Figure 8.3.5 presents the record sheet for one week that the resident had to fill in.

8.3.6 Results and Discussion

Figure 8.3.6 shows a representative 1 OU/m^3 isopleth estimated, in 2005, by reverse modelling, after field inspection by the sniffing team, and including at best the odour points as identified in the ambient air.

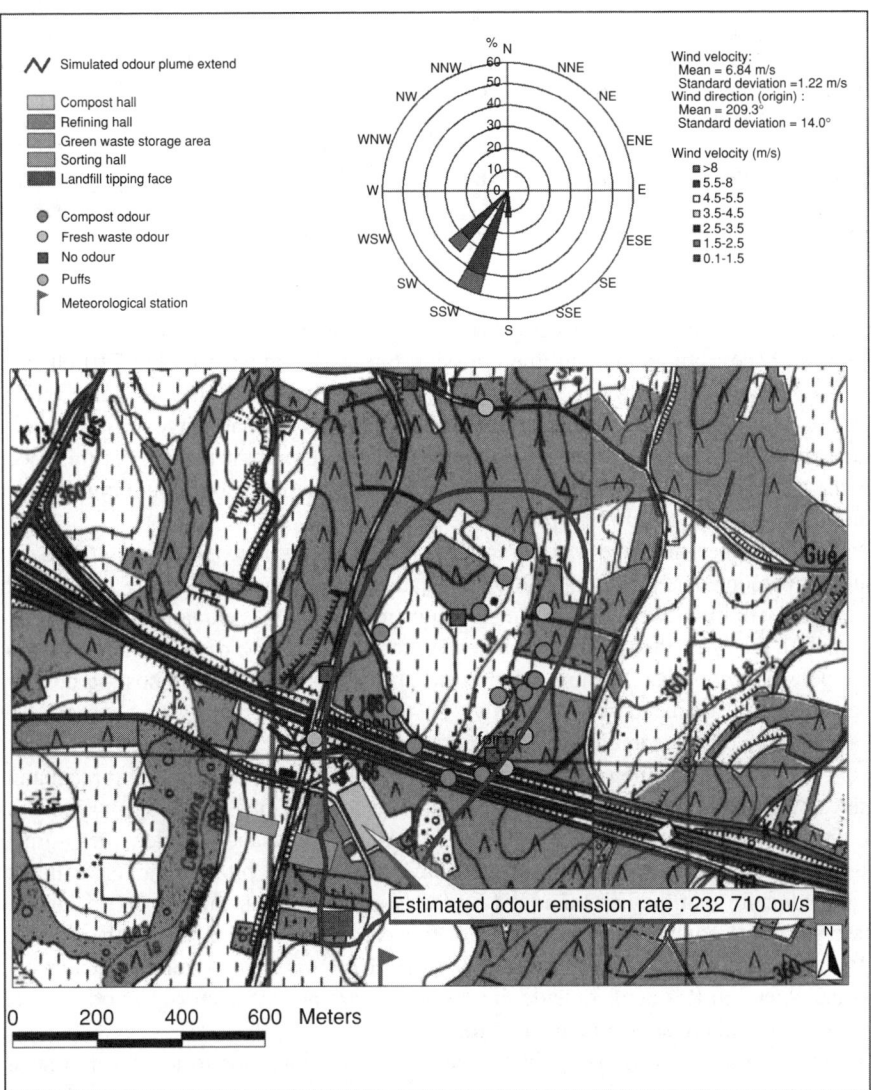

Figure 8.3.6 *1 OU/m^3 isopleth (in dark shading) estimated by Tropos model including at best the odour points identified by the sniffing team the 25 July, 2005.*

Table 8.3.1 *Emission rates obtained between July and November 2005 by the 'sniffing team' method.*

Date	Wind direction	Wind speed (m/s)	Stability class	Max distance (m)	Trucks number (trucks/h)	Emission rate (OU/s)
25/07/05	209°	6.8	D	850	14	232 710
29/07/05	224°	2.2	A	340	14	83 070
01/09/05	273°	4.4	C	680	11	134 500
02/09/05	354°	3.1	B	900	14	311 860
13/09/05	159°	1.5	B	–	–	–
14/09/05	261°	4.7	C	875	13	222 230
21/09/05	93°	1.9	B	950	10	330 560
22/09/05	84°	4.0	C	1475	11	405 810
23/09/05	196°	3.7	C	400	16	66 461
04/10/05	30°	4.6	D	3300	7	754 170
18/10/05	99°	2.8	B	1650	14	493 740
25/11/05	251°	6.1	D	3250	14	956 420

For this case, the maximum downwind distance of odour perception is about 850 m from the centre of the tipping area and the emission rate was evaluated to 232 710 OU/s. All the emission rates are shown in Table 8.3.1 with the respective values of wind variables and the stability class deduced from the Pasquill table. To evaluate the contribution of the waste arrivals, the number of trucks was considered to interpret the results. For this specific case, the truck traffic did not modify the emission. In fact, the major odour emissions were linked to the grinding and to the fresh waste deposal.

On 13 September, the wind was too weak to obtain a reliable result. During this day, the odour was perceived no further than 200 m from the composting hall. During 25 November, the compost windrows were turning and the high emission rate is explained by this activity. On 4 and 18 October, the quite high values are the results of the displacement of odorous municipal waste compost. Without these values, more typical of the composting process of garbage, the average mean would be only 223 400 OU/s. Nevertheless, the average of the 11 values has to be considered to represent the impact in the neighbourhood. This emission rate of 326 900 OU/s is then introduced into the dispersion model with the typical annual climate of the region. The percentiles 95, 98 and 99.5 for 1 OU/m^3 are shown in Figure 8.3.7

If the 98th percentile is considered to delineate the nuisance border, there are few residents in this zone. People within the percentile confirm by complaints or thank to the questionnaire that they are annoyed by the landfill emissions and in particular by the 'municipal waste compost' odour. Then, for this case (also confirmed for other landfills), the 98th-percentile for 1 OU/m^3 (C98, 1 h = 1 OU/m^3) is quite a good indicator of the nuisance potential. Residents outside this percentile are not annoyed by the odour even if the average global emission rate is quite high.

The measurements of the key compounds by the mobile laboratories during the same period highlight a local influence of the landfill for limonene, methane and hydrogen sulfide near the landfill but no significant influence was demonstrated for the first nearest habitations (ISSEP, 2005). For this specific study, the sensorial field measurements give

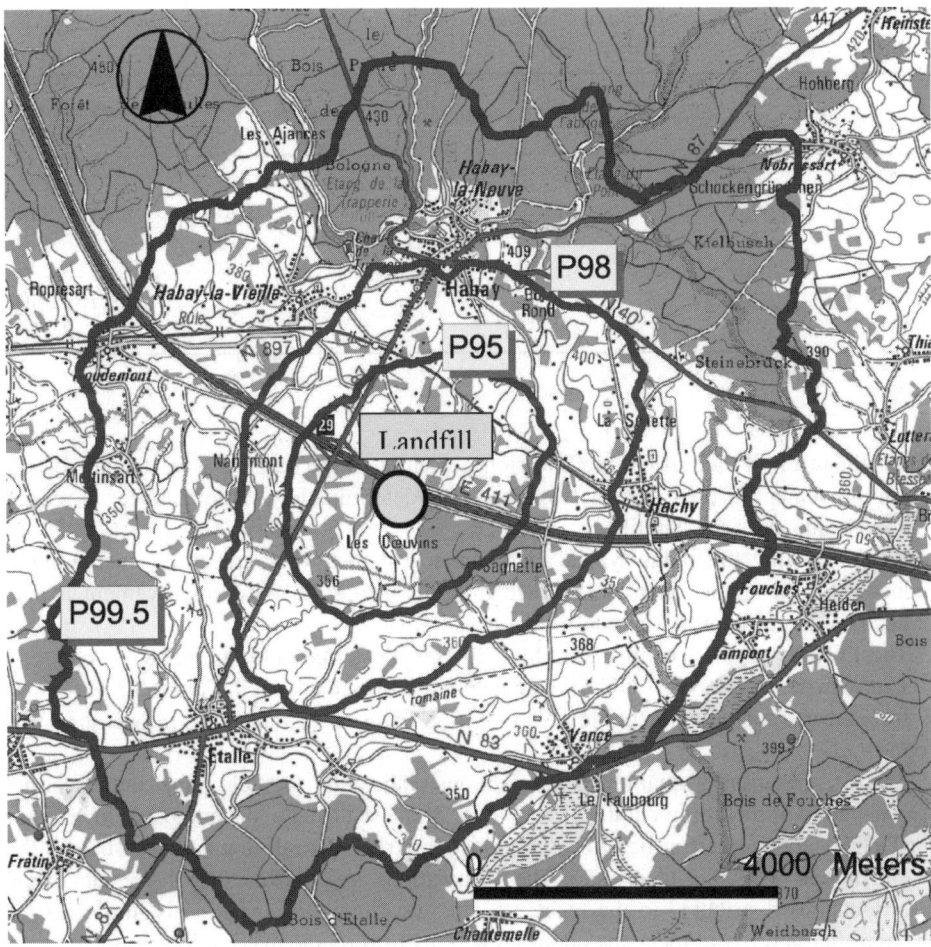

Figure 8.3.7   *Percentiles 95, 98 and 99.5 (from emission rate values determined in 2005).*

more representative results on the odour impact than the chemical analysers. Nonetheless, the mobile laboratory located on the border of a nearby residence can highlight a good correlation between the detection of important 'odour events' by the residents and recording of peak concentration for these tracers.

Four years after the 2005 campaign, a second odour impact assessment was carried out. The process had changed (see Section 8.3.2). Organic waste composting had stopped and replaced by the drying of inorganic residues (plastics, food packaging) of the municipal waste.

The emission rate obtained, by field inspection and reverse modelling are listed in Table 8.3.2 below.

The average value of the emission rates is 109 641 OU/s, a third of the 2005 value (326 900 OU/s). This value is also lower than in 2005 without considering the compost aeration by turning (223 400 OU/s). The significant diminution of the average emission

Table 8.3.2   *Emission rates obtained between March and July 2009 by the 'sniffing team' method.*

Date	Wind direction	Wind speed (m/s)	Stability class	Max distance (m)	Emission rate (OU/s)
30/03/09	82°	3.0	B	340	63 292
29/05/09	57°	4.5	B	500	159 930
04/06/09	350°	3.3	B-C	500	158 490
09/06/09	221°	3.1	B-C	600	119 270
12/06/09	261°	4.1	C	670	152 040
02/07/09	81°	2.8	B	300	59 714

rate is the result of the process modification (cessation of the municipal waste composting and improvement of the biogas network) and the determination of the manager to reduce the odour impact of the landfill.

By considering this value typical of a year to calculate the percentiles, an important reduction of the impact area is observed in relation to 2005. This improvement is also noticed by the residents in the questionnaires.

The 1 OU/m^3 isopleth estimated in May 2009 is illustrated in Figure 8.3.8.

The 98th percentiles for 1 OU/m^3 calculated in 2009 are shown in Figure 8.3.9.

In comparison to the percentile obtained in 2005, a reduction of the impact area was found.

The interpretation of qualitative observations realised by the field panel indicates that for the period of measurement the global odour of the site can be 'decomposed' in 40% for the LFG, in about 30% for the 'green' compost, in 20% for the drying hall and 10% for the waste storage (arrival of the waste). The observations are confirmed by the results of the questionnaires distributed to the residents. This qualitative information is important for the manager in order to focus the odour reduction effort on the major odour sources.

The hedonic tone is also an important dimension to consider. For instance, in this case, even if the green compost odour is more frequently perceived than the odour of waste coming from the bio-drying hall and from the waste storage, this odour is better appreciated by the residents. The impact of the green compost odour on the population should be considered less important.

A summary of the emission rate values for 2005 and 2009 is presented in Table 8.3.3, in comparison to the values obtained for the Wallonia landfills of the monitoring network managed by ISSEP during 10 years (Collard *et al.*, 2008).

The current study with the residents' diaries will give new information to better identify the specific conditions for the odour releasing in the neighbourhood as well as establish better dose-effect relationships including the hedonic tone of the odour source.

The field inspection was realized around from 11 am to 2 pm. It would be interesting to select other observation periods. For instance, in some questionnaires received by the residents, the early morning (from 7 am to 9 am) and the evening (from 6 pm to 8 pm) are the most critical periods. The thermal inversion but also the 'comeback' home periods are probably the principal explanations. A continuous monitoring of the odour is helpful

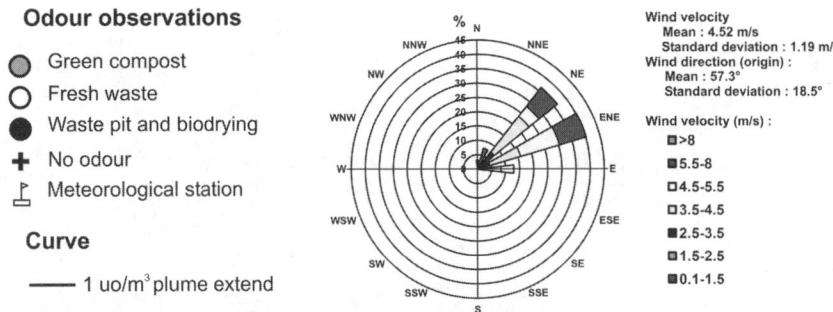

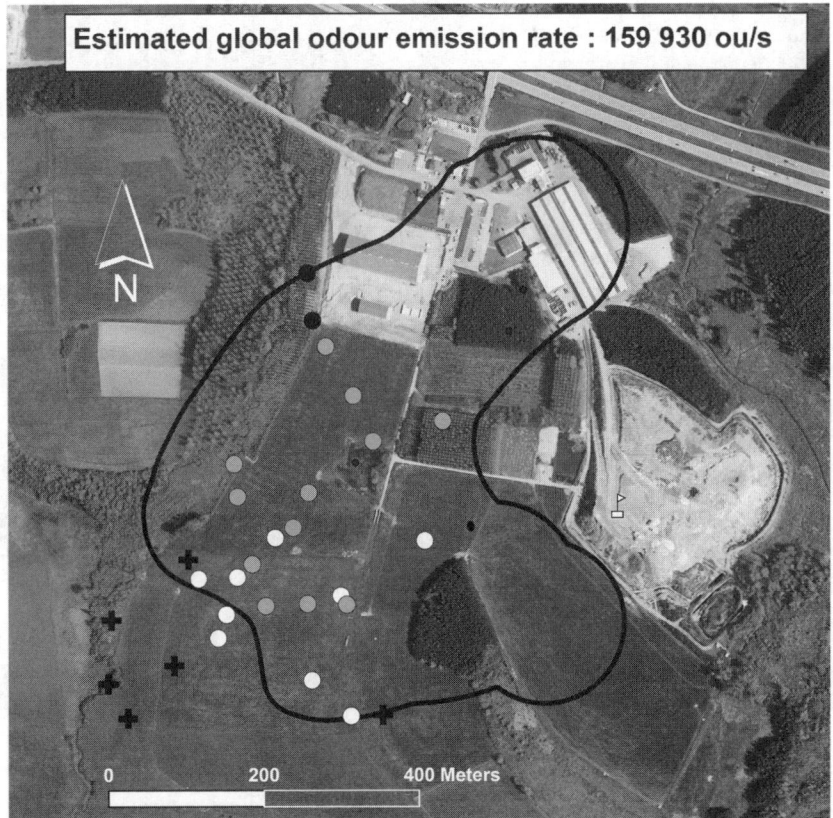

Figure 8.3.8 *1 OU/m³ isopleth (dark ring) estimated by Tropos model including at best the odour points identified by the sniffing team the 29 May, 2009.*

to better assess the odour impact in the neighbourhood. The resident diary is conducted in this way. It is also the reason why an e-noses network has been installed around the site.

The FIDOR® e-nose is configured to give continuous, different information to the manager. For instance, Figure 8.3.10 is a print screen of the user interface on which magnitude estimation of the odour is represented by odour thermometers.

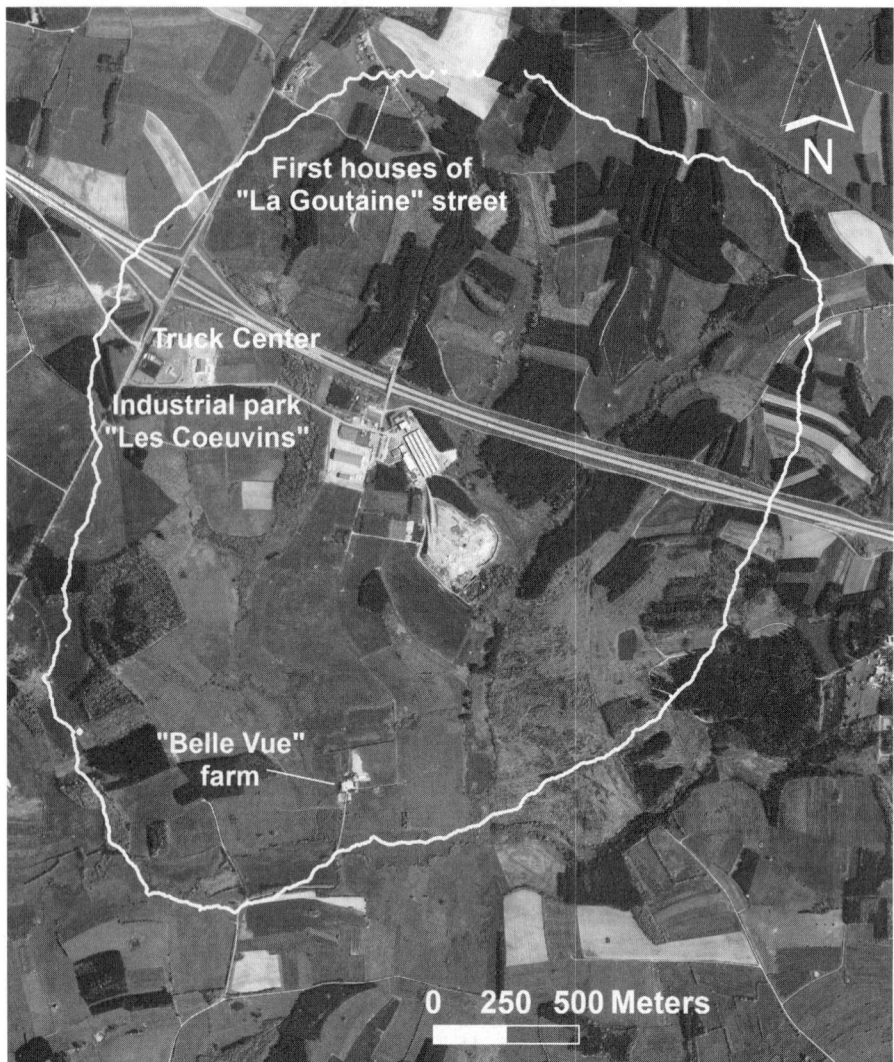

Figure 8.3.9   *The 98th percentile (for emission rate values determined in 2009).*

The calibration of the e-noses is realised either in regard to olfactometric values or by field observations. However, field observations are preferred. The current aim is to continually predict by E-noses the distance of perception in the neighbourhood, with field observations being more representative. To reach this goal, odour emission rates have to be evaluated by the E-noses. To determine this value, a FIDOR® algorithm has been developed to relate the answers of the E-noses with the results of various field inspections (by the sniffing team), considering the weather conditions and a bi-Gaussian model. After validation, again by field inspection, an E-nose can determine in continuous the maximum distance of perception. Even if this distance is still full of uncertainties, the system is able to give in continuous, an estimation of the impact area and its direction. Continual developments aim at defining

Table 8.3.3   *Emission rate values for 2005 and 2009 in comparison to landfill data acquired over 10 years in the monitoring network managed by ISSEP.*

		Network statistics (11 landfills)	HLC case study 2005	HLC case study 2009
	Survey number	15	1	1
	Field days	87	11	6
Odour	Min	8000	66 461	59 714
Emission	Mean	66 400	326 900	109 641
Rate (OU/s)	Max	405 810	956 420	159 930
P98 for 1 OU/m^3	Max distance (km)	3.4	3.3	0.7

and optimising the accuracy of this determination. Figure 8.3.11 represents a distance of perception given by the FIDOR® system, validated by a field observation.

Figure 8.3.12 shows the scatterplot of odour concentration estimated by partial least square model for one nose versus concentration as measured by dynamic olfactometry.

Furthermore, the emphasis is put on the performance evaluation of the FIDOR system when sensor arrays are placed at more distant locations with respect to the odour emitting facility. Two additional E-noses are placed at 100 (N6) and 2000 meters (N7) away to the first (N2) in the direction of the residents. Some interesting correlations have been found between the e-nose responses at different distances in the same direction. Figure 8.3.13 shows the comparison of the three E-nose responses as resulting of an odour-event.

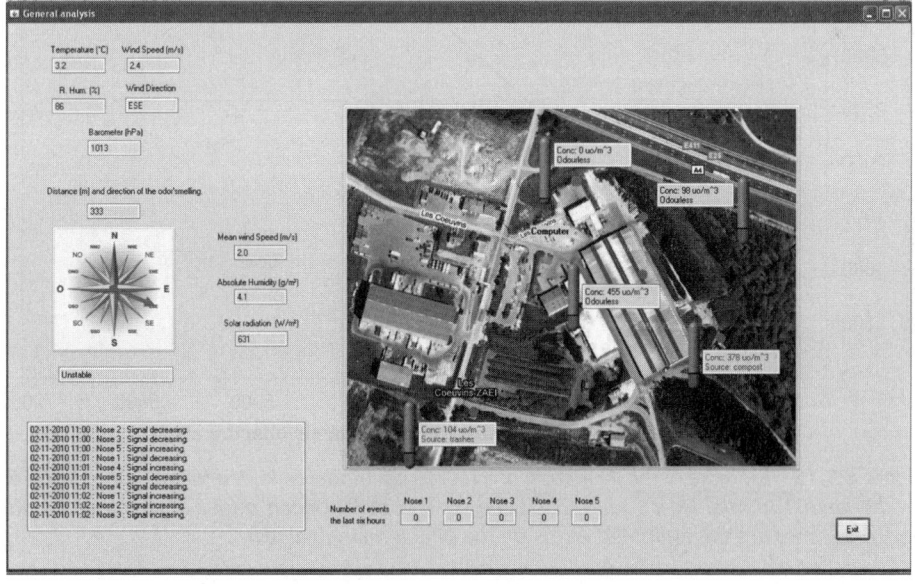

Figure 8.3.10   *Printscreen of the FIDOR user interface.*

Figure 8.3.11 *Maximum distance of perception given by the e-nose network, validated by filed inspection.*

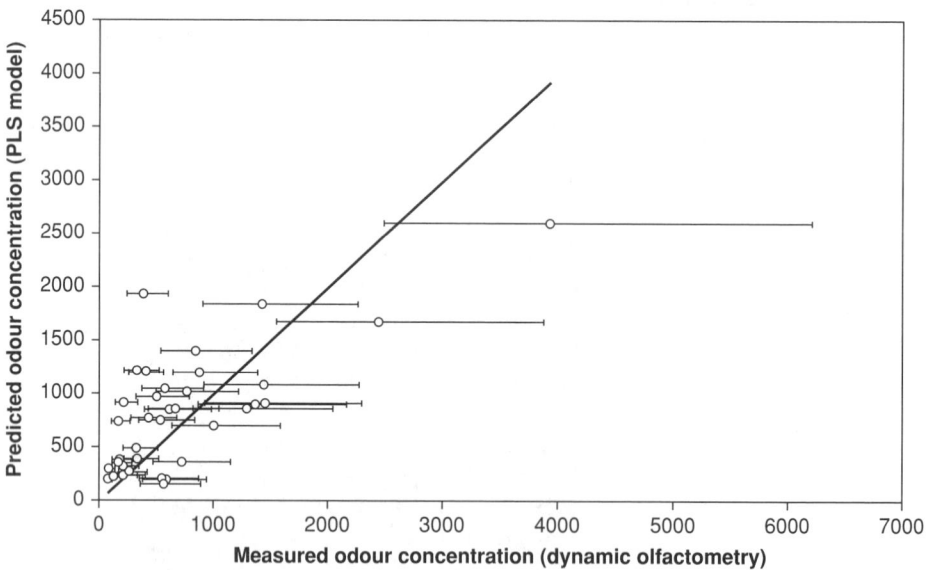

Figure 8.3.12 *Example of the validation of a regression model obtained for an e-nose installed on the landfill (partial least square has been used with European odour concentrations, the horizontal lines are the estimated errors on the olfactometric values).*

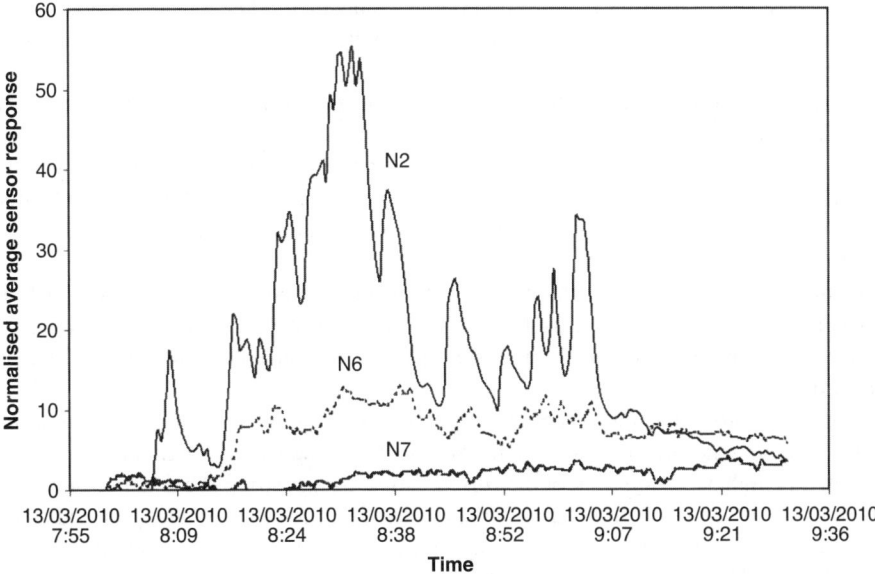

Figure 8.3.13 *Correlation of the signals given by the three E-noses located at different distances from the landfill: N2 is located on the site, N6, 100 m away and N7 2000 m.*

A correlation can be observed between the nose placed at the source level and the one placed 100 m away, while the most distant nose reacts less to the odour-event.

8.3.7 Conclusions, Recommendations and Outcomes

Today, in Wallonia, landfills are rather complex areas, with several installations of waste treatment. The HLC landfill, during its history, has progressed from a landfill to a waste treatment plant. This evolution linked to the modification of the regulations has resulted in not only the reduction of the odour impact but also in a diversification of the odours sources. To assess the odour impact, the different odours have to be taken into account. The global odour emission rate cannot be determined by simple emission sampling and olfactometric measurements in the laboratory, particularly for sites with windrows turning, fugitive emissions, vehicles gas emissions and heterogeneous emissions surface; typical of a landfill.

To assess the impact of a landfill area, different methods are available. Each one presents both advantages and drawbacks and none of them is sufficient. A landfill generates various odours in the atmosphere with a high variability rate. For this particular reason, the determination of the nuisance would find an optimum with a continuous monitoring in the neighbourhood. In Wallonia, the trend is to increase citizen participation. A resident-watchmen survey becomes more accepted and shows several advantages. The most important is probably the restoration of the dialogue between the stakeholders. However, this survey has to last at least one year in order to give all the variability as well as be assisted with frequent contacts and rigorous statistical data treatment. An alternative approach is the field inspection method combined with reverse modelling. This method has the advantages of using expert panels and determining the odours in the neighbourhood. Even, with some skill, it is possible to identify the different odours. This method is not carried out for more than one hour and has to be repeated several times in order to include all the conditions (weather and process).

Dose-effect relations need to be studied in greater detail in order to determine a level where 'no justified cause of annoyance' exists. In Europe, the regulation imposes exposure criteria without taking into account the hedonic tone of the odours. For this landfill, this percentile corresponds to the perception of the residents. The others campaigns organised for the Wallonia landfills indicate similarities but a more detailed evaluation is required in order to avoid to coming to easy conclusions.

Chemical analysers can continuously monitor some key compounds but their detection limit is rarely low enough to give useful results when they are located in the neighbourhood of a landfill. Moreover, chemical composition is not systematically related to the odour perception.

E-nose technology has the potential to continuously monitor a global odour as well as predict the emergence of a nuisance in the vicinity. Nevertheless, as for chemical analysers, the relatively high detection limit of the current gas sensors restricts the distance of use. The interest of these devices is that the relationships between E-nose signals, near and around the landfill area, and odour perception can be used to predict in continuous odour impact in the vicinity. The FIDOR® e-nose installed around HLC is able to predict a distance of perception but still with some inaccuracies.

The initiative of the Wallonia Ministry to organise a survey network for all the Wallonia landfills is a good start to the harmonization of the methods to assess the odour impact of the landfill sites as well as define standards dedicated to this activity. For instance, the validation of the 98th percentile, 1 OU/m^3 by the residents of HLC could be a first step to defining criteria for standards representative of the true perception of the population.

References

Collard, C., Lebrun, V., Fays, S., *et al.* Air survey around MSW landfills in Wallonia: feedback of 8 years' field measurements, *ORBIT 2008*, Wageningen, Netherlands, 13–15 October, (2008).

ISSEP, RÉSEAU DE CONTRÔLE DES C.E.T. EN RÉGION WALLONNE, C.E.T. de Habay, Troisième campagne de contrôle (2009), http://environnement. wallonie.be/data/dechets/cet/10hab/pdf/10_RapCMP_HAB_2009.pdf (accessed January, 2012).

ISSEP, RÉSEAU DE CONTRÔLE DES C.E.T. EN RÉGION WALLONNE, C.E.T. de Habay, Première campagne de contrôle (2005), http://environnement. wallonie.be/data/dechets/cet/10hab/pdf/10_RapCMP_HAB_2005.pdf (accessed January, 2012).

Muñoz, R., Sivret, E., Parcsi, G., *et al.* Monitoring techniques for odour abatement assessment, *Water Research*, **44**, (2010) 5129–5149.

Nicolas, J., Craffe, F. and Romain, A.C., Estimation of odor emission rate from landfill areas using the sniffing team method, *Waste Management*, **26** (11), (2006a) 1259–1269.

Nicolas, J., Romain, A.-C. and Ledent, C., The electronic nose as a warning device of the odour emergence in a compost hall, *Sensors and Actuators B: Chemical*, **116**, 1–2 (2006b) 95–99.

Nicolas, J., Cors, M., Romain, A.-C. and Delva, J., Identification of odour sources in an industrial park from resident diaries statistics, *Atmospheric Environment*, **44** (13), (2010) 1623–1631.

Romain, A.-C., Nicolas, J. and Andre, P., In situ measurement of olfactive pollution with inorganic semiconductors: Limitations due to humidity and temperature influence, *Seminars in Food Analysis*, **2**, (1997) 283–296.

Romain, A.-C., Godefroid, D. and Nicolas, J., Monitoring the exhaust air of a compost pile with an e-nose and comparison with GC-MS data, *Sensors and Actuators B: Chemical*, **106** (1), (2005) 317–324.

Romain, A.-C., Delva, J. and Nicolas, J., Complementary approaches to measure environmental odours emitted by landfill areas, *Sensors and Actuators B: Chemical*, **131**, 1 (2008) 18–23.

Van Langenhove, H. and Van Broeck, G., Applicability of sniffing team observations: experience of field measurements, *Water Science and Technology*, **44** (2001) 65–70.

8.4   Industrial Activities

I. Sówka

Institute of Environmental Protection Engineering, Wroclaw University of Technology, Wroclaw, Poland

8.4.1   Motivation for the Study

An important step in choosing the methodology to determine odour concentrations was the European standard EN 13725:2003 with the amendment introduced in 2006 and its Polish version PN-EN 13725:2005, subsequently amended by PN-EN 13725:2007 (Jakość powietrza, 2007).

The research methodology included in the above standards, refers to measurements of concentrations at the source: the emission. In the case of emission concentrations, several research proposals (Kośmider *et al.*, 1997, 1998, 2002) have been made and studies (Krysiak *et al.*, 2009; Sówka *et al.*, 2009; Sówka *et al.*, 2009; Kulig *et al.*, 2008; Kulig *et al.*, 2009; Kulig *et al.*, 2009) carried out on odour nuisance assessment in the areas around the odour source. In Europe, standards have been developed by the Association of German Engineers, Verein Deutscher Ingenieure (VDI) (VDI 3883, part I, 1997; VDI 3883, part II, 1997; VDI 3940, 1993; VDI 3940 B.1, 2006; VDI 3940 B.2, 2006; VDI 3940 B.3 (Draft), 2008; VDI 3940 B.4 (Draft), 2008). However, in Poland, there are no standards covering air quality and odour nuisance assessments methods in the form of field research as well as legislations covering standards of the odour air quality. These issues were partly included in subsequent drafts of the act on the prevention of odour nuisance.

Since 2007, three drafts of acts on the prevention of odour nuisance have been developed: (1) a draft of an act on the prevention of odour nuisance developed in February 2007, (2) a draft of an act on 17 October, 2008 on the prevention of odour nuisance and (3) a draft of an act on 27 February, 2009 on the prevention of odour nuisance.

Legislation is currently being drawn up, which includes a new draft of an act on the prevention of odour nuisance, developed by the Department of Environmental Protection, on 22 December, 2010, with subsequent developed assumptions to the draft of the act and in April 2011, they were subjected to public consultation.

Other projects on the 'odour topic' have also been realized in Poland, including (Kulig *et al.*, 2011):

(a) inventory survey of the main sources of odorants in Poland, carried out by the Institute of Environmental Management (1990s),
(b) developing materials for the national strategy to reduce odour nuisance by the Odour Air Quality Laboratory of the Technical University of Szczecin in cooperation with other national groups (1995–1997)
(c) research project PBZ-MEiN-5/2/2006: 'New methods and technologies of deodoriza-tion in industrial and agricultural production and municipal management' (2007–2010)

In September 2011, an elaboration by the Ministry of Environment was developed, entitled 'The proposition of methodology for assessing the content of olfactory active substances in the air'. The basis of the work was literature data on the subject of the elaboration – odour emissions sources, methods of reducing them, odorimetry and its applications as well as authors own research (Kulig, *et al.*, 2011). The result of the work was to be a proposal of a measurement methodology, representative of the Polish conditions. However, due to the complexity of the issue, the authors proposed a procedure to be followed on the assessment of odour nuisance. For the purposes of this expertise, the techniques of reducing odour nuisance were set together with the methods of assessing it for chosen potential odorants emission sources.

In this paper, a research methodology will be presented, aimed at identifying a group of industrial odour emission sources in Poland. Examples of odour concentration measure-ments (dynamic olfactometry) from chosen, industrial odour emission sources will also be presented as well as some suggestions related to the use of selected computational and spatial analysis methods.

The presented research methodology is aimed at identifying hazard odorous gases emis-sion from industrial plants, through the identification of potential technological odour emis-sion sources as well as complex odour air quality assessment on a chosen area, through the simultaneous use of several of these techniques.

8.4.2   Identification of Industrial Odour Emission Sources

An organizational scheme (Figure 8.4.1) associated with the complex identification of industrial odour emission sources, takes into account methods of identification and coop-eration areas in the proposed research field with administrative units, industry and local community representatives. In order to assess the impact of the chosen emission source, it is necessary to use various tools and methods (i.e. complaints analysis, analysis of market and statistical data, dynamic olfactometry, field research, survey and model tests), which make it possible to determine an area of odour nuisance and develop a strategy aimed at reducing odour emission from a selected plant (e.g. proposing optimal deodorization techniques) and improving the quality of life of the inhabitants living in the area subjected to the impact of odours.

For the purpose, a number of potential odour sources have been summarized in Table 8.4.1 The summary shows that the greatest influence on odorous air quality in Poland is munic-ipal management, agriculture and farming and industry. In the industrial sector, the most influential is the agro-food sector.

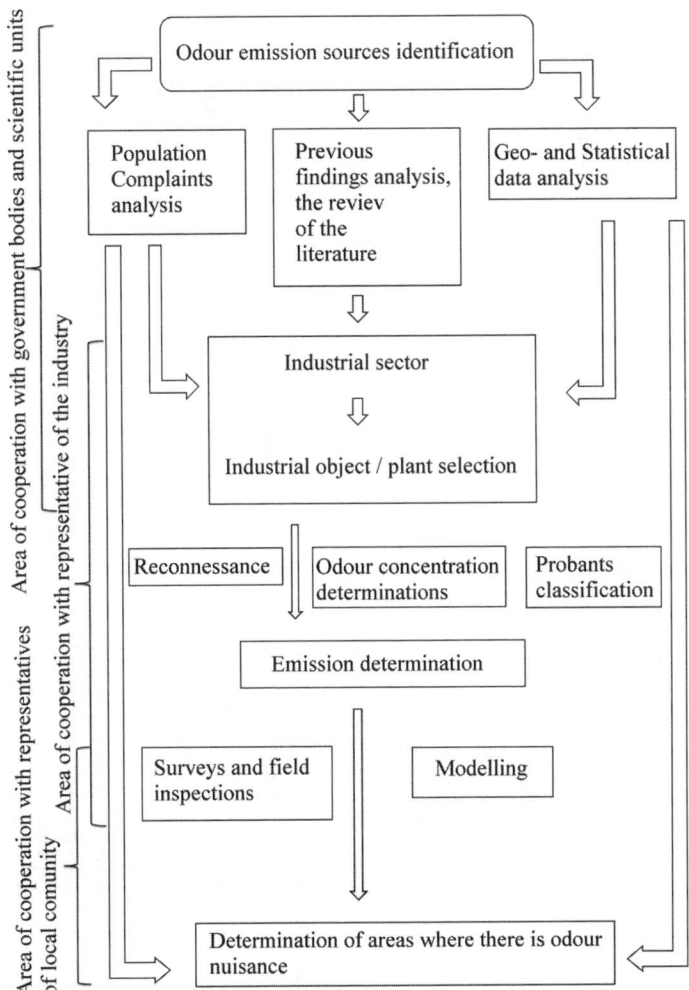

Figure 8.4.1 *An organizational scheme of the methodology of industrial odour emission sources identification (Reprinted with permission from Sówka, 2011, Copyright (2011) Institute of Environmental Engineering Technical University of Wroclaw).*

This is confirmed by the results of the summaries and complaints analysis carried out on the basis of information provided by the Regional Inspectorates of Environmental Protection and the Chief Inspectorate of Environmental Protection. Information can be also provided by the Provincial Sanitary-Epidemiological Stations and Environmental Protection Departments subjected to government units (city councils). These analysis show that in Poland the largest number of complaints are recorded in the vicinity of municipal management objects (39%), agriculture and farming (34%) and industry (11%) (Sówka, 2011).

In terms of space, complaints analysis in Poland can be used for the quantitative identification of odour emission sources, that is, the number of complaints will indicate areas or

Table 8.4.1 List of economic sectors and selected plants and facilities being potential sources of odours in Poland. (Reprinted with permission from Kulig et al., 2011, Copyright (2011) Ministry of the Environment, Poland).

	Economic sector	Type of a plant or facility	Number of potential sources of odours – scale of the problem
1.	Industry	Meat processing	about 400
2.		Fish processing (most located by seaside)	279 (2007)
3.		Alcohol beverages production distilleries	1000
4.		breweries	45
5.		wine production	>30
6.		Sugar production	18
7.		Vegetable oil processing	9 (rapeseed)
8.		Tobacco manufacture	7
9.		Dairy plants, different production range scale	>300
10.	Chemical industry, refineries		7
11.	Metallurgical industry, foundries		400
12.	Agriculture	animal production farms	about 2 million.
13.		large-scale farms	about 300
14.	Municipal sewage tratement plants		about 2900
15.	Municipal management	municipal waste management facilities	municipal waste sorting	173
16.		green waste composting	90
17.		mixed municipal waste biological and mechanical treatment plants	11
18.		municipal waste fermentation plants	3
19.		municipal waste incineration plants	1
20.		Landfills of non-hazardous and non-inert waste (excluding closed landfills)	520

regions 'exposed' to odour nuisance as well as types of sources that may cause the deterioration of air quality. Quantitative analysis of public complaints shows that such areas in Poland may be located in the territories of the Lower Silesia, Masovian and Silesian Regions.

Within the chosen economy (industrial) sector, another quantitative identification method of odour emission sources in Poland could be the statistical data analysis of the number of industrial plants from selected sectors located in a selected area. With data provided by the Main Statistical Office, it is possible to identify sources by sectors. Data on the number of plants that are a potential odour emission source can be classified and categorized according to the European Classification of Activities (ECA) code and according to the Polish Classification of Activities 2007 (PCA7).

Thanks to the classification of objects, mainly from the agro-food sector according to PCA7, it can be stated which type of activity is the most common in a selected region. On this basis, it is possible to conclude that, for example, in the Lower Silesia province, the most common are meat processing plants (PCA7 code: 10.11.Z), furskin article production plants (14.20.Z) and manufacturers of rusks and biscuits (10.72.Z) considered to be oppressive in terms of odour.

Figure 8.4.2: shows an example of a spatial location of companies and plants, mainly from the agro-food industrial sector, in the province of Lower Silesia where the biggest number of

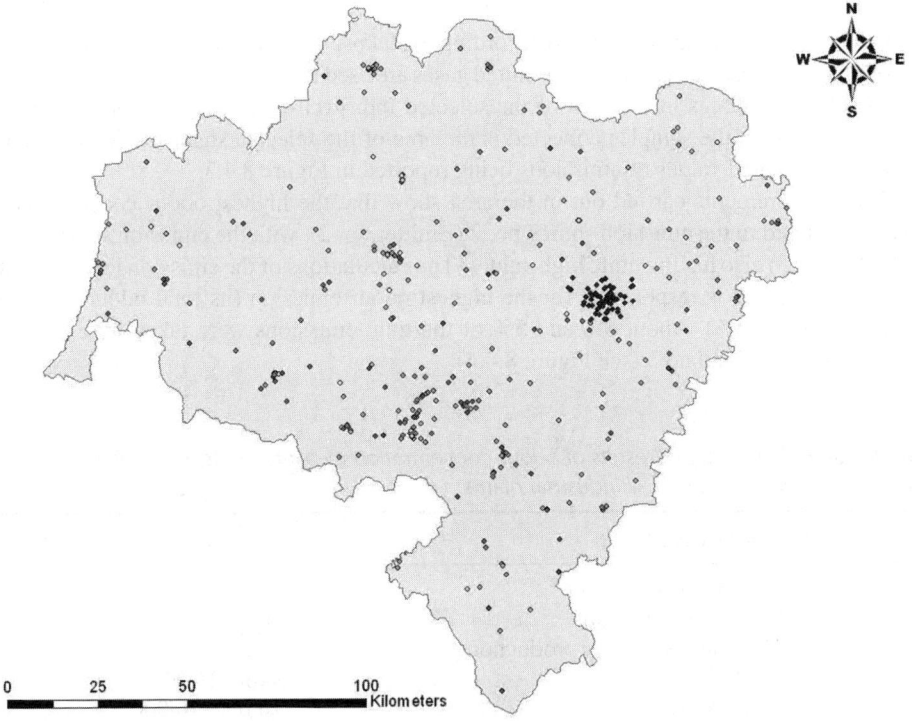

Figure 8.4.2 *Distribution of the agro-food industry plants that are potential sources of odour emissions in the province of Lower Silesia (Reprinted with permission from Sówka, 2011, Copyright (2011) Institute of Environmental Engineering Technical University of Wroclaw).*

public complaints have been reported. To complete an inventory project, GIS (*Geographic Information System*) tools have been used, which enable a visualisation of the spatial distribution of the industrial plants on the map as well as carrying out a spatial analysis.

Analysis carried out using GIS tools states that the quantitative inventory of industrial plants being potential odour emission source in a selected area is possible.

8.4.3   The Use of Selected Methods for Assessing the Odour Quality of Air

The following step, after the quantitative and spatial inventory of emission sources in a selected area, is the selection of a representative object (industrial plant) and determination of odour concentration (according to PN-EN 13725 standard), estimation of the odour emissions from particular technological processes, determination of potential range of impact of a selected plant using model tools as well as carrying out field research (Sówka *et al.*, 2011; Sówka *et al.*, 2011; Sówka *et al.*, 2011).

8.4.3.1   *Determination of Odour Concentration from Selected Industrial Plants*

Table 8.4.2 summarizes exemplary results of odour concentrations, determined using a method of dynamic olfactometry, in gas samples collected in selected industrial plants in Poland.

Odour concentrations determined according to (Jakość powietrza, 2007; Kośmider *et al.*, 2002), with volumetric flow rates of emitted gases are used to determine the odour emissions from selected sources in the area of the selected industrial plants. An example of odour concentrations in the samples collected is the area of the selected sugar production plant, with the estimated values of emissions being reported in Figure 8.4.3.

The measurements carried out in the area show that the highest odour concentrations were measured in the emission source no. 2 (Emitter no. 2), with the emission source no. 3 (Emitter no. 3) also having quite high values. The calculations of the emission level indicate that Emitter no. 2 is responsible for the highest odour emission (its total odour emission share is about 86%). About 8% and 5% of the total emissions were taken by emissions from source no.4 and no.3 (see Figure 8.4.4).

Table 8.4.2   *Summary of results of odour concentration measurements, determined in gas samples collected in selected industrial plants.*

	Type of activity	Odour concentration, ou_E/m^3
1.	meat processing plant	60–185 000
2.	Fish processing plant	10^3–10^5
3.	Alcoholic beverages production plant	
	– distilleries	9400–21 000
	– breweries	16 000–48 000
4.	Sugar production plant	30–56 700
5.	Rapeseed processing plant	2000–100 000
6.	Tobacco manufacture plant	500–70 000

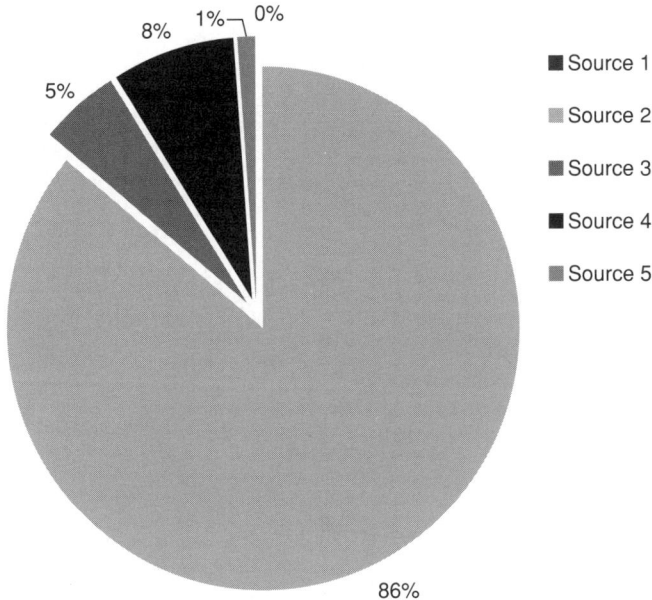

Figure 8.4.3 *The average odour concentrations and estimated odour emission for selected emitters (Reprinted with permission from Sówka et al., 2011, Copyright (2011) I. Sówka).*

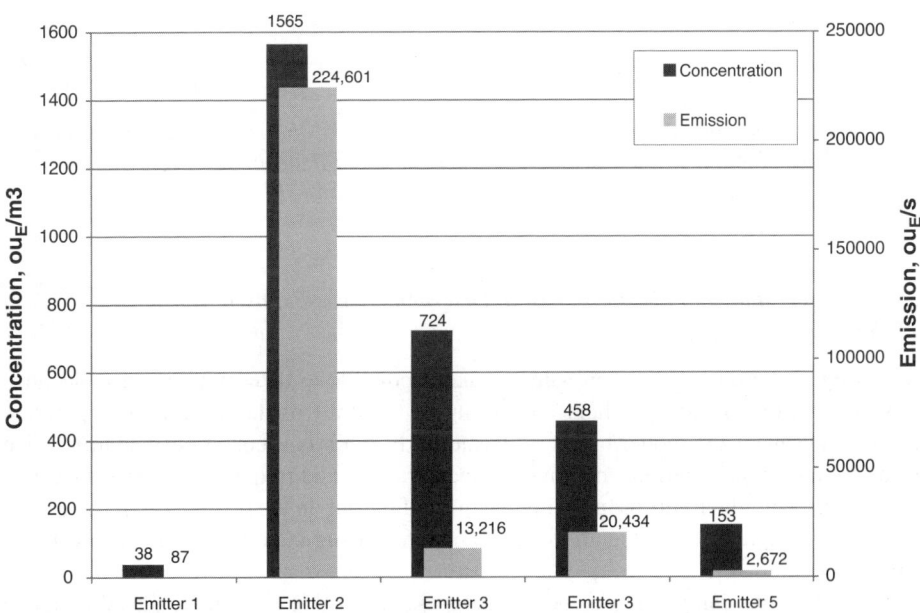

Figure 8.4.4 *The contribution of individual emission sources in the estimated total odour emission from the plant (Reprinted with permission from Sówka et al., 2011, Copyright (2011) I. Sówka).*

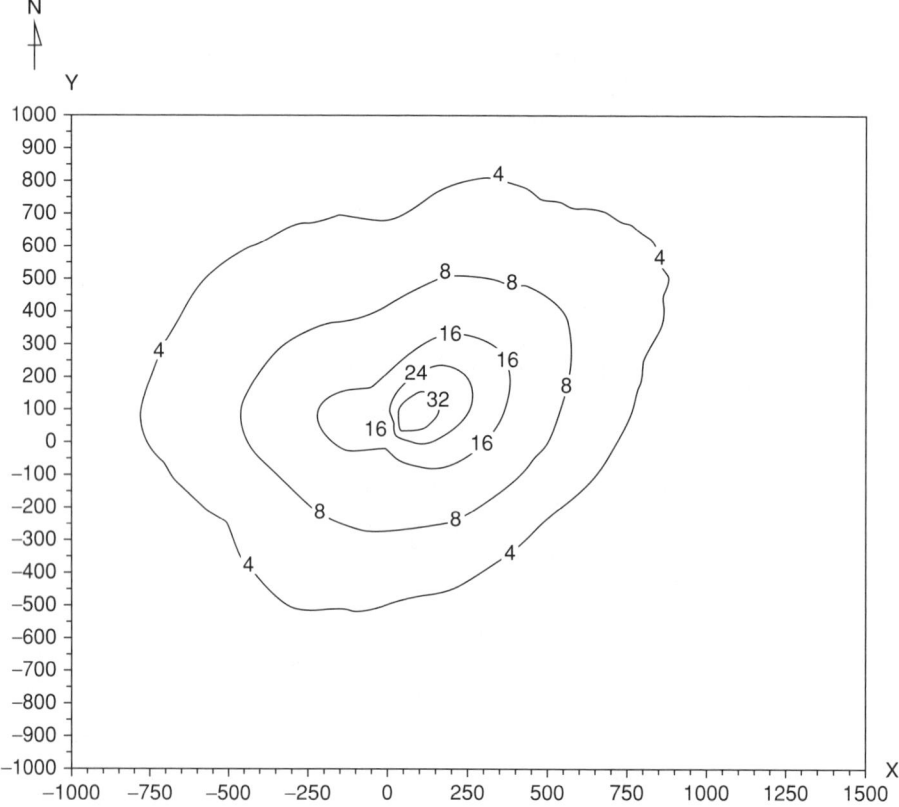

Figure 8.4.5 *Determined the frequency of exceedances of the odour concentration limit*
*,% . (*Proposed in Poland, the comparative level of odour concentration limit value for 1 h
averaging time, for residential areas, is 1 ou_E/m^3 with acceptable frequency of exceedances of
8% per annum.)

8.4.3.2   An Example of Methods Used to Determine Areas Exposed to Odour Nuisance

Estimation of the odour emission values make it possible to determine odour concentrations at receptor points; in Poland (according to the draft of the act on the prevention of odour nuisance) the model based on Pasquils' formula is used. The calculations give average annual concentrations, maximum concentrations and frequency of exceedances to be compared with the limits proposed in the draft of the act on the prevention of odour nuisance. An example of the calculation results obtained using the reference model is shown in Figure 8.4.5

The calculations using the reference model make it possible to estimate the odour range and point out zones in which the field inspections and survey research are carried out, according to VDI 3940 and 3883.

Due to the amount of time required and elevated costs of field research, which result in the estimation of the so called 'odour hours' (field inspections). Measuring sessions, lasting several days (research cycles in accordance to VDI 3940), were carried out in Poland, with the results leading to the introductory assessment of the usefulness of selected GIS techniques (IDW interpolation and kriging) used to interpolate the odour intensity values around the industrial plants and facilities (Sówka, 2011, Sówka *et al.*, 2011, Sówka *et al.*, 2011).

The results of studies using selected model tools, especially geostatistical, were complemented with results obtained in the sociological research described in (Sówka, 2011; Sówka *et al.*, 2011).

8.4.4 Conclusions, Recommendations and Outcomes

Industrial production in Poland, including the agro-food industry, is a reason for the deterioration of the air quality and quality of life in areas in the neighbourhoods of odour emission sources. Therefore, in order to improve the quality of life and the natural environment, it is necessary to estimate the range of odour impact of a selected industrial plant.

The proposed and tested research methodology leads to a spatial, quantitative and qualitative identification of the industrial odour sources. Using several identification techniques simultaneously, gives a trustworthy and complete assessment of the odour quality of the air around the selected odour emission source. The following are determinant: odour emission, range of odour source impact (model tests, field research), concentration of odour at the receptor point (model tests, field research), incidence of odours in the chosen area (field and survey research) as well as level of odour nuisance with an impact of odours on the quality of life of people living in the neighbourhood of the analysed industrial odour emission sources (survey research).

References

Kośmider, J., Mazur-Chrzanowska, B. and Wyszyński, B., *Odory*. Wydawnictwo Naukowe PWN, Warszawa (2002).

Kośmider, J., Ograniczanie emisji odorów. Proponowane procedury licencyjne i kontrolno-pomiarowe. *Materiały seminaryjne*, Wyd. Ekochem, Szczecin (1997).

Kośmider, J., Uciążliwość zapachowa. *Metodyka terenowych oznaczeń stężenia odorów w powietrzu na przykładzie fermy norek*, Archiwum Ochrony Środowiska 1-2, (1998) 19–32.

Krysiak, M., Mikołajczyk, M., Molińska, U. and Nych, A. *Badania ankietowe i analiza skarg ludności jako metody oceny jakości zapachowej powietrza, VII Konferencja Naukowa Studentów*, Tom 1 (2009) 49–53, Oficyna wydawnicza Politechniki Wrocławskiej, Wrocław.

Kulig, A., Sinicyn, G., Czyżkowski, B. *et al.* Identification and survey of the sources of potential olfactory impact in municipal management in Poland. *Chemistry for Agriculture*, 9 (2008) 549–563.

Kulig, A., Lelicińska-Serafin, K., Podedworna, J. *et al.* Inwentaryzacja źródeł odorantów w gospodarce ściekowej i odpadowej w Polsce oraz ocena ich uciążliwości na podstawie badań ankietowych. *Przemysł Chemiczny*, **88** (5) (2009) 484–492.

Kulig, A., Lelicińska-Serafin, K., Podedworna, J. *et al.* Charakterystyka i ocena oddziaływania zapachowego źródeł odorantów w gospodarce ściekowej i odpadowej w Polsce na podstawie badań ankietowych. *Chemik*, **11** (2009) 414–420.

Kulig, A., Zwoździak, J., Szklarczyk, M. and Sówka, I., Propozycja metodyki oceny zawartości substancji zapachowo czynnych w powietrzu, *Ministerstwo Środowiska*, Warszawa (2011).

PN-EN 13725, Jakość powietrza. *Oznaczanie stężenia zapachowego metodą olfaktometrii dynamicznej*, (2007).

Sówka, I., Nych, A., Zwoździak, J. and Szklarczyk, M. Zastosowanie badań ankietowych do oceny zapachowej jakości powietrza, *Polska inżynieria środowiska pięć lat po wstąpieniu do Unii Europejskiej*. Tom 1 / pod red. J. Ozonka, M. Pawłowskiej. Lublin: Komitet Inżynierii Środowiska PAN (2009) 299–303.

Sówka, I., Zwoździak, J., Nych, A. *et al.* Ocena uciążliwości zapachowej wybranego źródła przemysłowego na podstawie badań ankietowych, *Chemik*, **11** (2009) 400–402.

Sówka, I., Metody identyfikacji odorotwórczych gazów emitowanych z obiektów przemysłowych. Prace Naukowe Instytutu Inżynierii Ochrony Środowiska *Politechniki Wrocławskiej*. **90** (55). Wrocław (2011).

Sówka, I., Skrętowicz, M., Szklarczyk, M. and Zwoździak, J., Evaluation of nuisance of odour from food industry, *Environment Protection Engineering*, **37** (1), 5–12 (2011).

Sówka, I., Skrętowicz, M., Zwoździak, J. *et al.*, Ocena zapachowego oddziaływania wybranego zakładu chemicznego z zastosowaniem olfaktometrii dynamicznej, badań modelowych oraz ankietowych, *Przemysł Chemiczny*, **90** (6), 1000–1004 (2011).

Sówka, I., Skrętowicz, M., Nych, A. *et al.*, Zastosowanie metod geostatycznych do interpolacji intensywności zapachów emitowanych z zakładów przemysłowych, *Przemysł Chemiczny*, **90** (5), 174–178 (2011).

Sówka, I., Ocena zasięgu oddziaływania zapachowego zakładu przemysłowego na przykładzie wybranej cukrowni, *Ochrona Środowiska*, **33** (1), 31–34 (2011).

Sówka, I., Skrętowicz, M., Nych, A. *et al.*, The suitability assessment of ordinary kriging in GIS environment to interpolation of the odour intensity values in the area around the selected industrial plant, *PETrA 2011: Pollution and Environment – Treatment of Air: The 1st World Scientific Conference*, Prague, Czech Republic, May 17–20, 2011. Organizers: Odour, ČSCHI (2011).

VDI 3883, part I, *Effects and Assessment of odours. Psychometric Assessment of Odour Annoyance.* Questionnaires, Verein Deutscher Ingenieure, Berlin, Beuth Verlag (1997).

VDI 3883, part II, *Effects and Assessment Of Odours. Determination of Annoyance Parameters by Questioning.* Repeated brief questioning of neighbour panelists, Verein Deutscher Ingenieure, Berlin, Beuth Verlag (1997).

VDI 3940, *Determination of Odorants in Ambient Air by Field Inspections.* Verein Deutscher Ingenieure, Berlin, Beuth Verlag (1993).

VDI 3940 B.1, *Measurement of Odour Impact by Field Inspection* – Measurement of the impact frequency of recognizable odours – Grid measurement, Verein Deutscher Ingenieure, Berlin, Beuth Verlag (2006).

VDI 3940 B.2, *Measurement of Odour Impact by Field Inspection* – Measurement of the impact frequency of recognizable odours – Plume measurement, Verein Deutscher Ingenieure, Berlin, Beuth Verlag (2006).

VDI 3940 B.3 (Draft), *Measurement of Odour in Ambient Air by Field Inspections* – Determination of odour intensity and hedonic odour tone, Verein Deutscher Ingenieure, Berlin, Beuth Verlag (2008).

VDI 3940 B.4 (Draft), *Determination of the Hedonic Odour Tone* – Polarity profiles, Verein Deutscher Ingenieure, Berlin, Beuth Verlag (2008).

8.5 Concentrated Animal Feeding Operation (CAFO) Plants

K.Y. Wang

School of Biosystems Engineering and Food Science, Zhejiang University, Hangzhou, China

8.5.1 Motivation for the Study

With the rapid development of animal production, especially concentrated animal feeding operations (CAFOs) where animals are raised in confined facilities in great quantity (Copeland, 2010), while contributing to the human supplies of meat, eggs and milk, it has become one of the most common sources of odor emissions and complaint. CAFO odors come from three primary sources: animal housing, manure storage units, and land application of manure (Zhang *et al.*, 2005). There are more than 300 compounds in the CAFO odor mixture. Odors emitted from intensive livestock and poultry farms may compromise ambient air quality as well as pose threats to the health and welfare of the surrounding communities. Making CAFOs environmentally sustainable has become more and more challenging, and will likely depend upon addressing health and environmental concerns. The IPPC has published odor guidelines for odor management at intensive livestock installations (IPPC, 2005).

Odor concentration and odor intensity are the two most widely used measurements of CAFO odor. The sensory method using olfactometry is most widely used to quantify odor concentration. The CEN European Standard (EN13725:2003) has become the official olfactometry odor analysis approach for European countries as well as several other countries. This method was standardized in 2002. On the other hand, the Triangle Odor Bag Method was developed in 1973 and used in Japan as a regulation method. An inter-laboratory comparison of seven olfactometry laboratories was conducted in Japan in late 2000, it was reported that the measured results of odor concentration of the same samples by both measurement methods were almost similar (Higuchi, *et. al.*, 2002).

Currently, in China, the *Air Quality Determination of Odor-Triangle Odor Bag Method* (GB/T14675-93, in Chinese) has been adopted for sampling and measuring the odor concentration, with it being consistent with the Japanese standard of odor measurement method. The *Emission Standards for Odor Pollutants* (GB14554-93, In Chinese) is also used for assessing the acceptability of predicted odor impacts from quantitative odor impact assessments.

The most common approach to odor impact assessment is to use mathematical models to predict the downwind odor concentrations on the basis of odor emission rates, topography and meteorological data. Odor emission rate can be determined by odor concentration and airflow rate. The Gaussian plume models, puff models, fluctuating plume models, and other models that have been applied to livestock odor dispersion modeling, that is, the advanced Gaussian models (AERMOD and ADMS) should be considered (Yu *et al.*, 2010).

China is one of the major livestock producers in the world, according to WAICENT it has nearly 50% of pigs and close to 40% of eggs produced in the world. Among these, more than 50% of the meat and eggs are produced by CAFOs. During the 1990s, pork production in China increased by 70%, with much of the growth occurring in densely-populated coastal areas, with the farms being close to urban and rural-residential developments and the subsequent odor emissions having more serious adverse impacts. However, there is very little scientific information available about odor and airborne emissions from livestock operations in China.

Odor emissions have been estimated based on odor concentrations and emission rates. The estimated emission figures have been employed in an atmospheric dispersion model (AERMOD, version 2.2 in Chinese) and the results of this modeling have been used to assess odor impact in the area surrounding the monitored pig farm.

8.5.2   Methodology

8.5.2.1   The Monitoring Site

A commercial pig farm in Cixi city, China was selected to represent the Zhejiang province swine operations. Pigs raised on this farm were Long-White and fed with a corn-soybean ration diet. The pig farm has a total surface area of 68 034 m², a total of 8740 pigs raised in 16 growing-finishing buildings with natural ventilation, 2100 piglets in seven piglet buildings with natural ventilation, and 1400 sows in 12 sow buildings with mechanical ventilation, and 30 boars in a mechanical ventilation building. Swine manure was manually removed from all the buildings twice a day after feeding time, and transferred to a manure composting plant which has a total surface area of 1800 m² in order to produce organic fertilizer. The wastewater of all the pig buildings and management buildings was collected and dumped into 11 digesters with a total volume of 1100 m³, and the digester effluent was stored in a big lagoon with a volume of 10 000 m³, and was used as liquid fertilizer for the surrounding area where vegetables were grown.

The pig farm is situated 2500 m to the north-east of the village of Ximenwen, Cixi city. Ximenwen is in a rural area with a monsoon climate, on a coastal plain of south Hangzhou. In addition to the village of Ximenwen, the villages of Laotang (3.7 km to the west), Fulong (2.7 km to the west), Xiaoshishan (4 km to the northwest) and Dise (2 km to the southwest) are in the area of the proposed monitoring site (see Table 8.5.1 for main climactic factors). Fulong in the southwest rural areas is the closest to the pig farm under study. The nearest buildings to the pig farm are in Laotang, to the east, and Fulong, to the south-west. It is understood that neither of these buildings is residential. The land surrounding the pig farm is relatively flat, while to the east, there is a small river. A map of the surrounding area is presented in Figure 8.5.1

Among the 37 production houses, four pig buildings (see Figure 8.5.2) which were one piglets building, one growing-finishing building, one gestation building and one farrowing building, respectively, and downwind from composting and lagoon were selected for the monitoring study.

The monitoring study was conducted for one week from August 23–29, 2009 with higher ambient temperatures according to the requirements of the methods for ambient air quality monitoring (In Chinese).

Table 8.5.1   *Main climatic factors in a Cilong pig farm.*

Parameters		Parameters	
Annual average temperature	16.2°C	Prevailing wind direction in spring	E, NW
Max. temperature	38.5°C	Prevailing wind direction in summer	E, SE
Min. temperature	−9.3°C	Prevailing wind direction in fall	NE, E
Average Max./Min. temperature	28.8°C/4.2°C	Prevailing wind direction in winter	NW
Annual sunshine hours	2038 h	Annual wind speed	3m/s
Annual average humidity	79%	Mix wind speed	4m/s
Annual average precipitation	1272.8 mm	Frequency of clam	8.36%
Annual prevailing wind direction	East (13.15%)		

8.5.2.2   The Monitored Parameters

Measurement of Odor.   Air samples were collected from four monitoring swine buildings between 4–5 pm every afternoon before manure removal, due to the highest odor concentrations occurring around this period with higher indoor temperature. Air samples were collected from manure composting before and after turning activity every day. Due to odor complaints from open source being generally expressed when the wind comes down at night, two air samples from downwind from the lagoon and boundary were collected between 6–7 pm every evening, respectively. The samples were collected in 3 l sampling bottle, and double samples were collected from each sampling point.

Figure 8.5.1   *The area surrounding the monitored pig farm.*

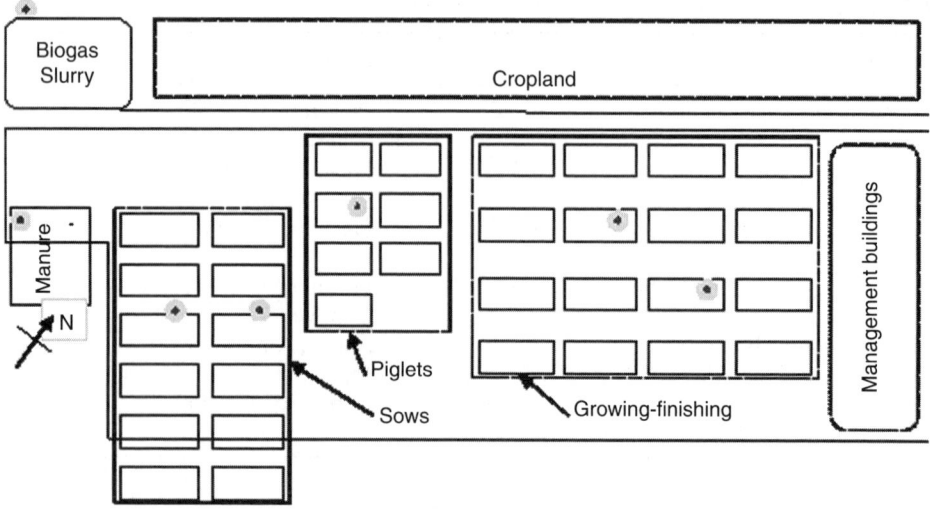

Figure 8.5.2   *Odor measuring positions in the monitored pig farm.*

The air samples were collected by two trained odor assessors, using the Triangle Odor Bag Method (GB/T14675-93), and were analyzed by six panelists who had passed both a written exam and aptitude test using five standard odorants within 24-h of sampling and in relation to the Chinese Odor Standard (GB14554-93).

Environmental parameter measurement.   In addition to odor concentration measurements, environmental variables that directly or indirectly influence emissions were also monitored at various locations. These variables included temperature, relative humidity, wind speed and wind direction (see Table 8.5.2).

8.5.2.3   Calculation of Odor Emission Rates
Estimation Of Ventilation Rate (Q): NV Monitoring Buildings.   Air velocities measured near the monitoring buildings were lower than 0.1 m/s, so the calculation of ventilation rate here was determined by thermal buoyancy ventilation without considering the wind pressure ventilation. The equation is as follows:

$$Q = \mu_b F_b \sqrt{\frac{2(1 - T_{out}/T_{in})gh}{(T_{out}/T_{in})^2 \left(\frac{\mu_b F_b}{\mu_a F_a}\right)^2 + T_{out}/T_{in}}} \qquad (8.5.1)$$

Where:

Q = Ventilation rate of the monitoring buildings, m³/s;
T_{in}/T_{out} = Indoor and outdoor air temperature, respectively, K;
h = Height between inlet and outlet middle point, m;
g = gravity acceleration, m/s²;
F_a/F_b = Area of air inlet and outlet, m²;
μ_a/μ_b = Discharge coefficient of air inlet and outlet.

Table 8.5.2 *Information from monitoring the pig buildings.*

Items	Pig buildings			
	Gestation	Farrowing	Piglets	Growing-finishing
Ventilation	MV	MV	NV	NV
Building Dimensions(m)	$62 \times 8 \times 3.5$	$46 \times 8.4 \times 3.5$	$32 \times 8.4 \times 3.5$	$50 \times 10.2 \times 3.5$
Floor type	concrete floor	concrete floor	semi-slatted floor	semi-slatted floor
Pen type	Orientation pen	High bed	Steel slatted pen	Steel slatted pen
Manure cleaning	2 times/day, 8:30 am, 16:30 pm			
Fans	3 (EM50, R = 0.7m)			
Wet-pad system	17 m²/building			
Inventory	186	48	480	480
Average weight of swine, kg/head	250	250(3–4 kg)	17	50
Feed type	Frequency and time: 2 times/day: 7:30 am, 3:30 pm			

MV monitoring buildings. The ventilation rate from a mechanical ventilated building can be calculated from the escape air velocity and cross sectional area of a side opening. It can be expressed by the equation:

$$Q = v \times A \tag{8.5.2}$$

Where:

v = air speed of the exhausting area, m³/s
A = the area of the exhausting, m².

Odor concentration (OC).

$$OC = (DT) = (Vo + Va)/Vo \tag{8.5.3}$$

DT = Dilution to threshold, V/V;
Vo = Volume of odorous air;
Va = Volume of fresh air.

Odor Emission Rate (OER). The Odor Emission Rate (OER) can be calculated from the odor concentration as measured by the panelists and ventilation rates in a building:

$$OER = OC \times Q \tag{8.5.4}$$

Odor emission rate based on per square meter of floor area (OU/m²/s):

$$OEF = OER/A$$

Odor emission rate based on per animal (OU/pig/s):

$$OEF = OER/N$$

Odor emission rate based on per 500kg live-weight (OU/AU/s):

$$OEF = OER/W/500$$

Where:

$A =$ floor area, m^2;
$N =$ number of pigs;
$W =$ average weight of pig, kg.

The odor emission rate can be entered into an AERMOD air dispersion model to predict the downwind odor concentration.

8.5.3   Dispersion Modeling Methodology

The model selected for this study is the AERMOD version 2.2 from Lakes Software. This model is widely used and accepted by the Environment Agency in China for undertaking such assessments and its predictions have been validated against real-time monitoring data by the USEPA. It is therefore considered a suitable model for this assessment.

8.5.4   Results and Discussion

8.5.4.1   *Ventilation Rates of Monitored Buildings*

The ventilation rates of the piglets and growing-finishing buildings with natural ventilation system were calculated by thermal buoyancy ventilation (Equation 8.5.1) while the ventilation rates of the sows building were calculated using Equation 8.5.2. The mean ventilation rates of the monitored buildings at sampling periods are reported in Table 8.5.3.

8.5.4.2   *Odor Concentrations and Emission Rates of Swine Buildings*

The odor concentration and emission factors on the samples from the monitored pig buildings are in Table 8.5.4.

The odor concentrations of all the monitoring pig buildings were higher than the limitation of Environmental quality standard for livestock and poultry farms (NY/T 388-1999). The highest odor concentration and odor emission rate based on per square meter of floor area were found in the growing-finishing pig building, the highest odor emission rate based on per pig was found in the farrowing building, and the highest odor emission rate based on per

Table 8.5.3   *Ventilation rates of the monitored pig buildings.*

Buildings type	Ventilation type	Q (m$^3 \cdot$ s^{-1})	Buildings type	Ventilation type	Q (m$^3 \cdot$ s^{-1})
Growing-Finishing	NV	4.07	Farrowing	MV	4.56
Piglets	NV	2.40	Gestation	MV	4.24

Table 8.5.4 *Odor concentration and emission factors of monitored pig buildings.*

Buildings type	Floor type	OC	OER OU/m²/s	OU/pig/s	OU/AU/s
Finishing	Part slatted floor	319.89	2.55	2.71	27.12
Piglets	Part slatted floor	209.46	1.87	1.05	30.80
Farrowing	High pen	106.82	1.26	10.15	20.30
Gestation	Concreted	171.91	1.47	3.92	7.84

500 kg live-weight was found in piglets building. The results were consistent with Hayes (2006).

8.5.4.3   Odor Concentrations of Manure Composting and Lagoons

Table 8.5.5 shows that the odor concentrations of the manure composting building and biogas effluent lagoon samples were higher than the standard (NY/T 388-1999).

The odor emission rates of the manure composting and lagoon were not determined in this study due to limited funds, they will be studied in the future.

8.5.4.4   Estimation of Total Odor Emission Rates of Pig Farm (TOER)

The overall odor emission rates of the monitored pig farm were obtained as the sum of the OEFs relevant to each of the odor sources, which are all pig buildings and animal inventory. The odor emission of the manure composting and lagoon was not included (see Table 8.5.6). The total odor emission rate was 22953.7 OU/s in this pig farm, which was used in odor dispersion simulation.

8.5.4.5   Odor Dispersion

Ground meteorological climate and sounding data.   The annual ground meteorological climate data of 2009 (58467) were provided by the Cixi weather station, the main index included hourly wind direction, wind speed, dry bulb temperature, relative humidity, total cloud cover and low cloud cover. Variations of temperature and wind speed in year around pig farm are in Figure 8.5.3.

Table 8.5.5 *Mean value of odor concentration of the composting yard and lagoon.*

Monitoring locals	Sampling time	OC (OU/m³)
Manure composting	Before turning	138.2
	After turning	486.3
Lagoon	Before effluent discharge	109.1
	After effluent discharge	288.4

Table 8.5.6   *Total odor emission rate for pigs in different growing stages.*

Phase of pigs	Number of pigs	OER/OU/pig/s	TOER/OU/s
Farrowing	280	10.15	2842
Gestation	1120	3.92	4390.4
Finishing	8470	2.71	22 953.7
Piglets	2100	1.05	2205

The annual sounding data of 2009 (58457) were collected by the Hangzhou sounding weather station which is the closest one to this pig farm. The main index included air pressure, dry bulb temperature, dew point temperature, wind direction, wind speed and ground level.

Gridding design.   The odor levels were predicted over an area of 5 × 5 km, a Cartesian grid with 500 m grid spacing (21 × 21) was set up for grid forecasting.

8.5.4.6   *Impact Assessment*

The dispersion model results are presented as contour plots in Figure 8.5.4.

It has been estimated that the population of the community around the pig farm is the town of Fulong which indicates that the appropriate odor criteria is 5 odor units. The highest odor concentration occurs around Laolongtang, secondly Fulong, the third Xiaosishan, and the lowest one was in the village of Dise. Therefore, the predicted daily mean odor concentration ≥ 5 OU/m^3 of the four villages accounted for 45.3%, 35.7%, 9.7%, 10.1%, respectively. Since the composting facility and lagoon were not considered, the total odor emission rate is relatively low

8.5.5   Conclusions, Recommendations and Outcomes

The results showed that the odor concentrations were in the range of 106–486 OU/m^3, with varying odor emission sources on the pig farm and all higher than the national standard. The maximum odor emission rate was calculated from the maximum ventilation rate and averaged odor concentrations measured inside the pig buildings. The odor emission rate

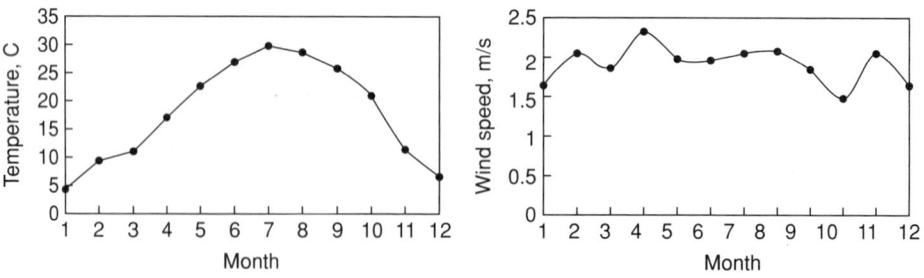

Figure 8.5.3   *Variation of annual temperature and wind speed in the monitored farm area.*

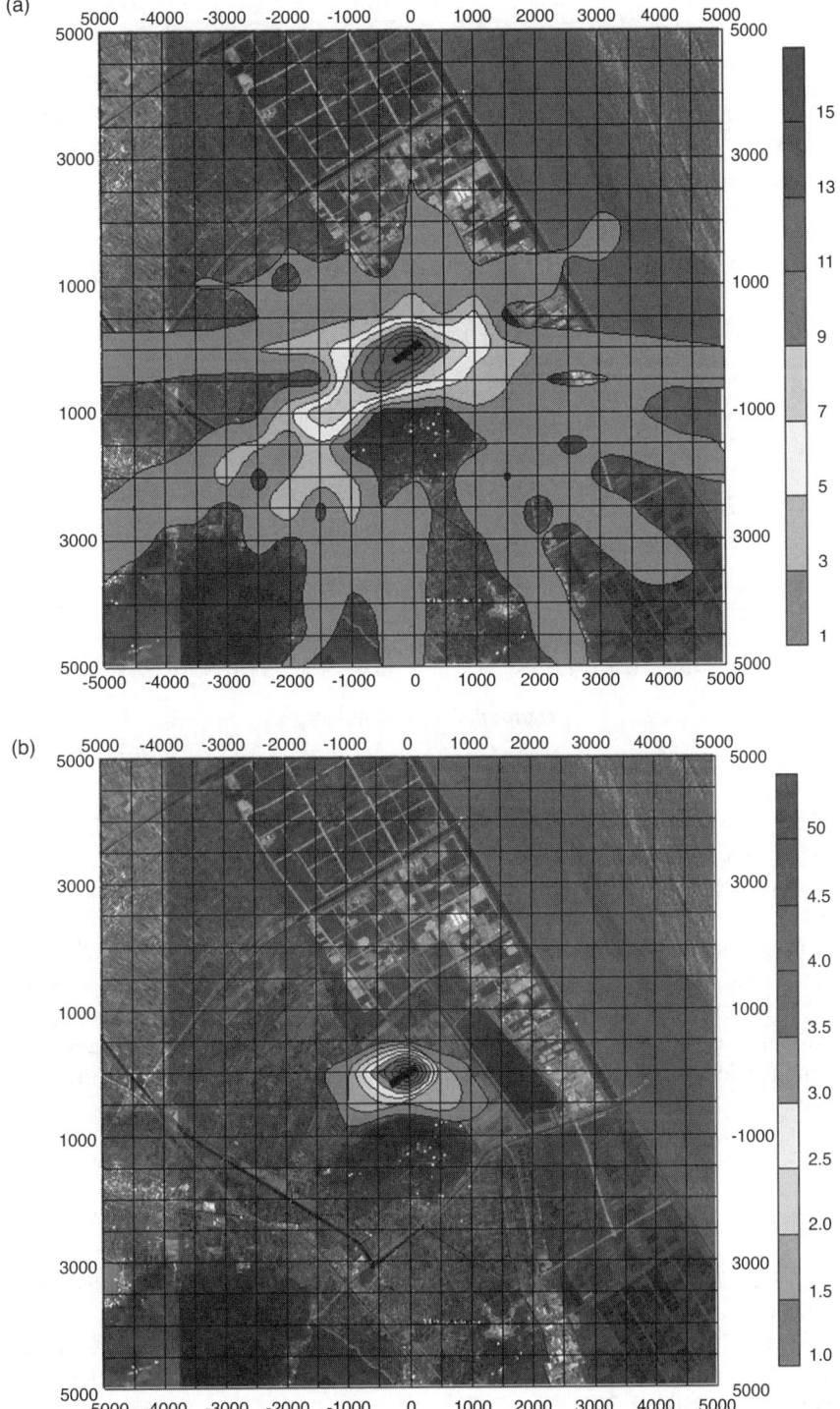

Figure 8.5.4 *Maximum odor concentration of grid points; (a) daily mean concentrations, (b) monthly mean concentrations, (c) yearly mean concentrations.*

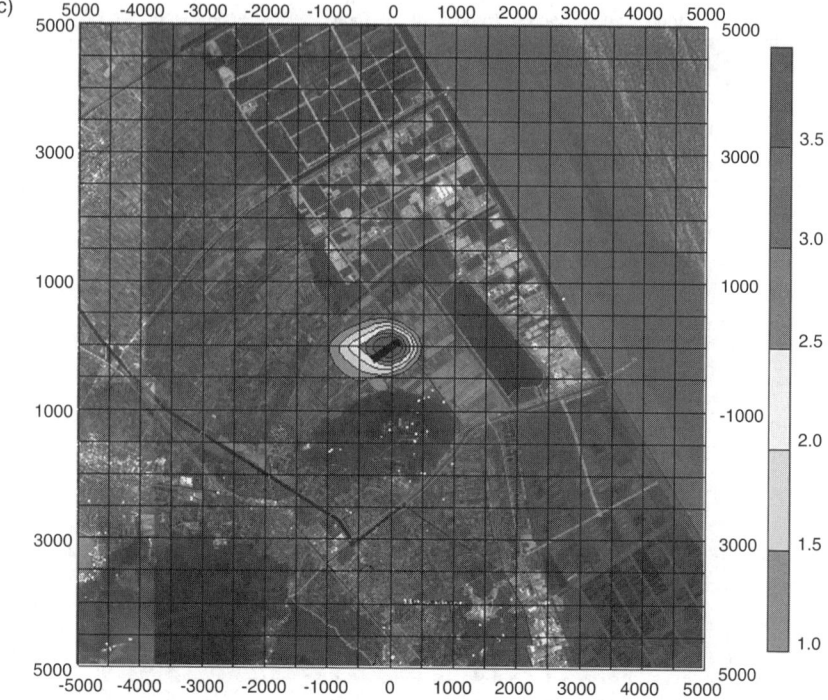

Figure 8.5.4 *(Continued)*

from the piglets, finishing, gestating and farrowing buildings were 1.05, 2.71, 3.92 and 10.15 OU/pig/s, respectively.

This study used AERMOD modeling to quantify the odor impacts around the monitored pig farm on the four neighboring villages. The dispersion modeling used local meteorological information and estimates of odor emissions from various pig buildings of the monitored farm in the study area. To assess the extent of the odor impacts, the dispersion model predictions were compared with the monitoring data, the predictions were lower than those measured, due to the emissions from the manure composting and lagoon having not been included. It has been recognized that there are some uncertainties in the emission estimates.

References

BS EN 13725:2003 Air Quality – Determination of Odor Concentration by Dynamic Olfactometry.

Copeland, C. (2010) *Air Quality Issues and Animal Agriculture: A Primer.* CRC Report RL32948. Available at http://ncseonline.org/NLE/CRSreports/10Jun/RL32948.pdf. (accessed on July 19, 2011).

Environment Agency (2005) *Odor Management at Intensive Livestock Installations*. Technical Guidance Note IPPC SRG 6.02 (Farming). Integrated Pollution Prevention and Control (IPPC).

Hayes, E.T., Curran, T.P. and Dodd, V.A. Odor and ammonia emissions from intensive pig units in Ireland. *Bioresource Technology*, **97** (2006), 940–948.

Higuchi, T., Masuda, J. and Hayano, A. (2002) Establishment of Quality Control Framework for Olfactometry in Japan, *Proceedings of the WEF Odors and Toxic Emissions 2002 Specialty Conference*.

Emission Standards for Odor Pollutants, GB14554-93 (In Chinese).

Environmental Quality Standards for livestock and poultry farms, NY/T 388-1999 (In Chinese).

Yu, Z., Guo, H. and Lague, C. (2010) Livestock odor dispersion modeling: a review, *Transactions of the ASABE*. **53** (4), 1231–1244.

Zhang, Q., Zhou, X.J., Guo, H.Q. *et al.* (2005) *Odor and greenhouse gas emissions from hog operations*. Project MLMMI 03−HERS−01. Winnipeg, Manitoba, Canada: Manitoba Livestock Manure Management Initiative.

8.6 Assessment, Control and Management of Odour in Sensitive Areas

N. Kalogerakis and M. Lazaridis

Department of Environmental Engineering, Technical University of Crete, Chania, Greece

8.6.1 Motivation for the Study

Municipal wastewater treatment plants (WWTP) can potentially emit odorous compounds that produce nuisance to the workers and nearby residents. Several chemical compounds contribute to odour problems, among them, sulfurous organic compounds, hydrogen sulfide, phenols and indoles, ammonia, volatile amines and volatile fatty acids; hydrogen sulfide (H_2S) and ammonia (NH_3) being most often the dominant odorant compounds. The detection and perception of odours by humans is an extremely complex process. WWTPs are often faced with complaints by nearby residents even if central odour control facilities have been installed. An integrated approach is required to successfully address the problem at the plant level. It involves the identification of odour sources within the WWTP, modelling the effect of mitigation actions, then using several alternative approaches (treatment technologies) to not only overcome the problem fully but also economically.

8.6.2 Description of the Situation

Odours emitted from wastewater treatment plants (WWTPs) are becoming a significant source of environmental annoyance. Often, it is the odours escaping from sewer manholes that cause complaints. More commonly, the odour source is the wastewater treatment plant. Odour-related complaints from communities surrounding WWTPs have increased constantly over the last decade. This is generally the result of new residential development near the plants.

Odorous compounds formed in sewer networks or during wastewater treatment are produced by the anaerobic decomposition of organic matter that contains sulfur or nitrogen.

Fresh wastewater has a distinctive, somewhat disagreeable odour, which is less objectionable than the odour of wastewater which has undergone anaerobic decomposition. The odorous compounds found in wastewater plants are usually generated by the microbial degradation of complex organic compounds. Under anaerobic conditions, the fermentation of fats, polysaccharides and proteins takes place. These compounds are hydrolysed first to fatty acids, shorter chain saccharides, amino acids and peptides and then to shorter chain compounds. Hydrolysis of proteinaceous material and organic sulfur compounds leads to the production of hydrogen sulfide (H_2S) and organic sulfides and disulfides. Domestic wastewater normally contains about 3–6 mg/l of organic sulfur in proteinaceous matter and additional organic sulfur in the form of sulfonates (about 4 mg/l) derived from household detergents (Boon, 1995).

The main odorous compounds emitted from WWTPs are sulfur containing substances such as hydrogen sulfide, methyl mercaptan, dimethyl sulfide, dimethyl disulfide, ethyl mercaptan, carbon disulfide and carbonyl sulfide. Nitrogen containing substances such as ammonia, amines, indole and skatole may also cause problems (Easter *et al.*, 2005; Nicell and Henshaw, 2007). The lower the molecular weight of a compound, the higher the volatility and potential for emission into the atmosphere. It has been noted that the human odour threshold for H_2S is very low, about 3–5 ppb.

The main impact from the existence of H_2S in the atmosphere is the annoyance caused to humans. Atmospheric dispersion modelling has been applied to the assessment of odour impacts in several cases using mainly Gaussian models such as the Industrial Source Complex (ISC) model (McIntyre, 2000; Henshaw *et al.*, 2006) and the CALPUFF model (Yu *et al.*, 2009). Dispersion modelling can effectively be used in order to estimate the dispersion of odours using available emission data as well as correlate it with the complaints and secondly to estimate the maximum odour emissions which can be permitted from a site in order to prevent odour complaints (McIntyre, 2000). However, the results of Gaussian models provide hourly concentrations which is not a representative scale to assess the annoyance caused from odours. Human annoyance can result after an odour exposure of few seconds and therefore a 'peak to mean' ratio is usually used to predict the maximum odour concentrations (Piringer *et al.*, 2007; Latos *et al.*, 2010a,b).

In addition, recent studies have given the relations between odour annoyance and odour exposure concentrations in order to express the odour impacts in terms of probability of detection and degree of annoyance (Nicell and Henshaw, 2007; Nicell, 2003; Henshaw *et al.*, 2006). Finally, a number of treatment technologies are currently being used to address the odour problem (Karageorgos *et al.*, 2010; Suffet *et al.*, 2004).

8.6.2.1   *Location, Local Conditions and Relevant Receptors*

The study was carried out at the WWTP of Chania (Greece). The WWTP of Chania is located close to the sea, 4 km North-East of the city of Chania. The wastewater treatment plant was designed for the treatment of domestic wastewater (105 500 equivalent inhabitants), industrial wastewater (5000 equivalent inhabitants) and sewage from septic tanks (7000 equivalent inhabitants), giving a total treatment capacity of 117 500 equivalent inhabitants and a flow rate of 26 000 m³/d. The average flow rate for the summer of 2007 was 19 500 m³/d. Sewage is treated with an Activated-Sludge method, while sludge with an Anaerobic Digestion method, utilizing the biogas which is produced to generate electricity. The plant is designed to give a 96% BOD5 removal and a 95% SS removal. The measurements focused on the summer period since the flow rate in the other seasons is lower (close to 17.000 m^3/d).

8.6.3   Specific Objectives of the Study

The main objective of the current study is to present an integrated approach to successfully address the odour problem in WWTPs. More specifically, the current study focus on the use of dose-response relationships of H_2S in conjunction with dispersion modelling and emission calculations for the determination of WWTPs odour impact in residential areas. The absolute goal is to estimate the probability of response and degree of annoyance of the communities surrounding wastewater treatment plants instead of the hourly mean concentrations.

In addition, another objective is a systematic search for all the odour sources within two WWTPs located on Crete and present results from different technologies (installation of covers and oxidation of collected gasses, addition of oxidants in the liquid phase, activated carbon/permanganate absorption, $FeCl_3$ addition in the liquid phase) that have been employed to address the odour-problem.

8.6.4   Methodology

The H_2S concentrations were measured by portable handheld devices at various locations in the municipal WWTP of Chania (Crete, Greece). In conjunction with the measurements, the Gaussian dispersion model AERMOD code was modified in order to estimate the maximum odour concentration for very short time steps, using peak-to mean ratios. In addition, relations between odour annoyance and odour exposure concentrations were embedded into the model, in order to express the odour impacts in terms of probability of detection and degree of annoyance of the population near the facilities. Furthermore, a case study is presented on how to address the odour problem from secondary sources within a municipal wastewater treatment plant (WWTP) by first identifying the locations of the problem and second by evaluating alternative treatment technologies. An evaluation of the effectiveness of the different technologies that were employed to address this problem are examined and comprehensively evaluated (cover installation, gas and liquid phase oxidation, activated carbon/permanganate absorption, $FeCl_3$ addition).

8.6.5   Data Collection

8.6.5.1   Odour Measurements

Hydrogen sulfide measurements were performed using a gold-film monitor. This instrument includes a thin gold film, which in the presence of hydrogen sulfide undergoes an increase in electrical resistance proportional to the mass of hydrogen sulfide contained in the sample taken. The instrument used in the current measurements is the Jerome 631-X H_2S analyser (Arizona Instruments, USA) which has a sensitivity of 3 ppb and can measure up to 50 ppm H_2S. Sample times vary depending on the H_2S concentration levels (Winegar and Schmidt, 1998).

8.6.5.2   Meteorological Data

Meteorological data are necessary in order to perform dispersion modelling of air pollutants. For this reason, meteorological data were collected during a period of over one year from the study area using a meteorological station (base year: 2007). The data collected were air temperature, relative humidity, wind speed and wind direction. The average annual temperature was found to be 18.6°C and the average annual relative humidity was 65.3% during 2008. The predominant wind direction during the measurement period was northerly.

8.6.5.3   Emission Data

Estimation of hydrogen sulfide emissions from area sources such as the sedimentation tanks is a rather complex procedure as a result of the variability of the emissions due to the effect of wind speed to the transportation of the pollutant from the tank to nearby areas. In the present study, several measurements were performed at the area close to the sources. In order to estimate the emissions of hydrogen sulfide, the Gaussian dispersion model AERMOD was used (USEPA, 2004; Latos *et al.*, 2010a,b). Several simulations of the model, with varying emission rates were performed in order to obtain the observed concentrations at the measurement points close to the sources. Furthermore, a minimization methodology is applied to determine the emission rate of hydrogen sulfide.

8.6.5.4   Model Description

The AERMOD Gaussian steady-state plume model was applied to study odour dispersion near the WWTP of Chania. Modifications in the AERMOD code were performed in order to be able to calculate peak concentration values instead of half-hour mean values. This is due to the fact that odour perception by humans is related proportionally to the instantaneous peak concentration of the odorant rather than to mean concentration values.

The widely used peak-to-mean ratio approach has been selected to resolve this modification. Equation 8.6.1 shows the methodology used to convert average concentrations to peak concentrations and therefore providing the ability to convert the 0.5-h modelled concentrations to 5-s peak concentration values:

$$C_p = C_m \left(\frac{t_m}{t_p}\right)^u \tag{8.6.1}$$

where C_p stands for peak concentration calculated for a short period t_p and C_m stands for the mean concentration calculated from the dispersion model for a longer period t_m. The exponent u depends on the atmospheric stability conditions and varies from 0.35 to 0.65.

The main impact of hydrogen sulfide at relatively low concentrations is the annoyance caused to the residents close to the emission source. This is the reason why providing contour plots of the H_2S concentrations are not giving the information of interest. However, the sense of smell is subjective and people react in a different way when being exposed to an odorant. Recently, dose-response relations have been developed that correlate odour concentrations to the probability of detection of the specific odour or probability of annoyance. The probability of detection of an odorous compound can be estimated (Nicell, 2003) as following:

$$P = \frac{100}{1 + \left(\dfrac{C_t}{C}\right)^{\frac{1-p}{p}}}$$

(8.6.2)

where, P (in %) stands for the probability of detection of an odour, C is the concentration of the odorous compound (in ppb), C_t is the threshold concentration of the specific odorous compound (in ppb) and p (dimensionless) is the 'persistence of response' of the specific odour. The 'persistence of response' varies from 0–1 depending on the compound. The value of p was set to 0.4 using observations about the percentage of people being able to identify the existence of hydrogen sulfide over a range of concentrations. The threshold concentration is the concentration at which 50% of people exposed to hydrogen sulfide can detect its presence, with it being set to 4.7 ppb (Nagata, 1990).

A similar relation has been proposed to estimate the annoyance caused by odours:

$$A = \frac{10}{1 + \left(\dfrac{C_{5AU}}{C}\right)^{\frac{1-a}{a}}}$$

(8.6.3)

where A (measured in annoyance units, AU) stands for the degree of annoyance of the population and ranges from 0–0, C is the concentration of the odorous compound (in ppm), C_{5AU} corresponds to the odorant concentration where the population annoyance has a value of 5 AU and the term a (dimensionless) is the 'persistence of annoyance' of the specific odour. The 'persistence of response' varies from 0– 1 depending on the compound (Nicell, 2003). Parameters C_{5AU} and were estimated to be 23 ppb and 0.68 respectively, based on the observations of people annoyed by the existence of H_2S in the atmosphere.

8.6.6   Results and Discussion

8.6.6.1   *Measurements*

Several multiple measurements took place in specific points within the whole area of the Chania WTP, using the Jerome 631-X instrument, in order to detect the main emission sources that cause odour problems. In total, 45 points were marked for the WTP and each measurement was taken three times in order to check the repeatability of the instrument. According to the measurement values of gaseous H_2S, the main emission sources with the highest effects on odours were the primary sedimentation tanks (ST-1 and ST-2) and two other tanks (T-1 and T-2) where the recycled activated sludge is already mixed with the sludge from the primary sedimentation tanks in an intermediate bacteria selection tank (BST) and is directed to the respective aeration tanks (see Figure 8.6.1).

The WTP at Chania is equipped with a chemical scrubber that utilizes hydrogen peroxide for the elimination of odorous gases. Specifically, scrubber collects odours coming from the pretreatment stages and ST-1, ST-2 but does not include the odorous gases from T-1 and T-2. Therefore, during the summer season of 2007, the installation of covers and fans

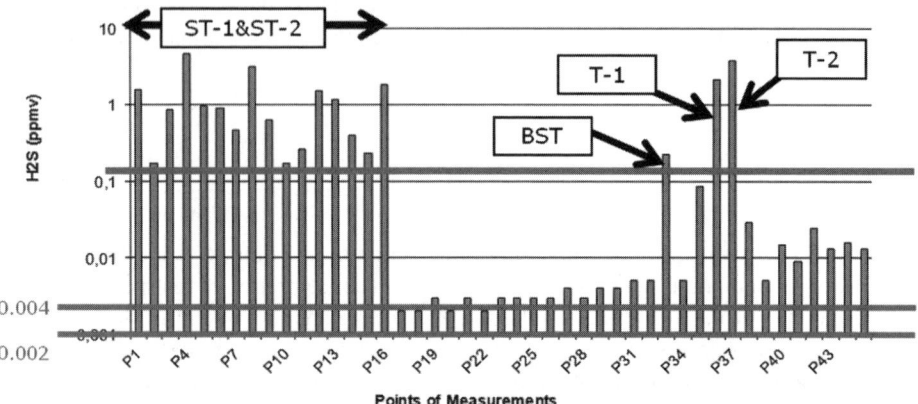

Figure 8.6.1   *Measurements of gaseous H₂S at several points of Chania's WTP.*

in T-1 and T-2 took place taking the odorous gases from these tanks to the central chemical scrubber.

These measurements indicate that high H₂S concentrations may be observed in the facility (reaching 30 ppm very close to the main H₂S sources) but these values decrease rapidly only a few meters away from the sources. The measurements conducted at the village of Koubeli, close to the WTP showed that the concentrations of H₂S were always below the threshold concentration.

8.6.6.2   Modelling

The model AERMOD was used to perform a study of the annoyance that may be caused due to the H₂S emissions from the wastewater treatment plant of Chania to the population living in the surrounding area. Therefore, the modified AERMOD model was then applied in order to estimate the peak values during a typical summer day (using available meteorological data). The model estimated the probability of detection of H₂S and degree of annoyance during a short time-interval (5 s).

The mean observed meteorological parameters and the mean H₂S emissions were used (average 2007 values) to calculate the ambient H₂S concentrations. Sensitivity calculations were also performed in order to study the effect of using different time steps of the peak concentration values. Figure 8.6.2 presents the results of the model for various short time steps (5s, 10s, 30s and 60s). The probability of detection of hydrogen sulfide at the main transport direction (North-South) was estimated in order to assess the impact of the choice of the short time step used in the proposed methodology. In Figure 8.6.2, the negative distance values correspond to a northern direction while positive distance values correspond to a southern one. Using the results that occurred for a time step of 5s as 'base-results', the change in the probability of detection of hydrogen sulfide has been estimated. Using a time step of 10 s (100% greater than the 'base-value of 5 s), the mean decrease of the probability of detection was 4%. Using 30s and 60s as a short time scale (600% and 1200% increased values according to the 'base value' of 5 s correspondingly) the mean decrease of the probability of detection of H₂S was 10% and 15% correspondingly. The main reason

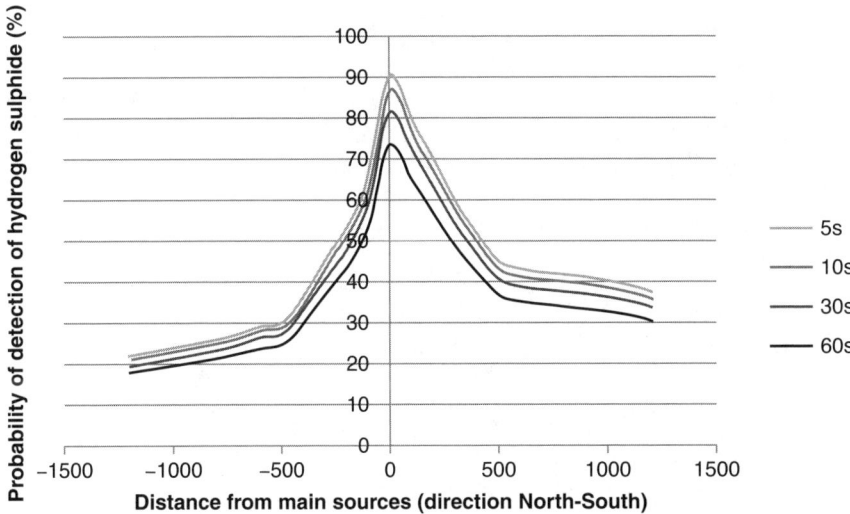

Figure 8.6.2 *Probability of detection of hydrogen sulfide (%) versus distance from the main sources (Direction North – South) for different time steps of the peak concentration values.*

for this is the sigmoid function that correlates H_2S concentrations to the probability of detection of the specific compound. The sensation of smell can be triggered at a very short time period exposure, and this is the reason why the value of 5 s was chosen to study the effects caused from the presence of hydrogen sulfide in the atmosphere.

In Figure 8.6.3, the equal probability curves of detection of hydrogen sulfide around the facility are presented. The main direction of the odour dispersion is south-east because of the prevailing wind direction and the complex landscape. The villages that are affected are Koumbeli and Profitis Ilias. The results indicate that there is a high degree of annoyance and probability of detection in the vicinity of the facility. The results indicate that the odours close to the sources have a high probability to be detected (90.1%) and this probability decreases as the distance from the source increases. Due to the low detection threshold of the specific compound, the probability of detection remains at high levels even 400 m from the main sources (Koumbeli).

The obtained results are in agreement with previous studies which have used the peak to mean ratios (e.g. Piringer *et al.*, 2007). Future work also has to examine the effect of different atmospheric stability conditions in conjunction with the surface topography (Piringer *et al.*, 2007; Nicell, 2009). The difficulty in obtaining an objective set of criteria for the assessment and regulation of odour impacts is pointed out by Nicell (2009). The current modified AERMOD model is used as an effective and easy to use modelling platform for the determination of human odour exposure.

8.6.6.3 Emission Data

The estimation of hydrogen sulfide emissions from area sources such as the sedimentation tanks is a rather complex procedure because of the variability of the emissions due to the effect of wind speed to the transportation of the pollutant from the tank to nearby areas. In the present study, several measurements were performed at the area close to the sources. In order

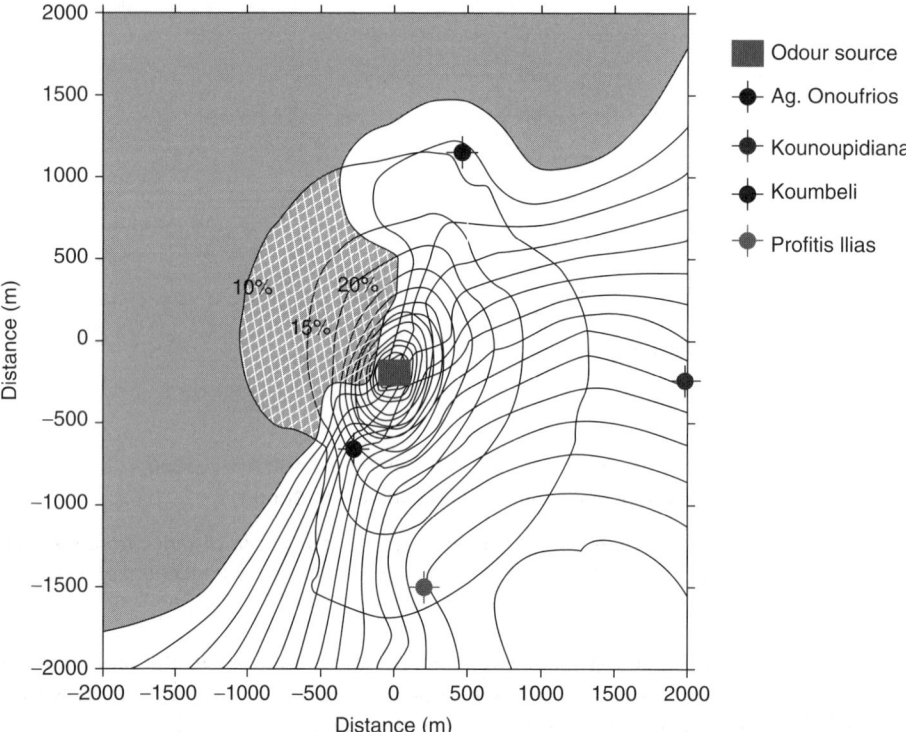

Figure 8.6.3 *Probability of detection of hydrogen sulfide in the vicinity of the facility based on the modified AERMOD model calculations. (Reprinted with permission from Karageorgos et al., 2010. Copyright (2010) IWA Publishing).*

to estimate the emissions of hydrogen sulfide, the Gaussian dispersion model AERMOD was used (USEPA, 2004). Several simulations of the model, with varying emission rates were performed in order to obtain the observed concentrations at the measurement points close to the sources. The minimization of the factor presented in Equation 8.6.1 provides the correct value of the emission rate of hydrogen sulfide.

$$\text{Minimize d} = \sum_{i=1}^{n} \left(C_i - \sum_{j=1}^{m} \left(Q_j \times C_{i,j,MOD} \right) \right)^2 \qquad (8.6.4)$$

where:

d = The parameter that was minimized
i = Point of measurement (from 1 to I = 5)
j = Source Index (1 for primary sedimentation tanks, 2 for tanks T-1 and T-2 and 3 for the intermediate bacteria selection tank (BST))
C_i = Observed concentration of H_2S at measurement point i ($\mu g/m^3$)
$C_{ij,MOD}$ = Concentration of H_2S at point i, from source j, using the dispersion model with emission rate of 1 g/m^2. ($\mu g/m^3$)
Q_j = The emission factor for each source.

Table 8.6.1 *Estimated emission rates (g/h) of hydrogen sulfide from the main facility sources.*

	Emissions (g/h)	Variance
S-T	32	±68%
T-1, T-2	26	±54%
BST	15	±34%

Using thos methodology, it is possible to both calibrate the model with realistic data sets obtained from the specific facility as well as estimate the emission factors from the main odour sources. Due to the high variability of the measurements obtained, it is also adequate to take into account the variance of the emission rates from each source. In Table 8.6.1, the estimated emission rates and their variance are presented.

Therefore, the modified AERMOD model was then applied in order to estimate the peak values during a typical summer day (using available meteorological data). The model estimated the probability of detection of H_2S and degree of annoyance during short time-interval (5 s). Three scenarios were studied. In scenario 1, the mean observed meteorological parameters and the mean H_2S emissions were used (average 2007 values). In scenario 2, the emissions were considered to have the maximum annual values during the study period and also the worst-case meteorological conditions were adopted (low wind speed, stable atmospheric conditions).

In scenario 3, a decrease of 50% of the peak observed H_2S emissions was used in combination with the mean observed meteorological parameters (average 2007 values). Scenario 2 was applied in order to examine the possible impact that can be caused by the presence of H_2S in the atmosphere under conditions that may occur especially during the summer. This is the reason why the emissions were considered to have higher values than the estimated mean values and the meteorological conditions adopted correspond to a minimum dispersion of the odorous compound. Scenario 3 was applied to study a possible solution to the problem caused by the presence of the odorous compound in the atmosphere reducing the emissions by 50%. The resulting probability of detection and degree of annoyance were calculated at four receptors. The first receptor was close to the sedimentation tanks (S-T) (points 1–16) since these points were identified as main emission sources. The second point was 100 m south of the sedimentation tanks (point 43). The other two points were at two villages close to the facility (Koumbeli and Profitis Ilias at 400 m and 1200 m respectively). The results from the simulations are presented in Table 8.6.2.

8.6.6.4 Treatment of Major Odour Sources

A case study on how to address the odour problem from secondary sources within a municipal wastewater treatment plant (WWTP) by first identifying the locations of the problem and second by evaluating alternative treatment technologies was performed for the WWTP of Chania (Crete, Greece).

The principal sources of odours in wastewater management facilities include the preliminary treatment operations (i.e. screening facilities, grit removal tanks, flow-equalization basins, primary clarifiers) and the solids processing facilities (anaerobic digestion, sludge

Table 8.6.2   *Model results at specific receptors in relation to the probability of detection of H_2S and degree of annoyance for the three scenarios. (S-T: (Sedimentation tank); T-1, T-2: (tanks); BST: (Bacteria selection tank).*

Location	Scenario 1		Scenario 2		Scenario 3	
	Probability of detection (%)	Degree of annoyance (AU)	Probability of detection (%)	Degree of annoyance (AU)	Probability of detection (%)	Degree of annoyance (AU)
S-T	90.1	4.8	98.2	6.2	72.1	3.8
Point 43*	79.0	4.2	90.7	4.9	62.8	3.5
Koumbeli	51.8	3.2	72.0	3.8	42.4	3.0
Profitis Ilias	37.4	3.0	29.2	2.7	32.4	2.7

*Outside the walls of the facility

storage basins, mechanical dewatering). The primary sources of odour have been identified by several researchers in the past. Locations with a high odour potential within a WWTP include headworks, screening facilities, flow equalization basins and primary clarifiers whereas aeration basins and secondary clarifiers constitute low/moderate potential odour sources. In addition, the locations of anaerobic digestion, sludge storage basins and mechanical dewatering exhibit a moderate/high odour potential.

The most common method to suppress odours originating from such locations includes the installation of covers, collection hoods, and air handling equipment for containing and directing odorous gases to treatment systems. In cases where the treatment facilities are close to residential areas, it has become a common practice to cover treatment units such as the headworks, primary clarifiers, and sludge processing facilities. Wherever covers are used, the trapped gases are collected by suitable vacuum pumps and treated.

Such a management policy has been adopted by the WWTP of Chania where all the facilities related to preliminary treatment (i.e. influent pumping station, fine screening, grit and oil removal tank, flow distributors of primary sedimentation tanks) and biosolids treatment (i.e. pre-thickening and post-thickening stages, digestion and dewatering buildings) have been covered and the odorous gases produced are directed to the central odour control building where a chemical scrubber uses hydrogen peroxide as the oxidizing agent. The installed unit treats most of the released H_2S in the gas phase.

8.6.6.5   *Treatment of Secondary Odour Sources*

In order to minimize and essentially eliminate the unpleasant odours emitted from the WWTP, it was decided to identify all the secondary sources of H_2S within the plant. The H_2S analyser was operated in the survey mode, resulting in the continuous recording of H_2S concentrations in the gas phase (Latos *et al.*, 2010a,b). According to the values of gaseous H_2S, the main open sources with the highest odour impact were the primary sedimentation tanks (ST-1 and ST-2) and the flow distribution tanks (T-1 and T-2) where the activated sludge from aeration tanks is mixed with sludge coming from the primary sedimentation tanks in the intermediate bacteria selection tank (BST) and then directed to the respective aeration tanks. In order to minimize odours originated from these identified secondary sources, several actions may be taken such as the installation of covers and the reduction of turbulence by changing the height of overflowing gates.

8.6.6.6   *Installation of Covers and Oxidation of Collected Gasses*

In sedimentation tanks ST-1, ST-2, the main odour source was at the overflow weirs where the wastewater from the effluent weirs is collected. The measured values of H_2S concentrations were even higher than 50 ppm which is the maximum detectable concentration of Jerome 631-X. In order to reduce the turbulent conditions several coverings of both effluent weirs around the tanks and overflow wells took place. After the installation of aluminium covers, a significant reduction in odours was noticed. The average concentration of gaseous H_2S became up to 50 times lower (Karageorgos *et al.*, 2010).

8.6.6.7   *Addition of Oxidants in the Liquid Phase*

The second source with the highest odour impact identified after the field measurements were the tanks T-1 and T-2. To check the efficiency of a rather strong oxidant such as H_2O_2, several experiments with different dosages of H_2O_2 solution (50% w/w, Degussa Co.) were carried out (Karageorgos *et al.*, 2010). The study concluded that instead of adding H_2O_2 into the liquid phase, it would be more economically effective to cover T-1 and T-2 and connect the gases to the vacuum pump directing the H_2S gases to the main chemical scrubber of the plant. This change was implemented and worked successfully.

8.6.6.8   *Main Chemical Scrubber*

When the detailed detection of the secondary odorous sources was completed, several new covers were installed in the sedimentation tanks (ST-1 and ST-2) and flow distributors T-1 and T-2. Odorous gases originating from these sources were collected and directed to the central odour treatment unit. In this unit, a cross-sectional flow chemical scrubber utilizing sodium hydroxide and hydrogen peroxide for the treatment of H_2S and ammonia odorous emissions is used. The scrubber maximum capacity is about 45 000 m^3/h and its H_2S removal efficiency should be at least 99% for corresponding H_2S inlet concentrations not higher than 15 ppmV. The work by Karageorgos *et al.* (2010) highlighted how when using a chemical scrubber all removal efficiencies exceeded 95% but only once achieved 99%.

8.6.6.9   *Auxiliary System for Treatment of Odorous Gases*

An auxiliary odour treatment unit was tested during the peak times of the summer months at the WWTP of Chania. Adsorbent systems for odour control in wastewater applications generally consist of static beds of granular materials in vertical cylindrical columns. Activated carbons are widely used as adsorbents, separation media and catalyst supports (Bansal *et al.* 1988). The commercially activated carbon adsorption device (DS-300 manufactured by Purafil, Inc., USA) was used to treat odorous emissions from the fine screening unit of the WWTP (Karageorgos *et al.*, 2010).

Several experiments were run to test the removal efficiencies of the activated carbon device under different H_2S loadings. During the initial set of experiments, H_2S inlet concentrations varied in the range of 20–45 ppm and the removal efficiencies were all above

Table 8.6.3 H_2S removal efficiencies for several experimental runs conducted with DS-300 adsorption system for an influent gas stream flow rate of 520 m^3/h.

Experimental run	Average H_2S inlet concentration, ppmv	Average H_2S outlet concentration, ppmv	H_2S removal efficiency, %
#1	30.07	0.105	99.64
#2	24.77	0.107	99.56
#3	29.69	0.087	99.70

99.5%. For inlet H_2S concentrations varying in the range of 20–45 ppm, the corresponding outlet concentrations were rather low and thus yielded a significant deodorizing profile. The average H_2S concentration values calculated for each experimental run corresponded to removal efficiencies higher than 99.5% as shown in Table 8.6.3 (Karageorgos *et al.*, 2010).

Despite the fact that more than 99.5% H_2S was consistently removed, the total deodorisation of influent gases could not be achieved since the H_2S human odour threshold level (OTL) is very low (i.e. 0.005 ppm) and all the outlet concentrations exceeded this value.

Therefore, the introduction of a chemical reagent in the main wastewater stream was performed. More precisely, an aqueous ferric chloride solution ($FeCl_3$, 40% w/v, density: 1.45 g/ml, Feri-Tri Co., commercial name: ferissol 140) in the inlet pumping station of the Chania WWTP took place. During the summer of 2008, the average wastewater flow rate was in the order of 26 000 m^3/d, and an injection of 7 l/h of ferissol 140 in the inlet pumping station was applied, resulting in a significant reduction of the H_2S emissions in the fine screening unit. Consequently, the influent H_2S concentrations of DS-300 were reduced and varied in the range of 1–12 ppm. As expected, the effluent gases were under 0.005 ppm resulting in a complete deodorisation as shown in Table 8.6.4

It should be mentioned that complete deodorisation of gas streams is often not required. Even for WWTPs that are very close to residential areas, the objective is to have H_2S concentrations near or just above the 0.005 ppm level at the WWTP boundary and not within the facility grounds. Further dilution by the air movement should alleviate completely any complaints by nearby residents. The unusually high level of control of odorous emissions

Table 8.6.4 H_2S removal efficiencies for several experimental runs conducted with DS_300 adsorption system for influent gas stream flow rate of 520 m^3/h, after the addition of $FeCl_3$.

Experimental run	Average H_2S inlet conc. ppmv	Average H_2S outlet conc. ppmv	H_2S removal efficiency, %
#4	6.00	0.002	99.96
#5	10.41	0.014	99.86
#6	1.07	0.003	99.71
#7	9.36	0.004	99.87
#8	3.77	0.002	99.95
#9	2.12	0.001	99.96

by the WWTP management of Chania was simply the result of political pressure from the Mayor's office.

8.6.7   Conclusions, Recommendations and Outcomes

To overcome the odour problem in WWTPs, it is not possible to rely only on a central gas treatment unit but rather take a systematic approach to addressing the problem. Namely, start with the measurements in all the potential locations within the plant; identify the secondary sources of odour; estimate the amounts released into the atmosphere, use a model to predict the reduction in odour nuisance in order to see the require magnitude of H_2S emission reductions. Subsequently, alternative technologies were considered, including simple low cost measures, to achieve the desired reduction in emissions. This procedure was followed and the problem addressed satisfactorily in the WWPT of the city of Chania using a combination of technologies (addition of FeCl3 and activated carbon DS-300 modules) and appropriate measures (caps primary sedimentation units).

In accordance to the proposed integrated methodology to address the odour problem in WWTPs measurements of hydrogen sulfide (H_2S) in and close to the municipal wastewater treatment plant of Chania, a study was carried out during the summer of 2007. The H_2S concentrations away from the source (WTP) were below the threshold concentrations.

Application of dispersion models showed that the hourly mean concentrations were also below the threshold during the same period. However, this does not explain the large number of complaints from the residents of the village especially during the summer. This highlights how the application of a standard dispersion model underestimates the impacts of odorous pollutants due to these impacts depending on peak concentrations rather than mean hourly values.

In order to estimate the impacts of the highly odorous compound H_2S to the surrounding community, several changes were carried out to the AERMOD model. The modified AERMOD model predicts peak concentrations of odorous compounds and estimates the probability of detection of a specific pollutant as well as the degree of annoyance of the surrounding community. The results of the model indicate that under typical summer period conditions, the odour can be detected at villages close to the facility. The probability of detection of H_2S exceeds 50% for the Koubeli village (400 m from the main emission sources) during a short time interval (5 s) with a relatively high degree of annoyance (3.2 AU). Finally, it was demonstrated that the meteorological conditions considerably affect both the probability of detection and degree of annoyance.

The most common practices used to suppress WWTPs emissions include the installation of covers, collection hoods, and air handling equipment to contain and direct odorous gases to central treatment systems. This methodology was adopted by the Chania WWTP where the sources with the highest odour potential (i.e. fine screening room, grit removal tanks, anaerobic digesters, mechanical dewatering facilities, etc.) are tightly covered and their odorous emissions are directed to the central chemical scrubber for treatment. In response, a systematic approach for the identification of all the secondary odour point sources was applied and several technological alternatives were evaluated, with the most economical of these being implemented. In addition, the efficacy of an auxiliary unit based on the adsorption on activated carbon and oxidation by permanganate was assessed as a backup whenever the H_2S emissions would increase beyond the capabilities of the main chemical scrubber unit. This work has shown that odour emissions from secondary sources within a WWTP can still be a serious problem that needs to be addressed. If the central chemical scrubber is near capacity, the adding of oxidants directly in the liquid phase should not be considered, since their use is rather expensive and requires the installation of an automatic control system in order for them to be successful. On the other hand, the adding of a chemical reagent such as $FeCl_3$ in the liquid phase can offer a desirable reduction of odorous emissions in the gas phase as well as further treatment with an activated carbon/oxidation device can ensure complete deodorising conditions. Such an auxiliary system can be operated only during peak periods when the central gas treatment facility operating at full capacity cannot cope with the total load.

References

Bansal, R.C., Donett, J.B. and Stoeckli, F. (1988) *Active Carbon*, Marcel Dekker, New York.

Boon, A.G. (1995) Septicity in sewers: causes, consequences and containment. *Water Sci. Technol.*, **31** (7), 237–253.

Easter, C., Quigley, C. Burrowes, P. *et al.* (2005) Odor and air emissions control using biotechnology for both collection and wastewater treatment systems. *Chemical Engineering Journal*, **113** (2–3), 93–104.

Henshaw, P., Nicell, J.A. and Sikdar, A. (2006) Parameters for the Assessment of Odor Impacts on Communities. *Atmospheric Environment* **40** (6), 1016–1029.

Karageorgos, P., Latos, M., Lazaridis, M. and Kalogerakis, N. (2010) Alternative approaches to treat secondary sources of unpleasant odors in municipal waste water treatment plants. *Water Science and Technology*, **61**, 2635–2644.

Latos, M., Karageorgos, P. and Mpasiakos, C.H. *et al.* (2010a) Dispersion modelling of odours emitted from pig farms: Winter Spring measurements. *Global Nest Journal*, **12**, 46–53.

Latos, M., Karageorgos, P., Kalogerakis, N. and Lazaridis, M. (2010b) Dispersion of odorous gaseous compounds emitted from wastewater treatment plants. *Water Air and Soil Pollution*, **215**, 667–677.

McIntyre, A. (2000) Application of dispersion modeling to odour assessment: a practical tool or a complex trap? *Water Science and Technology*, **41** (6), 81–88.

Nagata, Y. (1990) Measurement of Odor Threshold by Triangle Odor Bag Method, *Bull. Japan Environ. Sanit. Center.*, **17**, 77–89.

Nicell, J. and Henshaw P. (2007) Odor Impact Assessments Based on Dose-Response Relationships and Spatial Analyses of Population Response. *Water Practice*, **1** (2), 1–14.

Nicell, J. (2009) Assessment and regulation of odour impacts. *Atmospheric Environment*, **43** (1), 196–206.

Nicell, J. (2003) Expressions to Relate Population Responses to Odor Concentration. *Atmospheric Environment*, **37** (35), 4955–4964.

Piringer, M., Petz, E., Groehn, I. and Schauberger, G. (2007) A sensitivity study of separation distances calculated with the Austrian Odour Dispersion Model (AODM). *Atmospheric Environment*, **41** (8), 1725–1735.

Suffet, I.H., Burlingame, G.A., Rosenfeld, P.E. and Bruchet, A. (2004) The value of an odor-quality-wheel classification scheme for wastewater plants. *Water Sci. Technol.*, **50** (4), 25–32.

U.S. Environmental Protection Agency (2004) *AERMOD: Description of the model formulation*. EPA-454/R-03-004, U.S. Environmental Protection Agency, Research Triangle Park, NC.

Winegar, E.D. and Schmidt, C.C. (1998) Jerome 631-X portable hydrogen sulfide sensor: laboratory and field evaluation. *Report to Arizona Instrument Corporation*, 15p. 3375 N. Delaware Street, Chandler, AZ.

Yu, Z., Guo, H., Xing, Y. and Lague, C. (2009) Setting acceptable odour criteria using steady-state and annual hourly weather data. *Biosystems Engineering*, **103** (3), 329–337.

Index

Note: **Bold** indicates the page numbers where the term has been discussed in detail.

Odour Impact Assessment Handbook, First Edition. Edited by Vincenzo Belgiorno, Vincenzo Naddeo and Tiziano Zarra.
© 2013 John Wiley & Sons, Ltd. Published 2013 by John Wiley & Sons, Ltd.